T0210943

Communications in Computer and Information Science 834

Commenced Publication in 2007
Founding and Former Series Editors:
Alfredo Cuzzocrea, Xiaoyong Du, Orhun Kara, Ting Liu, Dominik Ślęzak,
and Xiaokang Yang

More information about this series at http://www.springer.com/series/7899

Debdas Ghosh · Debasis Giri
Ram N. Mohapatra · Ekrem Savas
Kouichi Sakurai · L. P. Singh (Eds.)

Mathematics and Computing

4th International Conference, ICMC 2018
Varanasi, India, January 9–11, 2018
Revised Selected Papers

 Springer

Editors
Debdas Ghosh
Department of Mathematical Sciences
Indian Institute of Technology BHU
Varanasi, Uttar Pradesh
India

Debasis Giri
Haldia Institute of Technology
Haldia
India

Ram N. Mohapatra
University of Central Florida
Orlando, FL
USA

Ekrem Savas
Istanbul Commerce University
Istanbul
Turkey

Kouichi Sakurai
Kyushu University
Fukuoka
Japan

L. P. Singh
Indian Institute of Technology (BHU)
Varanasi
India

ISSN 1865-0929   ISSN 1865-0937 (electronic)
Communications in Computer and Information Science
ISBN 978-981-13-0022-6   ISBN 978-981-13-0023-3 (eBook)
https://doi.org/10.1007/978-981-13-0023-3

Library of Congress Control Number: 2018940140

Printed on acid-free paper

This Springer imprint is published by the registered company Springer Nature Singapore Pte Ltd.
part of Springer Nature
The registered company address is: 152 Beach Road, #21-01/04 Gateway East, Singapore 189721, Singapore

Message from the General Chairs

It is our privilege and great pleasure to welcome you to the proceedings of the 4th International Conference on Mathematics and Computing 2018 (ICMC 2018). The scope of the conference is to provide an international forum for the exchange of ideas among interested researchers.

ICMC 2018 was supported by invited speakers giving talks on mathematical analysis, cryptology, approximation theory, graph theory, operations research, numerical methods, etc. Technical sessions on a variety of fields covering almost all aspects of mathematics were arranged. The conference addressed key topics and issues related to all aspects of computing.

The conference was held at the Indian Institute of Technology (Banaras Hindu University), which is situated in the oldest city of the world – Varanasi. Varanasi is well known for its heritage and culture, and the participants enjoyed the city by visiting many places of interests.

We hope the interactions and discussions during the conference provided the participants with new ideas and recommendations, useful to the research world as well as to society.

P. K. Saxena
P. D. Srivastava
U. C. Gupta
L. P. Singh
Debjani Chakraborty

Message from the Program Chairs

It was a great pleasure for us to organize the 4th International Conference on Mathematics and Computing 2018 held during January 9–11, 2018, at the Indian Institute of Technology, BHU, Varanasi, Uttar Pradesh, India. Our main goal in this conference is to provide an opportunity for participants to learn about contemporary research in cryptography, security, modeling, and different areas of mathematics and computing. In addition, we aim to promote the exchange of ideas among attendees and experts participating in the conference, both the plenary as well as the invited speakers. With this aim in mind, we carefully selected the invited speakers. It is our sincere hope that the conference helped participants in their research and training and opened new avenues for work for those who are either starting their research or are looking to extend their area of research to a new field of current research in mathematics and computing.

The inauguration ceremony of the conference was held on January 9, 2018, starting with the one-hour keynote talk of Prof. T. S. Ho, University of Surrey, UK, followed by 11 forty-five-minute invited talks by Prof. R. N. Mahapatra, University of Central Florida, Orlando, USA, Prof. Matti Vuorinen, University of Turku, Finland, Prof. Srinivas R. Chakravarthy, Kettering University, USA, Dr. Srinivas Pyda, Oracle's System's Technology, USA, Dr. Parisa Hariri, University of Turku, Finland, Prof. S. Ponnusamy, Indian Institute of Technology Madras, Prof. Debasis Giri, Haldia Institute of Technology, India, Prof. Kouichi Sakurai, Kyushu University, Fukuoka, Prof. Chris Rodger, Auburn University, Alabama, USA, Prof. S. K. Mishra, Banaras Hindu University, India, Prof. T. Som, IIT (BHU), and Dr. Arvind, SCUBE India. The speakers/contributors came from India, Japan, UK, and the USA.

After an initial call for papers, 116 papers were submitted for presentation at the conference. All the submitted papers were sent to external reviewers. After a thorough review process, 29 papers were recommended for publication for the conference proceedings published by Springer in its *Communications in Computer and Information Science* (CCIS) series.

We are truly thankful to the speakers, participants, reviewers, organizers, sponsors, and funding agencies for their support and help without which it would have been impossible to organize the conference. We owe our gratitude to the research scholars of the Department of Mathematical Sciences, IIT (BHU), who volunteered the conference and worked behind the scene tirelessly in taking care of the details to make the conference a success.

Debdas Ghosh
Debasis Giri
Ram N. Mohapatra
Ekrem Savas
Kouichi Sakurai
L. P. Singh

Preface

The 4th International Conference on Mathematics and Computing (ICMC 2018) was held at the Indian Institute of Technology (Banaras Hindu University) Varanasi, during January 9–11, 2018. Varanasi, located in the Indian state of Uttar Pradesh, is one of the oldest cities in the world and is well-known for its culture and heritage. The Indian Institute of Technology (BHU) Varanasi is an institution of national importance.

In response to the call for papers for ICMC 2018, 116 papers were submitted for presentation and publication through the proceedings of the conference. The papers were evaluated and ranked on the basis of their significance, novelty, and technical quality by at least two reviewers per paper. After a careful blind refereeing process, 29 papers were selected for inclusion in the conference proceedings. The papers cover current research in cryptography, security, abstract algebra, functional analysis, fluid dynamics, fuzzy modeling, and optimization. ICMC 2018 was supported by eminent researchers from India, USA, UK, Japan, and Finland, among others. The invited speakers from India are recognized leaders in government, industry, and academic institutions such as the Indian Statistical Institute Chennai, IIT Madras, University of Surrey, UK, University of Central Florida, Orlando, USA, University of Turku, Finland, Kettering University, USA, Oracle's Systems Technology, USA, University of Turku, Finland, Haldia Institute of Technology, India, Kyushu University, Fukuoka, Auburn University, Alabama, USA, Banaras Hindu University, India, IIT (BHU), and SCUBE India.

A conference of this kind would not be possible to organize without the full support of different people across different committees. All logistics and general organizational aspects are looked after by the Organizing Committee members, who spent their time and energy in making the conference a reality. We also thank all the Technical Program Committee members and external reviewers for thoroughly reviewing the papers submitted to the conference and sending their constructive suggestions within the deadlines. Our hearty thanks to Springer for agreeing to publish the proceedings in its *Communications in Computer and Information Science* (CCIS) series.

We are truly indebted to the Science and Engineering Research Board (Department of Science and Technology), Council of Scientific and Industrial Research (CSIR), Defense Research and Development Organization (DRDO), and Indian Institute of Technology (BHU) Varanasi and SCUBE India for their financial support, which significantly helped to raise the profile of the conference.

The Organizing Committee is grateful to the research students of the Department of Mathematical Sciences, IIT (BHU), for their tireless support in making the conference a success.

Last but not the least, our sincere thanks go to all the Technical Program Committee members and authors who submitted papers to ICMC 2018 and to all speakers and participants. We fervently hope that the readers will find the proceedings stimulating and inspiring.

March 2018

Debdas Ghosh
Debasis Giri
R. N. Mohapatra
Ekrem Savas
Kouichi Sakurai
L. P. Singh

Organization

Patron

Rajeev Sangal   IIT (BHU), Varanasi, India

General Chairs

P. K. Saxena   DRDO, Delhi, India
P. D. Srivastava   Department of Mathematics, IIT Kharagpur, India

General Co-chairs

U. C. Gupta   Department of Mathematics, IIT Kharagpur, India
L. P. Singh   IIT (BHU), Varanasi, India
Debjani Chakraborty   Department of Mathematics, IIT Kharagpur, India

Program Chairs

Debdas Ghosh   IIT (BHU), Varanasi, India
Ram N. Mahapatra   University of Central Florida, USA
Kouichi Sakurai   Kyushu University, Japan
Debasis Giri   Haldia Institute of Technology, Haldia, India
Ekram Savas   Istanbul Commerce University, Turkey

Organizing Chair

Debdas Ghosh   IIT (BHU), Varanasi, India

Organizing Co-chair

Anuradha Banerjee   IIT (BHU), Varanasi, India

Organizing Secretary

T. Som   IIT (BHU), Varanasi, India

Organizing Joint Secretary

S. Mukhopadhyay   IIT (BHU), Varanasi, India
Subir Das   IIT (BHU), Varanasi, India

Organizing Committee

L. P. Singh	IIT (BHU), Varanasi, India
Rekha Srivastava	IIT (BHU), Varanasi, India
K. N. Rai	IIT (BHU), Varanasi, India
T. Som	IIT (BHU), Varanasi, India
S. K. Pandey	IIT (BHU), Varanasi, India
Shri Ram	IIT (BHU), Varanasi, India
V. S. Pandey	IIT (BHU), Varanasi, India
S. Mukhopadhyay	IIT (BHU), Varanasi, India
S. Das	IIT (BHU), Varanasi, India
S. K. Upadhyay	IIT (BHU), Varanasi, India
Ashokji Gupta	IIT (BHU), Varanasi, India
Rajeev	IIT (BHU), Varanasi, India
Vineeth Kr. Singh	IIT (BHU), Varanasi, India
A. Banerjee	IIT (BHU), Varanasi, India
R. K. Pandey	IIT (BHU), Varanasi, India
D. Ghosh	IIT (BHU), Varanasi, India
Sunil Kumar	IIT (BHU), Varanasi, India
S. Lavanya	IIT (BHU), Varanasi, India

Technical Program Committee

TPC for Mathematics

Abdalah Rababah	Jordan University of Science and Technology, Jordan
Abdon Atangana	University of the Free State, South Africa
Alip Mohammed	The Petroleum Institute, Abu Dhabi
Ameeya Kumar Nayak	IIT Roorkee, India
Anuradha Banerjee	Indian Institute of Technology (BHU), Varanasi, India
Arya K. B. Chand	IIT Madras, India
Ashok Ji Gupta	Indian Institute of Technology (BHU), Varanasi, India
Atanu Manna	IICT Bhadhoi, India
A. Okay Celebi	Yediyepe University, Turkey
Bibaswan Dey	SRM University, India
Carmit Hazay	Bar-Ilan University, Israel
Chris Rodger	Auburn University, Alabama, USA
Conlisk A. Terrence	Ohio State University, USA
Debashree Guha Adhya	IIT Patna, India
Debdas Ghosh	Indian Institute of Technology (BHU), Varanasi, India
Debjani Chakraborty	Indian Institute of Technology, Kharagpur, India
Dipak Jana	Haldia Institute of Technology, India
Dina Sokol	Brooklyn College, USA
Ekrem Savas	Istanbul Commerce University, Turkey
Elena E. Berdysheva	Justus-Liebig University, Giessen, Germany
Emel Aşıcı	Karadeniz Technical University, Turkey

Fahreddin Abdullayev	Mersin University, Turkey
Gopal Chandra Shit	Jadavpur University, Kolkata, India
Gennadii Demidenko	Sobolev Institute of Mathematics, Siberian Branch of Russian Academy of Sciences, Novosibirsk, Russia
Heinrich Begehr	Freie University Berlin, Germany
Hemen Dutta	Gauhati University, Assam, India
Huseyin Cakalli	Maltepe University, Istanbul, Turkey
Huseyin Merdan	TOBB University of Economics and Technology, Turkey
Indiver Gupta	SAG, DRDO, Delhi, India
Kalyan Chakraborty	Harish-Chandra Research Institute, Allahabad, India
K. N. Rai	Indian Institute of Technology (BHU), Varanasi, India
Leopoldo Eduardo Cárdenas-Barrón	Tecnológico de Monterrey, Mexico
Ljubisa Kocinac	University of Nis, Serbia
L. P. Singh	Indian Institute of Technology (BHU), Varanasi, India
Madhumangal Pal	Vidyasagar University, India
Mahpeyker Öztürk	Sakarya University, Turkey
Manoranjan Maiti	Vidyasagar University, India
Margareta Heilmann	University of Wuppertal, Germany
Maria A. Navascues	University of Zaragoza, Spain
Mehmet Gurdal	Suleyman Demirel University, Turkey
Mujahid Abbas	University of Pretoria (UP), Pretoria, South Africa
Moshe Lewenstein	Bar-Ilan University, Israel
Naba Kumar Jana	IIT (ISM) Dhanbad, India
Narendra Govil	Auburn University, Auburn, Alabama, USA
Nita H. Shah	Gujarat University, Navrangpura, Ahmedabad, India
Okay Celebi	Yeditepe University, Istanbul, Turkey
P. D. Srivastava	Indian Institute of Technology Kharagpur, India
P. L. Sharma	Himachal Pradesh University, Shimla, India
Puhan Niladri Bihari	IIT Bhubaneswar, India
Partha Sarathi Roy	Kyushu University, Japan
Prakash Goswami	Indian Institute of Petroleum and Energy, India
Rajeev	Indian Institute of Technology (BHU), Varanasi, India
Rajesh Kumar Pandey	Indian Institute of Technology (BHU), Varanasi, India
Rajendra Pamula	IIT (ISM) Dhanbad, India
Rajesh Prasad	IIT Delhi, India
Ram N. Mohapatra	University of Central Florida, USA
Rekha Srivastava	Indian Institute of Technology (BHU), Varanasi, India
Sadek Bouroubi	University of Sciences and Technology Houari Boumediene, Algeria
S. Das	Indian Institute of Technology (BHU), Varanasi, India
S. Lavanya	Indian Institute of Technology (BHU), Varanasi, India
Shri Ram	Indian Institute of Technology (BHU), Varanasi, India
S. K. Pandey	Indian Institute of Technology (BHU), Varanasi, India
S. K. Upadhyay	Indian Institute of Technology (BHU), Varanasi, India
S. Mukhopadhyay	Indian Institute of Technology (BHU), Varanasi, India

Snehashish Kundu	IIIT Bhubaneswar, India
Somesh Kumar	Indian Institute of Technology Kharagpur, India
Srinivas Chakravarthy	Kettering University, USA
Subrata Bera	NIT Silchar, India
Suchandan Kayal	NIT Rourkela, India
Suneeta Agarwal	Motilal Nehru NIT Allahabad, India
Sunil Kumar	Indian Institute of Technology (BHU), Varanasi, India
Sushil Kumar Bhuiya	IIT Kharagaur, India
T. Som	Indian Institute of Technology (BHU), Varanasi, India
U. C. Gupta	Indian Institute of Technology Kharagpur, India
Valentina E. Balas	Aurel Vlaicu University of Arad, Romania
Vineeth Kr. Singh	Indian Institute of Technology (BHU), Varanasi, India
V. S. Pandey	Indian Institute of Technology (BHU), Varanasi, India

TPC for Computing

Ashok Kumar Das	IIIT Hyderabad, India
Athanasios V. Vasilakos	Luleå University of Technology, Sweden
Bart Mennink	Radboud University, The Netherlands
Bidyut Patra	NIT Rourkela, India
Bimal Roy	ISI Kolkata, India
Biswapati Jana	Vidyasagar University, India
Cheng Chen-Mou	National Taiwan University, Taiwan
Christina Boura	Université de Versailles Saint-Quentin-en-Yvelines, France
Chung-Huang Yang	National Kaohsiung Normal University, Taiwan
David Chadwick	University of Kent, UK
Debasis Giri	Haldia Institute of Technology, India
Debiao He	Wuhan University, China
Dipanwita Roy Chowdhury	IIT Kharagpur, India
Donghoon Chang	IIIT-Delhi, India
Dung Duong	Kyushu University, Japan
Elena Berdysheva	Mathematisches Institut
Fagen Li	University of Electronic Science and Technology, China
Gerardo Pelosi	Politecnico di Milano, Leonardo da Vinci, Italy
H. P. Gupta	IIT (BHU) Varanasi, India
Hafizul Islam	IIIT Kalyani, India
Hiroaki Kikuchi	Meiji University, Japan
Hung-Min SUN	National Tsing Hua University, Taiwan
Jaydeb Bhaumik	Haldia Institute of Technology, India
Joonsang Baek	University of Wollongong, Australia
Junwei Zhu	Wuhan University of Technology, China
Indivar Gupta	Scientific Analysis Group, Delhi, India
Kazuhiro Yokoyama	Rikkyo University, Japan

Contents

Security and Coding Theory

Achieving Better Security Using Nonlinear Cellular Automata as a Cryptographic Primitive

Swapan Maiti$^{(\boxtimes)}$ (iD) and Dipanwita Roy Chowdhury$^{(\boxtimes)}$

Indian Institute of Technology Kharagpur, Kharagpur, India
swapankumar_maiti@yahoo.co.in, drc@cse.iitkgp.ernet.in

Abstract. Nonlinear functions are essential in different crypto-primitives as they play an important role on the security of a cipher design. Wolfram identified Rule 30 as a powerful nonlinear function for cryptographic applications. However, Meier and Staffelbach mounted an attack (MS attack) against Rule 30 Cellular Automata (CA). MS attack is a real threat on a CA based system. Nonlinear rules as well as maximum period CA increase randomness property. In this work, nonlinear rules of maximum period nonlinear hybrid CA (M-NHCA) are studied and it is shown to be a better crypto-primitive than Rule 30 CA. It has also been analysed that the M-NHCA with single nonlinearity injection proposed in the literature is vulnerable against MS attack, whereas M-NHCA with multiple nonlinearity injections provide better cryptographic primitives and they are also secure against MS attack.

Keywords: Cellular Automata · Maximum period nonlinear CA
Meier and Staffelbach attack · Nonlinear functions

1 Introduction

Cellular Automata (CA) have long been of interest to researchers for their theoretical properties and practical applications. In 1986, Wolfram first applied CA in pseudorandom number generation [16]. In the last three decades, one-dimensional (1-D) CA based Pseudorandom Number Generators (PRNGs) have been extensively studied [2,14].

Maximum period linear CA (LCA) increase randomness property as well as provide security against different side channel attacks like power attack, timing attack etc., but a linear CA is known to be insecure. Therefore, nonlinearity is very essential in cryptographic applications. Wolfram proposed Rule 30 as a better cryptographic primitive and it was used in non-linear CA (NLCA) construction for cryptographic applications [15,16]. However, Meier and Staffelbach developed an algorithm (MS attack) and it has been shown in [12] that the NLCA based on Rule 30 is vulnerable. All the 256 elementary 3-neighborhood CA rules were analysed in [5,11], and it was found out that no nonlinear elementary CA

© Springer Nature Singapore Pte Ltd. 2018
D. Ghosh et al. (Eds.): ICMC 2018, CCIS 834, pp. 3–15, 2018.
https://doi.org/10.1007/978-981-13-0023-3_1

rule is correlation immune. In [7], 4-neighborhood nonlinear CA are introduced and their cryptographic properties have also been studied. However, because of left skewed rule, the diffusion rate of left neighbor cell and that of right neighbor cell with respect to every cell is not same. Moreover, this nonlinear CA does not provide a maximum length cycle. In [8], Lacharme et al. analysed all the 65536 CA rules with four variables to find 200 nonlinear balanced functions which are 1-resilient. In [9], nonlinear and resilient rules are selected from 5-neighborhood bipermutive CA rules.

In [6], maximum period nonlinear hybrid CA (M-NHCA) with single non-linearity injection is proposed, where nonlinear rule of the injected cell is balanced and 1-resilient (or 2-resilient). The M-NHCA may become a better crypto-primitive than Rule 30 CA and other nonlinear CA. The main contribution of this work can be summarized as below:

- Study of nonlinear rules of M-NHCA with single nonlinearity injection and their security analysis.
- Security analysis of M-NHCA with multiple nonlinearity injections.

This paper is organized as follows. Following the introduction, basics of CA, cryptographic terms and primitives are defined in Sect. 2. MS attack is also stated in this section as the pre-requisite of our work. Section 3 presents security analysis of M-NHCA [6] with single nonlinearity injection. In Sect. 4, M-NHCA is extended with multiple nonlinearity injections and their security analysis is shown. This section compares M-NHCA with Rule 30 CA with respect to non-linearity and other related work. Finally, the paper is concluded in Sect. 5.

2   Preliminaries

This section presents some basics of Cellular automata and some definitions involving cryptographic terms and primitives with examples, and MS attack on Rule 30 CA.

2.1   Basics of Cellular Automata

Cellular Automata (CA) are studied as mathematical model for self organizing statistical systems [13]. One-dimensional CA based random number generators have been extensively studied in the past [4,11,16]. One-dimensional CA can be considered as an array of 1-bit memory elements. Formally, for a 3-neighborhood CA, the neighbor set of ith cell is defined as $N(i) = \{s_{i-1}, s_i, s_{i+1}\}$ and the state transition function of ith cell is as follows: $s_i^{t+1} = f_i(s_{i-1}^t, s_i^t, s_{i+1}^t)$, where, s_i^t denotes the current state of the ith cell at time step t and s_i^{t+1} denotes the next state of the ith cell at time step $t+1$ and f_i denotes some combinatorial logic for i^{th} cell. Since, a 3-neighborhood CA having two states (0 or 1) in each cell, can have $2^3 = 8$ possible binary states, there are total $2^{2^3} = 256$ possible Boolean functions, called rules. Each rule can be represented as an decimal integer from 0 to 255 [4]. If the combinatorial logic contains only boolean XOR operation, then it is called linear or additive rule. Some of the additive rules are 0, 60, 90, 102, 150 etc. Moreover, if the combinatorial logic contains AND/OR operations, then it is called nonlinear rule. For example, Rule 30 is a nonlinear rule. An n-cell CA with cells $\{s_1, s_2, \cdots, s_n\}$ is called a null boundary CA if $s_{n+1} = 0$ and $s_0 = 0$, and a periodic boundary CA if $s_{n+1} = s_1$. A CA is called uniform, if all cells follow the same rule. Otherwise, it is called non-uniform or hybrid CA. The CA where all cells follow linear rules but not the same linear rules are called linear hybrid CA (LHCA). Similarly, the CA where some cell follows nonlinear rules are called nonlinear hybrid CA (NHCA). The sequence of corresponding rules of CA cells is called rule vector for the CA.

2.2   Cryptographic Terms and Primitives

Pseudorandom Sequence: A bit-sequence is pseudorandom if it cannot be distinguished from a truly random sequence by any efficient polynomial time algorithm.

Affine Function: A Boolean function which involves its input variables in linear combinations (i.e., combinations involving $\oplus$) only, is called an affine function. For example, $f(x_1, x_2) = x_1 \oplus x_2$ is an affine function, whereas the function, $f(x_1, x_2) = x_1 \oplus x_2 \oplus x_1 \cdot x_2$ is not an affine function, where $\cdot$ is the Boolean 'AND' operation.

Hamming Weight: Number of 1's in a Boolean function's truth table is called the Hamming weight of the function.

Balanced Boolean Function: If the Hamming weight of a Boolean function of n variables is 2^{n-1}, it is called a balanced Boolean function. Thus, $f(x_1, x_2) = x_1 \oplus x_2$ is balanced, whereas $f(x_1, x_2) = x_1 \cdot x_2$ is not balanced.

Hamming Distance: Hamming weight of $f_1 \oplus f_2$ is called the Hamming distance between f_1 and f_2. Thus, Hamming distance between $f_1(x_1, x_2) = x_1 \oplus x_2$ and $f_2(x_1, x_2) = x_1 \cdot x_2$ is 3.

Nonlinearity: The minimum of the Hamming distances between a Boolean function f and all affine functions involving its input variables is known as the nonlinearity of the function. Hence, nonlinearity of $f(x_1, x_2) = x_1 \cdot x_2$ is 1.

Resiliency: A Boolean function of n variables is called to have a resiliency t, if for all possible subsets of variables of size less than or equal to t, on fixing values of those variables in every possible subset, the resultant Boolean function still remains balanced. For example, resiliency of $f(x_1, x_2) = x_1 \oplus x_2$ is 1, but resiliency of $f(x_1, x_2) = x_1 \cdot x_2$ is 0.

Algebraic Degree: The algebraic degree of a Boolean function is the number of variables in the highest order term with non-zero coefficient. Thus, algebraic degree of $f(x_1, x_2) = x_1 \oplus x_2 \oplus x_1 \cdot x_2$ is 2.

Unicity Distance: The unicity distance of a cryptosystem is defined to be a value of n, denoted by n_0, at which the expected number of spurious keys (i.e. possible incorrect keys) becomes zero.

In the next subsection, Meier and Staffelbach attack (MS attack) [12] on Rule 30 CA is explained briefly as the pre-requisite of our work.

2.3   Meier and Staffelbach Attack (MS Attack)

In [12], the attack is a known plaintext attack where the keys are chosen as seed of the cellular automaton of size n (i.e. the size of the keys is n). The problem of cryptanalysis is in determining the seed (or the keys) from the produced output sequence. In [12], a nonlinear CA denoted by $\{s_1, s_2, \cdots, s_n\}$ of width $n = 2N + 1$ is considered. The site vector of the nonlinear CA (i.e. contents of the CA) at time step t is $\langle s_{i-N}^t, \cdots, s_{i-1}^t, s_i^t, s_{i+1}^t, \cdots, s_{i+N}^t \rangle$ as shown in Fig. 1. The bit-sequence of ith cell for N cycles, denoted by $\{s_i^t\}$ that is $\langle s_i^t, s_i^{t+1}, \cdots, s_i^{t+N} \rangle$, is the known output sequence, where $i = N + 1$. The site vector, which is the key of this attack, forms a triangle along with the temporal sequence column (i.e. $\{s_i^t\}$). From the knowledge of two adjacent columns in the triangle, that is, temporal sequence column (i.e. $\{s_i^t\}$) and right adjacent sequence column (i.e. $\{s_{i+1}^t\}$) or temporal sequence column (i.e. $\{s_i^t\}$) and left adjacent sequence column (i.e. $\{s_{i-1}^t\}$), one can determine the seed. Every cell of the null boundary nonlinear CA follows Rule 30. The state transition function of Rule 30 is as follows: $s_i^{t+1} = s_{i-1}^t \oplus (s_i^t + s_{i+1}^t)$, where s_i^t is the current state and s_i^{t+1} is the next state of the ith cell.

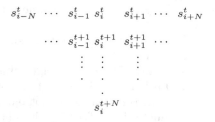

Fig. 1. Determination of the seed

First, a random seed $\langle s_{i+1}^t, \cdots, s_{i+N}^t \rangle$ is generated. In the completion forwards process, using the random seed and Rule 30 formula, $s_{i+1}^{t+1}, s_{i+2}^{t+1}, \cdots,$ s_{i+N-1}^{t+1} can be easily computed as it is only the unknown item in the expression of Rule 30. In this way, the random seed together with temporal sequence column forms the right triangle as shown in Fig. 1. The above formula can be written in another way: $s_{i-1}^t = s_i^{t+1} \oplus (s_i^t + s_{i+1}^t)$. The knowledge of right adjacent column and the temporal sequence column can compute the left triangle of the temporal sequence column and eventually, determine the seed $\langle s_{i-N}^t, \cdots, s_{i-1}^t \rangle$ (completion backwards process [12]).

Eventually, the CA is loaded with the computed seed $\langle s_{i-N}^t, \cdots, s_{i-1}^t, s_i^t, s_{i+1}^t, \cdots, s_{i+N}^t \rangle$ and produce the output sequence; the algorithm terminates if the produced sequence coincides with the known output sequence, otherwise, this process repeats for another choice of the random seed. There are 2^N ($\approx 2^{\frac{n}{2}}$) choices for random seed, so the required time complexity is $O(2^N)$ (i.e. $O(2^{\frac{n}{2}})$).

3   Security Analysis of Synthesized M-NHCA with Single Nonlinearity Injection

In this section, cryptographic properties of nonlinear functions of synthesized M-NHCA introduced in [6] are studied and the security analysis of the synthesized M-NHCA is presented. Before presenting our work on the security analysis, the synthesis of M-NHCA is briefly described with an example shown below.

3.1   Synthesis of M-NHCA

The algorithm [6] to synthesize a maximum period NHCA (M-NHCA) is explained briefly as the pre-requisite of our work. The following example clearly illustrates how a M-NHCA can be synthesized by injecting nonlinearity into a selected position of a maximum period LHCA.

Example 1. Let us consider a 3-neighborhood 7-bit maximum period null-boundary LHCA $\mathcal{L}'$ denoted by $\{x_0, x_1, x_2, x_3, x_4, x_5, x_6\}$ of a characteristic polynomial (primitive polynomial [4]) $x^7 + x + 1$ with rule vector $[1, 0, 1, 1, 0, 0, 1]$, where $0 \equiv$ Rule 90 and $1 \equiv$ Rule 150. Let nonlinearity be injected at position 3 (i.e. on the cell x_3) with the nonlinear function $f_N(x_1^t, x_5^t) = (x_1^t \cdot x_5^t)$. The updated state transition function (nonlinear) is $x_3^{t+1} = x_2^t \oplus x_3^t \oplus x_4^t \oplus (x_1^t \cdot x_5^t)$ and other functions x_i^{t+1}, for $i = 0, 1, 2, 4, 5, 6$, can be generated by 90/150 rules.

However, as mentioned in [6], to ensure maximum periodicity the neighboring transition functions need to be updated with the same nonlinear function $f_N(x_1^t, x_5^t) = (x_1^t \cdot x_5^t)$ by applying one cell shifting operations and an additional Boolean function $f_N(x_1^{t+1}, x_5^{t+1}) = ((x_0^t \oplus x_2^t) \cdot (x_4^t \oplus x_6^t))$ needs to be injected to the same inject position 3. Thus, the functions x_i^{t+1}, for $i = 0, 1, 5, 6$, can be generated by 90/150 rules and the updated state transition functions (nonlinear) of M-NHCA $\mathcal{N}'$ can be generated as follows:

$$x_2^{t+1} = x_1^t \oplus x_2^t \oplus x_3^t \oplus (x_1^t \cdot x_5^t)$$
$$x_3^{t+1} = x_2^t \oplus x_3^t \oplus x_4^t \oplus (x_1^t \cdot x_5^t) \oplus ((x_0^t \oplus x_2^t) \cdot (x_4^t \oplus x_6^t))$$
$$x_4^{t+1} = x_3^t \oplus x_5^t \oplus (x_1^t \cdot x_5^t)$$

3.2   Cryptographic Properties of Non-linear Rules

Balancedness is an important property of cryptographic Boolean functions. Indeed, resiliency is a balancedness test for certain functions obtained from the target cryptographic Boolean functions. Lack of resiliency implies correlation among input and output bits. CA with linear rules provide best resiliency. But this kind of CA can be trivially crytanalyzed by linearization. Nonlinearity is another important property of cryptographic Boolean functions. Like resiliency, nonlinearity should also increase with each iteration for a cryptographically suitable CA. It is difficult to have a balance between them in CA designs. It turns out that, only hybrid CA can be employed in providing both good nonlinearity and resiliency.

Table 1. Cryptographic properties of M-NHCA with iterations

Itr#	Nonlinearity							Balancedness							Resiliency						
	x_0	x_1	x_2	x_3	x_4	x_5	x_6	x_0	x_1	x_2	x_3	x_4	x_5	x_6	x_0	x_1	x_2	x_3	x_4	x_5	x_6
1	0	0	4	48	2	0	0	True	True	True	True	True	True	True	1	1	1	2	0	1	1
2	0	8	0	32	8	4	0	True	True	True	True	True	True	True	1	2	3	1	2	1	1
3	4	0	0	32	0	16	8	True	True	True	True	True	True	True	1	1	1	1	3	3	2
4	4	4	0	8	16	2	0	True	True	True	True	True	True	True	1	1	1	1	3	3	3
5	0	4	4	32	8	8	4	True	True	True	True	True	True	True	2	1	1	2	2	2	1
6	16	16	16	32	0	0	0	True	True	True	True	True	True	True	3	3	3	1	4	4	3

Table 1 shows the cryptographic properties of all rules of the M-NHCA shown in Example 1 with iterations. The M-NHCA generates balanced outputs, but the increase of resiliency and nonlinearity with iterations is not regular.

3.3   Vulnerability Against MS Attack

In this section, we show that the synthesized M-NHCA with single nonlinearity injection is not secure against MS attack. In this work, we consider a 3-neighborhood n-bit maximum period null-boundary LHCA $\mathcal{L}'$ denoted by $\{x_0, x_1, \cdots, x_{n-1}\}$ with rule vector $[d_0, d_1, \cdots, d_{n-1}]$, where $d_i = 0$ if x_i follows Rule 90 and $d_i = 1$ if x_i follows Rule 150. Let nonlinearity be injected at position j with the nonlinear function $f_N(x_{j-2}^t, x_{j+2}^t) = (x_{j-2}^t \cdot x_{j+2}^t)$. In the synthesized M-NHCA $\mathcal{N}'$, the state transition functions (nonlinear) of neighboring cells around the non-linearity position j are as follows:

$$x_{j-1}^{t+1} = x_{j-2}^t \oplus d_{j-1} \cdot x_{j-1}^t \oplus x_j^t \oplus (x_{j-2}^t \cdot x_{j+2}^t) \tag{1}$$

$$x_j^{t+1} = x_{j-1}^t \oplus d_j \cdot x_j^t \oplus x_{j+1}^t \oplus d_j \cdot (x_{j-2}^t \cdot x_{j+2}^t)$$
$$\oplus ((x_{j-3}^t \oplus d_{j-2} \cdot x_{j-2}^t \oplus x_{j-1}^t) \cdot (x_{j+1}^t \oplus d_{j+2} \cdot x_{j+2}^t \oplus x_{j+3}^t)) \tag{2}$$

$$x_{j+1}^{t+1} = x_j^t \oplus d_{j+1} \cdot x_{j+1}^t \oplus x_{j+2}^t \oplus (x_{j-2}^t \cdot x_{j+2}^t) \tag{3}$$

where $\langle x_0^t, x_1^t, \cdots, x_{n-1}^t \rangle$ is the site vector of $\mathcal{N}'$ at time step t and all other cells x_i, $0 \leq i \leq j-2$ and $j+2 \leq i \leq n-1$, of synthesized NHCA follow Rule 90/150 as the corresponding cells of $\mathcal{L}'$ follow. The attack is a known plaintext attack. The output sequence $\{x_i^t\}$ (i.e. the temporal sequence $\{x_{j-1}^t\}$) is known upto the unicity distance N shown in Table 2, where $i = j - 1$.

Our aim is to determine the seed $\langle x_0^t, x_1^t, \cdots, x_{i-1}^t, x_i^t, x_{i+1}^t, \cdots, x_{n-1}^t \rangle$ from the knowledge of given output sequence $\{x_i^t\}$. A random seed $\langle x_{i+1}^t, \cdots, x_{n-1}^t \rangle$ is generated out of $2^{n-(i+1)}$ possibilities. Now, x_{j-2}^t can be determined from the Eq. (1) with probability $\frac{1}{2}$. In the completion forwards process (i.e. left to right approach), x_j^{t+1}, x_{j+1}^{t+1} can be computed using the Eqs. (2) and (3) respectively, since in every expression only one item is unknown like Rule 30. x_{j+2}^{t+1}, x_{j+3}^{t+1}, $\cdots$, x_{n-1}^{t+1} can be computed as per 3-neighborhood 90/150 rule. For next time step (i.e. at time step $t+2$) we can compute all above values in the similar way. In this way, right triangle of the temporal sequence column (i.e. $\{x_i^t\}$), shown in Table 2,

Table 2. Determination of the seed for M-NHCA $\mathcal{N}'$

x_0^t	$\cdots$	x_{i-6}^t	x_{i-5}^t	$\cdots$	x_{i-1}^t	x_i^t	x_{i+1}^t *	$\cdots$ ***	x_{n-1}^t *
x_0^{t+1}	$\cdots$	$\cdots$	$\cdots$	$\cdots$	x_{i-1}^{t+1}	x_i^{t+1}	x_{i+1}^{t+1}	$\cdots$	x_{n-1}^{t+1}
	$\cdots$	$\cdots$	$\cdots$	$\cdots$	x_{i-1}^{t+2}	x_i^{t+2}	x_{i+1}^{t+2}	$\cdots$	
	$\cdots$	$\cdots$	$\cdots$	$\cdots$	x_{i-1}^{t+3}	x_i^{t+3}	x_{i+1}^{t+3}	$\cdots$	
	$\vdots$	$\vdots$	$\vdots$	$\vdots$	$\vdots$	$\vdots$	$\vdots$	$\vdots$	
	$\cdots$	$\cdots$	$\cdots$		$.$	$.$	$.$	$\cdots$	
		$\cdots$	$\cdots$		$.$	$.$	$.$	$\cdots$	
			$\cdots$		$.$	$.$	$.$	$\cdots$	
					x_{i-1}^{t+N-1}	$.$	x_{i+1}^{t+N-1}		
						x_i^{t+N}			

'*' represents "guess" value

can be determined. Because of single nonlinearity injection and since all CA cells in the opposite side of injection point in respect of temporal sequence column follow Rule 90/150 as in LHCA, the only knowledge of right adjacent column (i.e. $\{x_{i+1}^t\}$) in the right triangle together with temporal sequence column can determine the seed $\langle x_0^t, \cdots, x_{i-1}^t \rangle$. The columns $\{x_{j-3}^t\}, \{x_{j-4}^t\}, \cdots, \{x_0^t\}$ can be computed as per 3-neighborhood 90/150 rule. Here, each column is computed by bottom-up approach. In this way left triangle of the temporal sequence column (i.e. $\{x_i^t\}$) can be formed (completion backwards process) and hence, the seed $\langle x_0^t, \cdots, x_{i-1}^t \rangle$ can be determined.

Eventually, the CA is loaded with the computed seed $\langle x_0^t, \cdots, x_{i-1}^t, x_i^t, x_{i+1}^t, \cdots, x_{n-1}^t \rangle$ and produce the output sequence; the algorithm terminates if the produced sequence coincides with the given temporal sequence, otherwise, this process repeats for another choice of random seed $\langle x_{i+1}^t, \cdots, x_{n-1}^t \rangle$.

The random seed $\langle x_{i+1}^t, \cdots, x_{n-1}^t \rangle$ can be chosen with $2^{n-(i+1)}$ possibilities. Since, x_{j-2}^t is determined from the Eq. (1) with probability $\frac{1}{2}$, therefore, for the column $j - 2$, $\frac{n-(i+1)}{2}$ values can be computed deterministically and other $\frac{n-(i+1)}{2}$ values can be chosen randomly with $2^{\frac{n-(i+1)}{2}}$ possibilities. The required time complexity is: $2^{n-(i+1)} \cdot 2^{\frac{n-(i+1)}{2}} = 2^{\frac{3}{2}(n-1-i)} = 2^{n-\frac{n+3}{4}}$, where $i = j - 1$ and $i = \frac{n-1}{2}$, the middle cell position of the CA. Hence, the required time is less than 2^n (reqd. for exhaustive search).

4   M-NHCA with Multiple Nonlinearity Injections

M-NHCA with single nonlinearity injection described in Sect. 3 is not secure against MS attack. In this section, we extend M-NHCA with multiple nonlinearity injections and study their cryptographic properties, and it is also shown that M-NHCA with multiple nonlinearity injections is secure against MS attack. Here, we consider an n-cell maximum period LHCA denoted by

Table 3. Nonlinearity comparison w.r.t. injection points

LHCA polynomial	Nonlinearity inject position(s)	CA cell for nonlinearity	Nonlinearity with iterations						
			1	2	3	4	5	6	7
7, 1, 0	3	x_3	48	32	32	8	32	32	16
10, 3, 0	3	x_3	48	8	64	128	128	256	256
	3, 7	x_3	48	16	64	192	256	256	384
12, 7, 4, 3, 0	3	x_3	48	64	64	32	32	512	512
	3, 8	x_3	48	64	128	32	48	768	768
16, 5, 3, 2, 0	5	x_5	48	32	512	512	512	1024	1024
	5, 9	x_5	48	64	1024	512	1024	1024	1024
32, 28, 27, 1, 0	11	x_{11}	16	64	512	2048	2048	3072	3072
	7, 11, 15, 19	x_{11}	16	256	2048	3072	4096	4096	4096

$\{x_0, x_1, \cdots, x_{n-2}, x_{n-1}\}$. For multiple nonlinearity injections, we follow the following two criteria: (1) Non-linearity can be injected in cell position i, $2 \leq i \leq n - 3$ such that the injected nonlinear function $f_N(x_{i-2}^t, x_{i+2}^t) = (x_{i-2}^t \cdot x_{i+2}^t)$ can be formed properly. (2) To retain the maximum length cycle, there must be at least three cells in between any two non-linearity inject positions; that is, if i and j be two inject positions then there must be $|i - j| \geq 4$.

4.1   Achieving Better Nonlinearity

In this section, we compute nonlinearity of some synthesized M-NHCA with single and multiple nonlinearity injection(s). The result is shown in Table 3. The underlying maximum period LHCA is synthesized [3] from a primitive polynomial represented as a listing of non-zero coefficients. For example, the set (7, 1, 0) represents the CA polynomial $x^7 + x + 1$. The set (i, j, k) in the 2nd column of Table 3 represents that nonlinearity is injected in i^{th}, j^{th} and k^{th} cell positions simultaneously. Table 3 clearly illustrates that the nonlinearity of M-NHCA increases more in multiple injections than single injection.

4.2   Diffusion and Randomness Properties

Nonlinear function of the nonlinearity injected cell of synthesized M-NHCA is a 7-neighborhood rule as described in Subsect. 3.1. Therefore, the diffusion rate of cell contents of M-NHCA is more than that of 3-neighborhood CA. To test the randomness property of the M-NHCA, 100 bit-streams with each stream of 10,00,000 bits are generated from each cell of a 32-bit M-NHCA which is synthesized from a 32-bit 90/150 LHCA of CA polynomial (primitive polynomial [4]) $x^{32} + x^{28} + x^{27} + x + 1$, and are tested by NIST test suite [1]. Table 4 shows high randomness property of the generated bit-streams.

Table 4. Results of NIST-statistical test suite for randomness of M-NHCA

Test name	Status	Test name	Status
Frequency test	Pass	Cumulative sums	Pass
Block frequency (block len. = 128)	Pass	Runs	Pass
Non-overlapping template (block len. = 9)	Pass	Longest run	Pass
Overlapping template (block len. = 9)	Pass	FFT	Pass
Approximate entropy (block len. = 10)	Pass	Universal	Pass
Random excursions test	Pass	Serial	Pass
Random excursions variant test	Pass		

4.3 Resistance Against MS Attack

A new design construction of a stream cipher is presented in [10] based on CA, and the authors have shown its security analysis including MS attack resistance of the cipher. MS attack is a real threat on a CA based system. In this work, the detailed proof of MS attack resistance of a synthesized M-NHCA is shown.

Let us consider a 3-neighborhood n-bit maximum period null-boundary LHCA $\mathcal{L}'$ denoted by $\{x_0, x_1, \cdots, x_{n-1}\}$ with rule vector $[d_0, d_1, \cdots, d_{n-1}]$, where $d_i = 0$ if x_i follows Rule 90 and $d_i = 1$ if x_i follows Rule 150. Let nonlinearity be injected at positions j and k with the nonlinear functions $f_N(x_{j-2}^t, x_{j+2}^t) = (x_{j-2}^t \cdot x_{j+2}^t)$ and $f_N(x_{k-2}^t, x_{k+2}^t) = (x_{k-2}^t \cdot x_{k+2}^t)$ respectively, where $k - j = 4$ which is the 2nd criteria for multiple nonlinearity injections. The state transition functions (nonlinear) of neighboring cells of synthesized M-NHCA $\mathcal{N}'$ around the non-linearity positions j and k respectively, are as follows: for j^{th} position,

$$x_{j-1}^{t+1} = x_{j-2}^t \oplus d_{j-1} \cdot x_{j-1}^t \oplus x_j^t \oplus (x_{j-2}^t \cdot x_{j+2}^t) \tag{4}$$

$$x_j^{t+1} = x_{j-1}^t \oplus d_j \cdot x_j^t \oplus x_{j+1}^t \oplus d_j \cdot (x_{j-2}^t \cdot x_{j+2}^t)$$
$$\oplus ((x_{j-3}^t \oplus d_{j-2} \cdot x_{j-2}^t \oplus x_{j-1}^t) \cdot (x_{j+1}^t \oplus d_{j+2} \cdot x_{j+2}^t \oplus x_{j+3}^t)) \tag{5}$$

$$x_{j+1}^{t+1} = x_j^t \oplus d_{j+1} \cdot x_{j+1}^t \oplus x_{j+2}^t \oplus (x_{j-2}^t \cdot x_{j+2}^t) \tag{6}$$

Similarly, for k^{th} position, the expressions (nonlinear) for x_{k-1}^{t+1}, x_k^{t+1} and x_{k+1}^{t+1} can be generated as 2nd rule set, where $(x_0^t, x_1^t, \cdots, x_{n-1}^t)$ is the site vector of $\mathcal{N}'$ at time step t. Now, this 2nd rule set can be stated with $k = j+4$ as follows:

$$x_{j+3}^{t+1} = x_{j+2}^t \oplus d_{j+3} \cdot x_{j+3}^t \oplus x_{j+4}^t \oplus (x_{j+2}^t \cdot x_{j+6}^t) \tag{7}$$

$$x_{j+4}^{t+1} = x_{j+3}^t \oplus d_{j+4} \cdot x_{j+4}^t \oplus x_{j+5}^t \oplus d_{j+4} \cdot (x_{j+2}^t \cdot x_{j+6}^t)$$
$$\oplus ((x_{j+1}^t \oplus d_{j+2} \cdot x_{j+2}^t \oplus x_{j+3}^t) \cdot (x_{j+5}^t \oplus d_{j+6} \cdot x_{j+6}^t \oplus x_{j+7}^t)) \tag{8}$$

$$x_{j+5}^{t+1} = x_{j+4}^t \oplus d_{j+5} \cdot x_{j+5}^t \oplus x_{j+6}^t \oplus (x_{j+2}^t \cdot x_{j+6}^t) \tag{9}$$

All other cells x_i, for $0 \leq i \leq j-2$, $i = j+2$ and $j+6 \leq i \leq n-1$, of $\mathcal{N}'$ follow Rule 90/150 as corresponding cells of $\mathcal{L}'$ follow. Our aim is to determine the

Table 5. Determination of the seed for M-NHCA $\mathcal{N}'$

x_0^t	⋯	x_{i-6}^t	x_{i-5}^t	⋯	x_{i-1}^t	x_i^t	x_{i+1}^t	⋯	x_{n-1}^t
		*	*				*	***	*
x_0^{t+1}	⋯	*	*	⋯	x_{i-1}^{t+1}	x_i^{t+1}	x_{i+1}^{t+1}	⋯	x_{n-1}^{t+1}
	⋯	*	*	⋯	x_{i-1}^{t+2}	x_i^{t+2}	x_{i+1}^{t+2}	⋯	
	⋯	*	*	⋯	x_{i-1}^{t+3}	x_i^{t+3}	x_{i+1}^{t+3}	⋯	
⋮	⋮	⋮	⋮	⋮	⋮	⋮	⋮	⋮	
		*	*	⋯	.	.	.	⋯	
			*	⋯	.	.	.	⋯	
				⋯	.	.	.	⋯	
					x_{i-1}^{t+N-1}	.	x_{i+1}^{t+N-1}		
						x_i^{t+N}			

'*' represents "guess" value

seed $\langle x_0^t, x_1^t, \cdots, x_{i-1}^t, x_i^t, x_{i+1}^t, \cdots, x_{n-1}^t \rangle$ from the knowledge of given output sequence $\{x_i^t\}$ (i.e. the temporal sequence $\{x_{j+3}^t\}$) upto the unicity distance N shown in Table 5, where $i = j + 3$ and $i = k - 1$ since $k - j = 4$.

We choose a random seed $\langle x_{i+1}^t, \cdots, x_{n-1}^t \rangle$ out of $2^{n-(i+1)}$ possibilities. Now, x_{j+2}^t can be determined from the Eq. (7) with probability $\frac{1}{2}$. In the completion forwards process (i.e. left to right approach), x_{j+4}^{t+1}, x_{j+5}^{t+1} can be computed using the Eqs. (8) and (9) respectively, in the 2nd rule set. $x_{j+6}^{t+1}, x_{j+7}^{t+1}, \cdots, x_{n-1}^{t+1}$ can be computed as per 3-neighborhood 90/150 rule. For next time step (i.e. at time step $t + 2$) we can compute all above values again using the 2nd rule set. In this way, right triangle of the temporal sequence column (i.e. $\{x_i^t\}$), shown in Table 5, can be determined. Here, the only knowledge of right adjacent column in the right triangle together with temporal sequence column can not determine the seed $\langle x_0^t, \cdots, x_{i-1}^t \rangle$. The column $\{x_{j+1}^t\}$ can be computed using the state transition function of x_{j+2}^{t+1}. The column $\{x_j^t\}$ can only be computed from the Eq. (6) if the column $\{x_{j-2}^t\}$ (i.e. $\{x_{i-5}^t\}$) is chosen as random out of 2^{j+1} possibilities, because $\{x_{j-2}^t\}$ is unknown. The column $\{x_{j-1}^t\}$ can only be computed from Eq. (5) of the 1st rule set if the column $\{x_{j-3}^t\}$ (i.e. $\{x_{i-6}^t\}$) is chosen as random out of 2^j possibilities, because $\{x_{j-3}^t\}$ is unknown. The column $\{x_{j-4}^t\}, \{x_{j-5}^t\}, \cdots, \{x_0^t\}$ can be computed as per 3-neighborhood 90/150 rule. Here, each column is computed by bottom-up approach. In this way left triangle of the temporal sequence column (i.e. $\{x_i^t\}$) can be formed (completion backwards process) and hence, the seed $\langle x_0^t, \cdots, x_{i-1}^t \rangle$ can be determined.

Eventually, the CA is loaded with the computed seed $\langle x_0^t, \cdots, x_{i-1}^t, x_i^t, x_{i+1}^t, \cdots, x_{n-1}^t \rangle$ and produce the output sequence; the algorithm terminates if the produced sequence coincides with the given temporal sequence, otherwise, this process repeats for another choice of random seed $\langle x_{i+1}^t, \cdots, x_{n-1}^t \rangle$.

The random seed $\langle x_{i+1}^t, \cdots, x_{n-1}^t \rangle$ can be chosen with $2^{n-(i+1)}$ possibilities. Since, x_{j+2}^t is determined from the Eq. (7) with probability $\frac{1}{2}$, therefore, for the column $j + 2$, $\frac{n-(i+1)}{2}$ values can be computed deterministically and other $\frac{n-(i+1)}{2}$ values can be chosen randomly with $2^{\frac{n-(i+1)}{2}}$ possibilities. The column $j - 2$ is chosen as random out of 2^{j+1} possibilities. The column $j - 3$ is chosen as random out of 2^j possibilities. Therefore, the required time complexity is:

$$2^{n-(i+1)} \cdot 2^{\frac{n-(i+1)}{2}} \cdot 2^{j+1} \cdot 2^j = 2^{\frac{3}{2}(n-i-1)} \cdot 2^{2j+1} = 2^{n+\frac{3}{4}(n-9)}$$

where $j = i - 3$ and $i = \frac{n-1}{2}$, the middle cell position of the CA. Hence, the required time is greater than 2^n (reqd. for exhaustive search) for $n > 9$.

Following the similar approach, we can determine the seed $\langle x_{i+1}^t, \cdots, x_{n-1}^t \rangle$ from the given output sequence $\{x_i^t\}$ (i.e. the temporal sequence $\{x_{j+3}^t\}$) upto the unicity distance N, by guessing the seed $\langle x_0^t, \cdots, x_{i-1}^t \rangle$ out of 2^i possibilities. In the completion forwards process, the left triangle of the temporal sequence column (i.e. $\{x_i^t\}$) can be determined. In the completion backwards process, the right triangle of the temporal sequence column (i.e. $\{x_i^t\}$) can be formed. The random seed $\langle x_0^t, \cdots, x_{i-1}^t \rangle$ can be chosen with 2^i possibilities. The column $j + 6$ (i.e. $k + 2$) is chosen as random out of 2^{n-k} possibilities. The column $j + 7$ (i.e. $k + 3$) is chosen as random out of 2^{n-k-1} possibilities. Therefore, the required time complexity is:

$$2^i \cdot 2^{n-k} \cdot 2^{n-k-1} = 2^{i+2n-2k-1} = 2^{n+\frac{n-5}{2}}$$

where $k = i + 1$ and $i = \frac{n-1}{2}$, the middle cell position of the CA. Hence, the required time is greater than 2^n (reqd. for exhaustive search) for $n > 5$.

4.4 Comparison with Rule 30 CA

The comparison of M-NHCA with Rule 30 CA is shown in Table 6. Nonlinearity of M-NHCA synthesized from LHCA of CA polynomial $x^{32} + x^{28} + x^{27} + x + 1$ is shown for 3 iterations, which is already shown in Table 3. Nonlinearity of M-NHCA increases very fast with iterations than that of Rule 30 CA. M-NHCA with multiple nonlinearity injections is secure against MS attack. Although, hardware requirement of this M-NHCA is slightly more than that of Rule 30 CA, yet this M-NHCA is fair with respect to the security features.

Table 6. Comparison of M-NHCA with Rule 30 CA

Nonlinear CA	Nonlinearity			Maximum period CA	MS attack resistant
	Itr#1	Itr#2	Itr#3		
Rule 30 CA	2	4	36	No	No
M-NHCA with single nonlinearity injection	16	64	512	Yes	No
M-NHCA with multiple nonlinearity injection	16	256	2048	Yes	Yes

5   Conclusion

MS attack is a real threat on a CA based system. M-NHCA with single nonlinearity injection is vulnerable against MS attack, whereas M-NHCA with multiple nonlinearity injections can resist MS attack and are better cryptographic primitive with respect to the security features. The M-NHCA can be used in designing a CA based cipher. Other applications of M-NHCA than pseudorandom sequence generation (e.g. hash functions) can also be investigated.

References

1. NIST SP 800-22: A statistical test suite for random and pseudorandom number generators for cryptographic applications. U.S. Department of Commerce (2010)
2. Bardell, P.: Analysis of cellular automata used as pseudorandom pattern generators. In: Proceedings of the IEEE International Test Conference 1990, Washington, D.C., 10–14 September 1990, pp. 762–768 (1990)
3. Cattell, K., Muzio, J.C.: Synthesis of one-dimensional linear hybrid cellular automata. IEEE Trans. CAD Integr. Circuits Syst. $15(3)$, 325–335 (1996)
4. Chaudhuri, P.P., Roy Chowdhury, D., Nandi, S., Chattopadhyay, S.: Additive Cellular Automata: Theory and Applications. IEEE Computer Socity Press, New York (1997)
5. Formenti, E., Imai, K., Martin, B., Yunès, J.-B.: Advances on random sequence generation by uniform cellular automata. In: Calude, C.S., Freivalds, R., Kazuo, I. (eds.) Computing with New Resources. LNCS, vol. 8808, pp. 56–70. Springer, Cham (2014). https://doi.org/10.1007/978-3-319-13350-8_5
6. Ghosh, S., Sengupta, A., Saha, D., Roy Chowdhury, D.: A scalable method for constructing non-linear cellular automata with period $2^n - 1$. In: Cellular Automata: Proceedings of the 11th International Conference on Cellular Automata for Research and Industry, ACRI 2014, Krakow, Poland, 22–25 September 2014, pp. 65–74 (2014)
7. Jose, J., Roy Chowdhury, D.: Four neighbourhood cellular automata as better cryptographic primitives. IACR Cryptology ePrint Archive 2015, 700 (2015)
8. Lacharme, P., Martin, B., Sole, P.: Pseudo-random sequences, Boolean functions and cellular automata. In: Proceedings of Boolean Functions and Cryptographic Applications, pp. 80–95 (2008)
9. Leporati, A., Mariot, L.: 1-resiliency of bipermutive cellular automata rules. In: Kari, J., Kutrib, M., Malcher, A. (eds.) AUTOMATA 2013. LNCS, vol. 8155, pp. 110–123. Springer, Heidelberg (2013). https://doi.org/10.1007/978-3-642-40867-0_8
10. Maiti, S., Ghosh, S., Roy Chowdhury, D.: On the security of designing a cellular automata based stream cipher. In: Pieprzyk, J., Suriadi, S. (eds.) ACISP 2017, Part II. LNCS, vol. 10343, pp. 406–413. Springer, Cham (2017). https://doi.org/10.1007/978-3-319-59870-3_25
11. Martin, B.: A Walsh exploration of elementary CA rules. Cell. Autom. $3(2)$, 145–156 (2008)
12. Meier, W., Staffelbach, O.: Analysis of pseudo random sequences generated by cellular automata. In: Davies, D.W. (ed.) EUROCRYPT 1991. LNCS, vol. 547, pp. 186–199. Springer, Heidelberg (1991). https://doi.org/10.1007/3-540-46416-6_17

13. Neumann, J.V.: The Theory of Self-reproducing Automata. (Edited by A.W. Burks) University of Illinois Press, Urbana (1966)
14. Serra, M., Slater, T., Muzio, J.C., Miller, D.M.: The analysis of one-dimensional linear cellular automata and their aliasing properties. IEEE Trans. CAD Integr. Circuits Syst. **9**(7), 767–778 (1990)
15. Wolfram, S.: Cryptography with cellular automata. In: Williams, H.C. (ed.) CRYPTO 1985. LNCS, vol. 218, pp. 429–432. Springer, Heidelberg (1986). https:// doi.org/10.1007/3-540-39799-X_32
16. Wolfram, S.: Random sequence generation by cellular automata. In: Advances in Applied Mathematics, vol. 7, pp. 123–169 (1986)

Context Sensitive Steganography on Hexagonal Interactive System

T. Nancy Dora, S. M. Saroja T. Kalavathy$^{(\boxtimes)}$, and P. Helen Chandra

Jayaraj Annapackiam College for Women (Autonomous), Theni District,
Periyakulam, Tamilnadu, India
nancydora.t@gmail.com, kalaoliver@gmail.com, chanrajac@yahoo.com

Abstract. Cryptography is the science of using mathematics to encrypt and decrypt data and Steganography is the art and science of hiding communication. A steganographic system thus embeds hidden content in unremarkable cover media so as not to arouse an eavesdropper's suspicion. The Steganography hides the message so it cannot be seen. In this paper, a new method is proposed to embed data in images. The security is provided through context sensitive rules. Hexagonal Finite Interactive System is taken as a base to choose the Carrier. Experimental results show that the method is very efficient especially when gluing is done by matching the border label so that the tiling is done uniformly.

Keywords: Image steganography · Secret image
Context sensitive stego technique · Stego image · Secret key

1 Introduction

Data security is a challenging issue of data communications today that touches many areas including secure communication channel, strong data encryption technique and trusted third party to maintain the database. The rapid development in information technology, the secure transmission of confidential data herewith gets a great deal of attention. The conventional methods of encryption can only maintain the data security. The information could be accessed by the unauthorized user for malicious purpose. Therefore, it is necessary to apply effective encryption/decryption methods to enhance data security. The two important techniques for providing security are cryptography and steganography [Mi1]. Both are well known and widely used methods in information security.

Steganography has been progressively becoming one of the popular technique to be used for secret communication between two parties or more. The term of steganography originated from two Greek words which were stegano and graphos. Stegano could be described as cover or secret and graphos defined the meaning of writing or drawing. The combination of both words delineated the meaning of "covered writing" [SK1].

D. Ghosh et al. (Eds.): ICMC 2018, CCIS 834, pp. 16–26, 2018.
https://doi.org/10.1007/978-981-13-0023-3_2

Several steganography methods began to propose embedded secret message in multimedia objects such as images. Images could be a powerful host to hide information because of the spacious spaces it offers. Moreover, the changes in digital images are usually unnoticeable to naked eye. Nowadays, computer technology has given a new life to the ancient steganography. Computer technology introduces digital steganography and makes the steganography easier to execute but harder to crack. These facts stimulate to propose a new model for hiding text in image. A tiling system generates a grid of tiles using the set R of production. Another type of steganographic scheme based on context-sensitive tilings is explored in [PR1] rules.

A novel method of hiding text in image is defined and implemented through tiling. A context sensitive rule is used to generate the Stego image. The images thus produced can be said to have both structure and semantics. A new algorithm has been proposed that would satisfy all the principles of security i.e. confidentiality, authentication, integrity and non-repudiation and also satisfy the requirements of steganography i.e. capacity, undetectability and robustness.

2   Preliminaries

2.1   Steganography [KK1, MB1]

Steganography is the technique of embedding hidden messages in such a way that no one, except the sender and intended receiver(s) can detect the existence of the messages. The main goal of steganography is to hide the secret message or information in such a way that eavesdroppers are not able to detect it [NS1]. Other goal of steganography is to communicate securely in a completely undetectable manner. The various forms of data in steganography can be audio, video, text and images etc.

The basic model of Steganography consists of three components: The Carrier image: The carrier image is also called the cover object that will carry the message that is to be hidden. The Message: A message can be anything like data, file or image etc. The Key: A key is used to decode/decipher/discover the hidden message.

Steganography can be achieved when the user can retrieve a secret message unnoticeably. This involves two main processes. The first process is embedding process, where a secret message is embedded in the host. The host and a secret message can be an image, a video, an audio or text. The second process is involving the extraction of the secret message that has been embedded. Generally steganography concepts can be represented by a basic model of steganography as in Fig. 1.

Various types of Steganography include Image Steganography, Audio Steganography, Video Steganography, Text files Steganography, etc. The image steganography is the process in which we hide the data within an image so that there will not be any perceivable change in the original image. Different techniques of Steganography like Least Significant Bit and Bitmap Steganography are available. Recently, image has been used in steganography as a carrier to

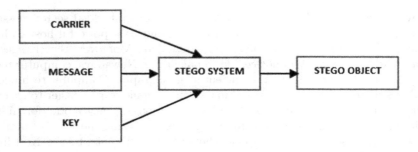

Fig. 1. Basic model of steganography

transmit or send the secret message from a sender to a receiver. The reason is because a huge amount of information can be hidden without noticeable impact to the image that is used as carrier. In addition, the usage of image in information hiding is an ideal technique to have a secured steganography because digital image is insensitive to human visual system.

We are also looking back the definition of hexagonal tiles and scenarios designed by tiling hexagonal unit cells with colors representing two dimensional hexagonal pictures and the structure of a complete 3×3 Hexagonal Finite Interactive System ($HFIS$) [NA1].

2.2 Hexagonal Grid

Hexagonal grid is an alternative representation of pixel tessellation scheme for the conventional square grid for sampling and representing discredited images. Each pixel is represented by a horizontal deflection followed by a deflection upward and to the right. These directions are represented by a pair of unit vectors u and v and this coordinate system is referred as the "h_2" system. Given a pixel with coordinates (u, v) (assumed integer), the coordinates of the neighbors are illustrated in Fig. 2.

2.3 Hexagonal Tiles and Scenarios

Let Σ be a finite alphabet. A hexagonal tile is a hexagonal cell labelled with symbol from the given alphabet and enriched with additional information on each border. This information is represented abstractly as an element from a finite set and is called a border label. The role of border labels is to impose local gluing constraints on self-assembling tiles: two neighbouring cells, sharing a side border (east-west or north east-south west or north west-south east) should agree on the label on that border. A hexagonal scenario is similar to a two-dimensional hexagonal picture, but: (1) each hexagonal cell is replaced by a tile; and (2) east-west or north east-south west or north west-south east neighbouring cells have the same label on the common border.

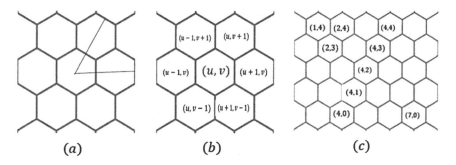

Fig. 2. (*a*) A coordinate system based on unit vectors u and v, (*b*) the neighborhood of a hexagonal pixel and (*c*) labeling of a hexagonal pixel

Graphically, a hexagonal scenario is obtained using the hexagonal tiles representing the transitions and identifying the matching classes or states of the neighbouring cells. The labels on the north east and north west borders represent north memory states, while the south east and south west borders represent south memory states and the ones on the west and east borders represent interaction classes. The selected labels on the external borders are called initial for south west, west and north west borders and final for north east, east and south east borders.

To construct the hexagonal scenarios by assembling hexagonal tiles, we make use of the following. N is the set of natural numbers $\{0, 1, 2, ...\}$, $Z = N \cup -N$ is the set of integers and R is the set of real numbers. We will be working in the two-dimensional hexagonal grid of integer positions $Z \times Z$. The directions $\mathcal{D} = \{EE, WW, NE, NW, SE, SW\}$ will be used as functions from $Z \times Z$ to $Z \times Z$: A point on the side borders in a unit cell is specified by its middle points such that $EE(x, y) = (x + 1, y), WW(x, y) = (x - 1, y), NE(x, y) = (x, y + 1), NW(x, y) = (x - 1, y + 1), SE(x, y) = (x + 1, y - 1)$ and $SW(x, y) = (x, y - 1)$. We say that (x, y) and (x', y') are neighbors if $(x', y') \in \{EE(x, y)/WW(x, y)/NE(x, y)/NW(x, y)/SE(x, y)/SW(x, y)\}$. Note that $EE = WW^{-1}, NE = SW^{-1}$ and $NW = SE^{-1}$. Examples of tiles and scenarios are presented in Fig. 3.

2.4   Hexagonal Finite Interactive System (HFIS)

Let Σ be a finite alphabet. A Hexagonal Finite Interactive System **(HFIS)** over Σ is defined by: a set $S = s_1, s_2, s_3, s_4$ of states and a set $C = c_1, c_2$ of classes; a set T of transitions of the form: $(s_1, c_1, s_2) \rightarrow a \rightarrow (s_3, c_2, s_4)$ where a is a symbol of a given alphabet Σ; specification of the initial/final states and classes. Let s_a be the scenario of direct transition where a is the labelled symbol over the alphabet. The set of all scenarios representing hexagonal picture is denoted by $\Sigma_{s_2}^{**}$.

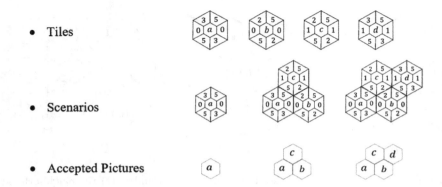

Fig. 3. Tiles and scenarios

A *HFIS* is complete if it specifies a transition $(s_1, c_1, s_2) \to t \to (s_3, c_2, s_4)$ for any pair $((s_1, c_1, s_2), (s_3, c_2, s_4))$ in $((S \times C \times S) \times (S \times C \times S))$. A tile representation is used which is based on showing the transitions and stating which states and classes are initial/final. The states/classes of this *HFIS* is denoted by the initials of the colors: The classes c_1 and c_2 are g (*green*) and b (*blue*) while the swne memory states s_1 and s_3 are p (*purple*) and m (*magenta*) and the nwse memory states s_2 and s_4 are r (*red*) and o (*orange*). A scenario is called indecomposable if all its south west and north east borders are labelled with s_1 and s_3 west and east borders with c_1 and c_2 and the north west and south east borders with s_2 and s_4 respectively and it doesn't contain any sub-scenarios with this property. A complete 3×3 *HFIS* is specified by the 64 transitions shown in Fig. 4.

3   Steganography Through *HFIS*

A New Model for Hiding Text in an image using Image Steganography through Hexagonal Finite Interactive System is proposed. An interactive system *HFIS* recognizing the Hexagonal grid consisting of a parallelogram array with empty tokens is considered as a carrier to carry the image. A simple and an efficient model based on context sensitive classes/states (CSC/CSS) replacement technique is stretched out for calculating secret message that can be embedded in an image. The embedding process distribute the secret message inside a shared colored images.

3.1   Context Sensitive Classes/States (CSC/CSS) Replacement Technique

In image steganography techniques, this proposed model uses substitution technique. CSC replaces the color of the class in the image cell and CSS replaces the color of the state in the image cell.

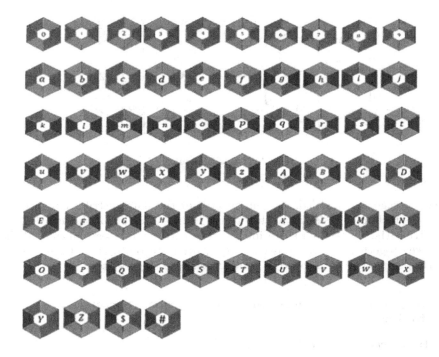

Fig. 4. Transitions of 3×3 $HFIS$ (Color figure online)

Each character of the original text is represented by a tile with border rule representing the action/transition of the unit cell. Each tile is a unit cell surrounded by two classes and four states The transition is directed by a context sensitive rule $(x_1 \alpha_1 \gamma_1 y_1 \beta_1 \delta_1) \Rightarrow (x_2 \alpha_2 \gamma_2 y_2 \beta_2 \delta_2)$. The graphical representation of the rule is shown in Fig. 5.

$\{x_i, y_i\}$ represents the classes either b ($blue$) or g ($green$); $\{\alpha_i, \beta_i\}$ represents the swne memory states either m ($magenda$) or p ($pink$); $\{\gamma_i, \delta_i\}$ represents the nwse memory states either o ($orange$) or r (red).

The $HFIS$ depends on the constraints of the transitions identifying the matching classes or states of the neighboring cells. The three sensitive class/states of each character in the secret message is represented by three bits of zeros. If there is a mismatching of class or states in a character, the particular character is sensitive. Usually the class representing x_1 and the states representing α_1 and δ_1 are more sensitive than the others. These sensitive class and states are modified to hide the character of that cell. In that case, the corresponding color is modified so that the original character is concealed in the cell and the identified bit is changed from zero to one. The collective bits are converted into decimal value to find the secret key. The conversion is done by choice of taking bits of length less than twenty one at a time. We may develop a secret key for each row. The assembled scenario is considered as the stego image in which the

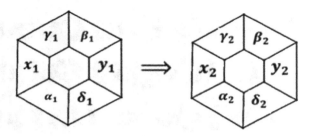

Fig. 5. Transition rule

secret message is concealed. The uniqueness of the data distribution process, made this technique resistant to the attacks as it is difficult for the attackers to reconstruct the shape from stego images. The carrier, the stego image and the secret key together are called as context sensitive stego object and the process of getting stego object is called as context sensitive stego system.

3.2 Steganography Algorithm for Encryption: (Secret Image to Stego Image)

Input: $SecretText, SecretImage, Carrier$
Output: $Secretkey, StegoImage$
Algorithm:

1. Fix the number of rows as I and identify the row as $Row(I)$
2. Calculate the length of the $SecretText$ in each row and fix it as $textlength(Row(I))$
3. Let $StegoImage = EmptyGrid(Carrier)$
4. For each $Row(I)$ do the following
5. Place the SecretImage of each characters one by one in the Grid as follows
6. Fix the sensitive class and states of the Current Image
7. Check the sensitive class/ states with the StegoImage
8. If $maching = ok$ assemble with StegoImage and goto step 5
9. If not, modify the unmatched class and states and assemble the image with StegoImage
10. Modify $x\alpha\delta$ and goto step 5
11. Convert $x\alpha\delta$ of $row(I)$ into decimal ; fix it as $Secretkey(I)$ and goto step 4
12. Get final StegoImage and Secret Key as output

3.3 Implementation of Context Sensitive Steganography

The following example elucidates the algorithm. Secret Image of the 64 characters are stored in a stack. The authors wish to send 'GOOD, VERY GOOD, and EXCELLENT' as a secret message and choose the Hexagonal grid consisting of a

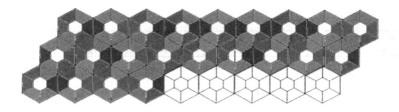

Fig. 6. Stego image

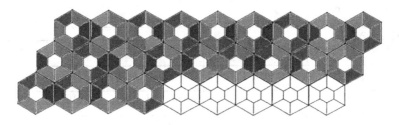

Fig. 7. Secret image

parallelogram array with empty tokens on a hexagonal finite interactive system as a Carrier to carry the image as a cover object. "GOOD" will be assembled by four transition tiles in the first row, "VERY GOOD" will be assembled by nine transition tiles in the second row and "EXCELLENT" will be assembled by nine transition tiles in the third row. The minimum required size of the secret image is 3×9. Choose a cover object of size 3×9.

Now the characters in the Secret Text is given as input and the Context Sensitive Classes/States (CSC/CSS) Replacement Technique is applied. The transitions for "GOOD", "VERY GOOD" and "EXCELLENT" are modified. The process of modification is given in table 1 (Fig. 8). The modified tiles are assembled in the cover object as a hexagonal scenario. The scenario is considered as the Stego Image. The identified bits of the sensitive classes and states are modified and collected row wise. The modified bits for the first, second and third row are respectively 000 000 100 000, 011 000 010 100 100 000 000 100 000 and 010 011 001 100 110 100 100 000 100. They are converted into decimal value and the corresponding Secret keys are obtained. The Secret keys are 32, 51003424, 40265988. The final Stego Image is shown in Fig. 6.

Now, Stego image and secret keys are sent to the receiver. The receiver recognize the character of the Stego Image row wise. The Secret keys are converted into binary digits. The receiver now identify the modified bits and the corresponding modified state or class of each tile. The identified class or states are remodified and the corresponding Secret image is received as shown in Fig. 7. The equivalent secret text is obtained.

secret Character	Secret Transition	secret Image	CSS/CSC Rule	Modified Transition	Stego Image	Modified Bits($x\alpha\delta$)	Stego Character
First row		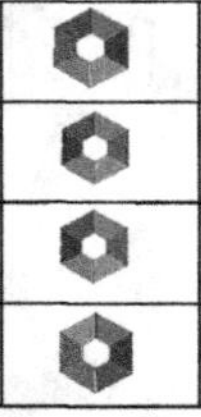			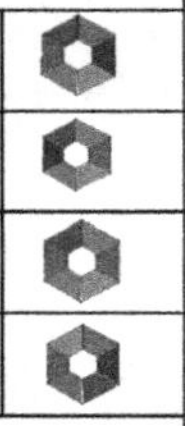		
G	pgrmbo		-	-		000	G
O	pbrpgo		-	-		000	O
O	pbrpgo		b → g	pgrpgo		100	2
D	mgombr		-	-		000	D
Second row		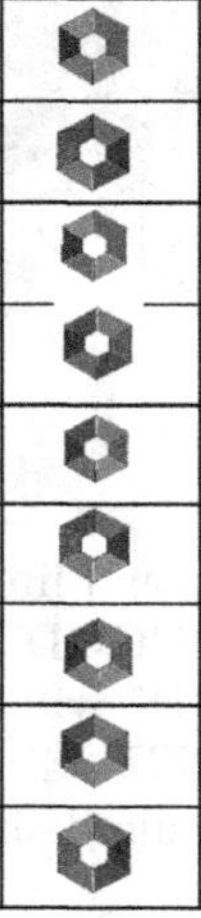			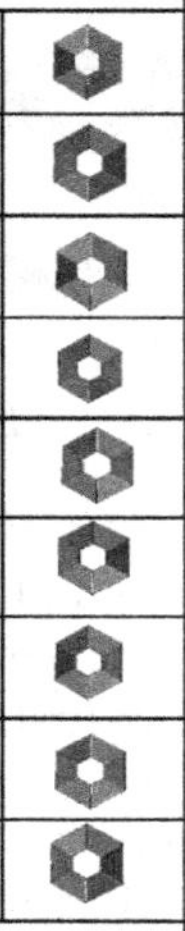		
V	pbomgo		p → m o → r	mbomgr		011	T
E	pgrmbr		-	-		000	E
R	mbomgo		m → p	pbomgo		010	V
Y	mbrpgr		b → g	mgrpgr		100	c
space	mbrpgo		b → g	mgrpgo		100	e
G	pgrmbo		-	-		000	G
O	pbrpgo		-	-		000	O
O	pbrpgo		b → g	pgrpgo		100	2
D	mgombr		-	-		000	D
Third row		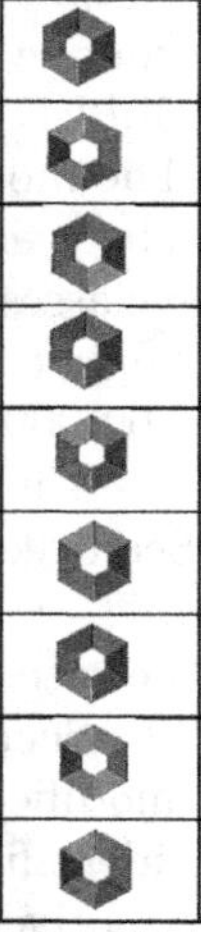			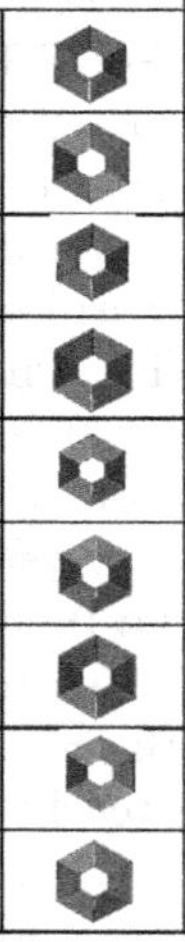		
E	pgrmbr		p → m	mgrmbr		010	A
X	pbomgr		p → m r → o	mbomgo		011	R
C	mgrmbo		o → r	mgrmbr		001	A
E	pgrmbr		g → b	pbrmbr		100	o
L	mgopbr		g → b m → p	pbopbr		110	j
L	mgopbr		g → b	mbopbr		100	v
E	pgrmbr		g → b	pbrmbr		100	o
N	pbopgo		-	mbopgo		000	N
T	mbomgr		b → g	mgomgr		100	7

Fig. 8. Process of secret image to stego image

3.4 Steganography Algorithm for Decryption: (Stego Image to Secret Image)

Input: $Secret key, StegoImage, Carrier$
Output: $Secret key, SecretImage$
Algorithm:

1. Fix the number of rows as I and identify the row as $Row(I)$
2. Calculate the length of the $StegoText$ in each row and fix it as $textlength(Row(I))$
3. Let $SecretImage = EmptyGrid(Carrier)$
4. For each $Row(I)$ do the following
5. convert the Secret key of Row(I) to binary digit
6. Extract the binary digit in bits of length 3 and store it in an array BIN
7. For each StegoImage of Row(I) do the following
8. Let $J = J^{th}$ tile in $Row(I)$
9. If $BIN(J) = 000$, place the Stego Image in the carrier.
10. If not, Identify the modified bits and corresponding modified states or class
11. Modify the stego transition into secret transition and place it in the carrier and goto step 7.
12. Goto Step 4 and do the process of next Row:
13. Get final SecretImage as output from the Carrier.

4 Conclusion

We have presented a stego-system which generates stego-objects using context sensitive tiling. A new steganographic algorithm for hiding text in images is proposed. This new steganographic approach is robust and very efficient for hiding text in images. We have further planned to develop the system in java based on the proposed algorithm. Steganography will continue to increase in popularity over cryptography. The system would be tested on the basis of various illustrations and the results would be compared with those of existing algorithms.

References

[KK1] Rahmani, M.K.I., Arora, K., Pal, N.: A crypto-steganography: a survey. Int. J. Adv. Comput. Sci. Appl. **5**(7), 149–155 (2014)
[MB1] Maiti, C., Baksi, D., Zamider, I., Gorai, P., Kisku, D.R.: Data hiding in images using some efficient steganography techniques. In: Kim, T., Adeli, H., Ramos, C., Kang, B.-H. (eds.) SIP 2011. CCIS, vol. 260, pp. 195–203. Springer, Heidelberg (2011). https://doi.org/10.1007/978-3-642-27183-0_21
[Mi1] Rajyaguru, M.H.: Crystography - combination of cryptography and steganography with rapidly changing keys. Int. J. Emerg. Technol. Adv. Eng. **2**, 329–332 (2012)
[NA1] Nancy Dora, T., Athisaya Ponmani, S., Helen Chandra, P., Kalavathy, S.M.S.T.: Generation of hexagonal patterns in finite interactive system and scenarios. Glob. J. Pure Appl. Math. **13**(5), 17–26 (2017)

[NS1] Johnson, N.F., Jajodia, S.: Exploring steganography: seeing the unseen. Computer **31**, 26–34 (1998)
[PR1] Ritchey, P.C., Rego, V.J.: A context sensitive tiling system for information hiding. J. Inf. Hiding Multimedia Sig. Process. **3**(3), 212–226 (2012)
[SK1] Sharma, V., Kumar, S.: A new approach to hide text in images using steganography. Int. J. Adv. Res. Comput. Sci. Soft. Eng. **3**(4), 701–708 (2013)

A Novel Steganographic Scheme Using Weighted Matrix in Transform Domain

Partha Chowdhuri[1]([✉]), Biswapati Jana[1], and Debasis Giri[2]

[1] Department of Computer Science, Vidyasagar University,
Midnapore 721102, West Bengal, India
prc.email@gmail.com, biswapatijana@gmail.com
[2] Department of Computer Science and Engineering, Haldia Institute of Technology,
Haldia 721657, West Bengal, India
debasis_giri@hotmail.com

Abstract. In this paper a weighted matrix based steganographic scheme has been developed based on Discrete Cosine Transform (DCT). First, (8×8) quantized DCT coefficient blocks are obtained from the cover image. Instead of hiding the data directory to the quantized DCT coefficient blocks, a different approach has been taken here. The AC coefficients, except 0 coefficients, are used to form a series of 3×3 temporary matrices. Then, each four bits secret data is converted into an integer value. An user defined weighted matrix is used to select the position in the temporary matrix where the data will be embedded. The integer value is then embedded into that particular position of the selected temporary matrix. The proposed method is tested using different steganographic attacks like RS analysis and NCC to show that the scheme is undetectable under these analysis and more robust that other schemes. This scheme provides good embedding capacity with high visual quality of stego images.

Keywords: Steganography · Weighted matrix
Quantized DCT coefficient · PSNR · NCC

1 Introduction

Due to the rapid development of computer technology and Internet, hiding data within the digital format of multimedia became very popular. Many schemes for hiding data have been proposed till date. These approaches are classified mainly into two categories: the spatial-domain and the frequency-domain. In spatial domain, the pixel values are directly manipulated to hide data. Noticeable distortion in any position of the image is a common case in spatial domain. Therefore, different approaches have been developed to increase the embedding capacity and to adjust the position to minimise the distortion noticeable to human eye. Some inherent problem of spatial-domain data hiding is there. For an example, for lossy compression it is very difficult to find the redundant portion of the

© Springer Nature Singapore Pte Ltd. 2018
D. Ghosh et al. (Eds.): ICMC 2018, CCIS 834, pp. 27–35, 2018.
https://doi.org/10.1007/978-981-13-0023-3_3

image for hiding data. But this redundant portion can be easily detected when we transform the image to frequency domain. The sharp transitions and edges of an image contribute the high-frequency content to its discrete cosine transform. Thus, to find the appropriate pixels to hide the data, transform domain scheme is a better approach. In transform domain schemes, some frequency oriented mechanisms like Discrete Cosine Transform (DCT) is used to transform the cover image. The secret data are then embedded into the cover image by modifying the frequency coefficients.

The Joint Photographic Experts Group (JPEG) digital image format is most popularly used image format nowadays. But, as JPEG is a compressed image format and uses transform domain principles, a slight modification in the transform domain may cause more distortion to the cover image. Most of the existing steganographic schemes changes one coefficient for hiding one bit of data. So, hiding more data causes changing more coefficients that gradually degrade the quality of the cover image which makes it prone to suspicion. We have used weighted matrix for our scheme to embed more data by changing one DCT coefficient. This increases the robustness as well as the quality of the image.

The remainder of the paper is constructed as follows: In Sect. 2 Motivation and objectives of our proposed techniques is described. In Sect. 3 the proposed scheme is presented. Experimental results and comparisons are shown in Sect. 4. In Sect. 5 Steganalysis and evolution are discussed Finally conclusions are depicted in Sect. 6.

2   Motivation and Objective

The main motivation and objective of the proposed work are listed below:

 (i) Embedding Capacity: In the literature, it is observed that payload is limited when using weighted matrix in the steganographic scheme. So our motivation is to increase data embedding capacity.
 (ii) Imperceptibility: It is well known that imperceptibility is the main requirement in any steganographic scheme. So first and foremost objective is to maintain imperceptibility in the proposed scheme.
(iii) Robustness: From the literature, it is seen that till now there exist some security loop hole in any steganographic scheme in real life. So our objective is to develop a steganographic scheme using weighted matrix through predefined integer sequence to enhance security and robustness.

3   Proposed Method

In this section, we proposed a novel data hiding technique based on discrete cosine transformation (DCT) using weighted matrix. We first transform a given cover image into a sequence of 8×8 blocks of DCT coefficients. The schematic diagram of secret data embedding and extraction are depicted in Figs. 1 and 2 respectively.

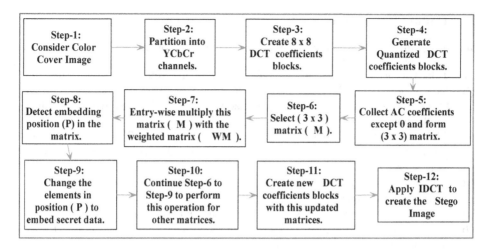

Fig. 1. Details of embedding process

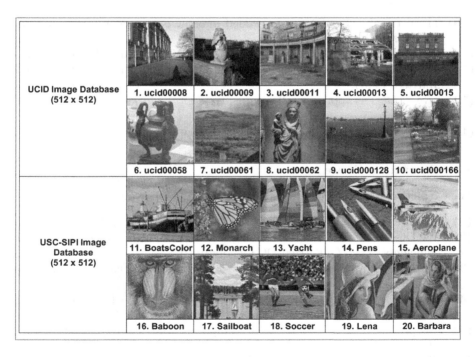

Fig. 2. Cover images with (512 × 512) pixel collected from the standard benchmark databases like UCID and USC-SIPI

3.1   Embedding Phase

The embedding technique can be explained by the following steps:

(1) Take the original cover image $CI_{m \times n}$ and secret image (SI) as input.

(2) Now to create DCT coefficient matrices DCT is applied to all 8×8 image blocks of the $CI_{m \times n}$. Then Quantized DCT is obtained from the DCT coefficient matrix.

(3) The AC coefficients, except 0 coefficients, from the DCT coefficient matrices are taken to form a series of 3×3 Temporary Matrices (TM) for embedding secret data.

(4) The secret image (SI) is also converted to binary form to create binary secret message (SM).

(5) First, a Temporary Matrix (TM) is taken for processing.

(6) 4 bits data is taken from the secret message (SM) and then converted to its equivalent decimal value (DV).

(7) Elementary multiplication is done between the Temporary Matrix (TM) and the weighted matrix (WM).

(8) The embedding position in the Temporary Matrix (TM) is selected using the standard weight matrix rule.

(9) The decimal value (DV) is embedded into the particular position in Temporary Matrix (TM).

(10) Step 5 to Step 9 is applied for the rest of the Temporary Matrices (TM).

(11) The quantized DCT block coefficients are updated using the Temporary Matrices (TM).

(12) Inverse DCT is applied to all the quantized DCT blocks to form the YCbCr channels.

(13) Finally, using the YCbCr channels, the stego image is created.

3.2   Extraction Phase

The extraction technique can be explained with the following steps:

(1) Take the stego image $SI_{m \times n}$ and secret Weighted Matrix (3×3) as input.

(2) Now DCT is applied in all 8×8 image blocks of the $SI_{m \times n}$ to get the DCT coefficients. Then Quantized DCT is obtained.

(3) Temporary Matrices ($TM_{3 \times 3}$) are created from the AC coefficients (except 0 coefficient).

(4) Now 4 bit secret data is extracted from Temporary Matrix (TM) by applying entry-wise multiplication with the weighted matrix (WM).

(5) This process is done with all other Temporary Matrices to extract all 4 bit data.

(6) All 4 bit data, extracted from all the Temporary Matrices, are concatenated to form the secret binary bits.

(7) The original secret image is generated from this secret binary bits.

4   Experimental Results and Comparisons

The proposed method is discussed in this paper is implemented in Java 9 windows 10 (operating system) environment. The computational platform was a Intel Core i5-6200U processor with a speed of 2.40 GHz and 4 GB RAM. A set of standard colour test images with size (512 × 512) are chosen to evaluate the performance of the proposed scheme. Figure 2 shows some of the standard colour test images which were collected form "USC-SIPI" image database [9] collected from the University of Southern California, "UCID" image dataset [5] consist of 1338 uncompressed color image collected from the Nottingham Trent University, UK. We have used secret message as logo image of size (54×54). The performance evaluation of the proposed scheme is evaluated using Mean Square Error (MSE), Peak-Signal-to- Noise-Ratio ($PSNR$) Normalized Cross Correlation (NCC) and Structural similarity ($SSIM$) index.

The imperceptibility of the stego image from the original image is indicated by the PSNR defined as:

$$MSE = \frac{\sum\limits_{i=1}^{Row} \sum\limits_{j=1}^{Col} [X(i,j) - Y(i,j)]^2}{(Row \times Col)} \tag{1}$$

$$PSNR = 10 \; log_{20} \frac{I_{max}}{MSE} dB \tag{2}$$

Table 1. Comparison of proposed scheme with Chang et al. and Weng et al's scheme in terms of visual quality (PSNR).

Image	Chang et al. [1]		Weng et al. [11]		Proposed scheme	
	Capacity	PSNR	Capacity	PSNR	Capacity	PSNR
Lena	12288	35.15	28.364	42.42	36864	49.92
Baboon	12288	31.34	6708	48.17	36864	47.25
Airplane	12288	35.22	27694	45.73	36864	46.33
Boat	12288	34.92	15564	45.72	36864	47.25
Zelda	12288	38.04	22504	43.05	36864	46.89
Pepper	12288	36.32	20191	48.15	36864	46.92
Average	12288	35.17	20170	45.54	36864	47.37

where Row and Col is the size of cover image ($X(i,j)$) and stego image ($Y(i,j)$). Where I_{max} is the peak signal value of the cover image which is equal to 255 for 8 bit images. High PSNR value assure better image quality and low PSNR implies poor image quality. Table 1 represent the experimental value of average PSNR which is greater than 47 dB and the maximum embedding capacity is = 36,864 bits. The bpp in this scheme = $\frac{Total\ embedded\ bits}{(Row \times Col)}$ = 0.14. Structural similarity (SSIM) index is a parameter for measuring the similarity between two images. Its value lies between -1 and $+1$. Its value approaches to $+1$ when two images are identical. The following formula is used to find the SSIM value of the Cover and Stego images.

$$\text{SSIM}(x, y) = \frac{(2\mu_x\mu_y + c_1)(2\sigma_{xy} + c_2)}{(\mu_x^2 + \mu_y^2 + c_1)(\sigma_x^2 + \sigma_y^2 + c_2)} \tag{3}$$

where, μ_x is the average of x, μ_y is the average of y;
$\sigma^2{}_x$ is the variance of x, $\sigma^2{}_y$ is the variance of y;
σ_{xy} is the covariance of x and y
$c_1 = (k_1L)^2$ and $c_2 = (k_2L)^2$, two variables to stabilize the division with weak denominator.
L is the dynamic range of the pixel-values.
$k1 = 0.01$ and $k2 = 0.03$ by default.

The Normalized correlation coefficient (NCC) is used to measure robustness. It calculates the difference between the original and stego image. It may be defined as:

$$NCC = \frac{\displaystyle\sum_{i=1}^{Row}\sum_{j=1}^{Col} x(i,j)y(i,j)}{\displaystyle\sum_{i=1}^{Row}\sum_{j=1}^{Col} |x(i,j)|^2} \tag{4}$$

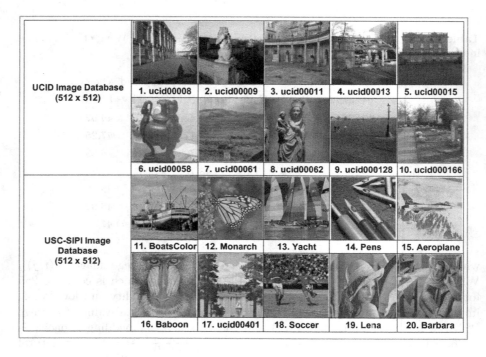

Fig. 3. Stego images with (512×512) pixel after embedding data.

where Row and Col are the number of row and column in the images, respectively, $x(i, j)$ and $y(i, j)$ are the original cover image and the stego image respectively. Figure 3 shows the stego images after embedding secret data. Experimental results of our proposed scheme in terms of different comparison metrics are shown in Tables 2, 3, 4 and 5.

Table 2. The comparison table of our proposed scheme with respect to PSNR, SSIM, NCC

Image database	Cover image	PSNR	SSIM	NCC
USC - SIPI image database (512 × 512)	BoatsColor	45.32	0.996	0.996
	Monarch	42.3	0.996	0.996
	Yatch	41.96	0.996	0.996
	Pen	43.25	0.996	0.996
	Aeroplane	44.32	0.996	0.996
	Baboon	45.32	0.996	0.996
	Sailboat	42.3	0.996	0.996
	Soccer	41.96	0.996	0.996
	Lena	43.25	0.996	0.996
	Barbara	44.32	0.996	0.996
UCID image database (512 × 512)	Ucid00008	48.3	0.996	0.996
	Ucid00009	47.6	0.996	0.998
	Ucid00011	45.3	0.996	0.996
	Ucid00013	48.6	0.996	0.946
	Ucid00015	48.3	0.996	0.996
	Ucid00058	47.6	0.996	0.998
	Ucid00061	45.3	0.996	0.996
	Ucid00062	48.6	0.996	0.946
	Ucid00128	48.6	0.996	0.946
	Ucid00166	44.7	0.996	0.996

Table 3. Comparison with other schemes with respect to execution time

Techniques	Execution time for embedding
Singh and Singh [8]	19.7653 s
Kim and Lee [4]	11.82 s
Lutovac [3]	2.36 s
Proposed scheme	2.25 s

Table 4. Comparison table with respect to PSNR, NCC and execution time for embedding

Image	Singh and Singh [8]			Rahman et al. [6]			Proposed scheme		
	PSNR	NCC	Execution time	PSNR	NCC	Execution time	PSNR	NCC	Execution time
Lena	39.7928	0.9973	19.6405	31.4550	0.9976	...	49.92	0.9998	2.2365
Boat	39.9285	0.9984	19.7653	30.0704	0.9980	...	47.25	0.9996	2.1565
Baboon	39.5467	0.9980	19.8433	30.4195	0.9906	...	48.46	0.9989	2.2641
House	39.1570	0.9981	19.9525	31.4114	0.9959	...	48.39	0.9984	2.2741
Pepper	40.1907	0.9980	19.9993	30.5127	0.9945	...	46.92	0.9985	2.2947
Zelda	39.7255	0.9980	19.6405			...	47.89	0.9991	2.1851
Aeroplane	...	...	...	...	...	...	46.33	0.9979	2.1894

Table 5. Comparison scheme with the other scheme in terms of PSNR and payload

Cover image	Lin and Shiu scheme [2]		Saidi scheme [7]		Wei et al.'s scheme [10]		Our proposed scheme	
	Capacity	PSNR	Capacity	PSNR	Capacity	PSNR	Capacity	PSNR
Lena	90,112	35.28	163,840	36.3762	64,008	33.971	36864	49.92
Airplane	90,112	34.53	163,840	36.7870	64,008	31.949	36864	46.33
Boat	90,112	33.05	163,840	34.4317	64,008	31.147	36864	47.25
Baboon	90,112	28.22	163,840	26.4209	64,008	25.995	36864	48.46
Peppers	90,112	35.09	163,840	35.6165	64,008	33.400	36864	46.92

5   Steganalysis and Evolution

The proposed data scheme is highly robust and only authorized person can extract the secret image by using proper weighted matrix. Six different numbers are required to form the weighted matrix. Also the sequence of numbers in the weighted matrix should be proper. Even if a single number is different in weight matrix or the sequence of these numbers are different, the recovery of the secret image is not possible. This increases the robustness of the proposed scheme.

6   Conclusion

In this paper, a secured stenographic scheme using weighted matrix in discrete cosine transform domain is proposed. Here, to embed confidential messages a shared secret key in the form of a matrix is used. The PSNR is reasonable and higher than 47 dB. In this scheme it is possible to embed more bits per block by applying weight matrix more than once to get higher bits per pixel. But in that case the PSNR will be as low as around 30 dB which makes the image vulnerable for attacking.

For future research, some authentication features can be included into the recent technique. This not only can be used for authentication but also can introduce some randomness and other extra security features into the proposed scheme. Also the idea of using weighted matrix may be used in other transform domain in order to improve results of steganography.

References

1. Chang, C.C., Lin, C.C., Tseng, C.S., Tai, W.L.: Reversible hiding in DCT-based compressed images. Inf. Sci. **177**, 2768–2786 (2007)
2. Lin, C.C., Shiu, P.F.: High capacity data hiding scheme for DCT-based images. J. Inf. Hiding Multimed. Signal Process. **1**(3), 220–240 (2010)
3. Lutovac, B., et al.: An algorithm for robust image watermarking based on the DCT and Zernike moments. Multimed. Tools Appl. **76**(22), 23333–23352 (2017)
4. Kim, H.S., Lee, H.-K.: Invariant image watermark using Zernike moments. IEEE Trans. Circuits Syst. Video Technol. **13**(8), 766–775 (2003)
5. UCID Image Database, Nottingham Trent University, UK. http://jasoncantarella.com/downloads/ucid.v2.tar.gz
6. Rahman, M.M., et al.: A semi blind watermarking technique for copyright protection of image based on DCT and SVD domain. Global J. Res. Eng. (2017)
7. Saidi, M., et al.: A new adaptive image steganography scheme based on DCT and chaotic map. Multimed. Tools Appl. **76**(11), 13493–13510 (2017)
8. Singh, D., Singh, S.K.: DCT based efficient fragile watermarking scheme for image authentication and restoration. Multimed. Tools Appl. **76**(1), 953–977 (2017)
9. University of Southern California: The USC-SIPI Image Database. http://sipi.usc.edu/database/database.php
10. Wei, S., Zhe-Ming, L., Yu-Chun, W., Fa-Xin, Y., Rong-Jun, S.: High performance reversible data hiding for block truncation coding compressed images. Signal Image Video Process. **7**(2), 297–306 (2013)
11. Weng, C.-Y., Huang, C.-T., Kao, H.-W.: DCT-based compressed image with reversibility using modified quantization. In: Pan, J.-S., Tsai, P.-W., Watada, J., Jain, L.C. (eds.) IIH-MSP 2017. SIST, vol. 81, pp. 214–221. Springer, Cham (2018). https://doi.org/10.1007/978-3-319-63856-0_27

Repeated Burst Error Correcting Linear Codes Over GF(q); q = 3

Vinod Tyagi[1] and Subodh Kumar[2(✉)]

[1] Shyam Lal College (Eve.), University of Delhi, Delhi, India
vinodtyagi@hotmail.com
[2] Shyam Lal College, University of Delhi, Delhi, India
subodh05031981@gmail.com

Abstract. In this paper, we develop a simple matrix method of constructing a parity check matrix for non binary (5k, k; b, q, m) linear codes capable of correcting m repeated burst errors of length b or less.

Keywords: Repeated burst · Burst error · Open loop and closed loop bursts
Parity check digits · Error patterns and syndrome

1 Introduction

At a very early stage in the history of coding theory, codes were meant for detecting and correcting single errors only. But later on, it was noticed that in almost all communication channels errors occur more in adjacent positions and quite less in random manner. Adjacent error correcting codes were introduced by Abramson [1]. The generalization of this idea was put in the category of errors that is now known as burst error in the literature of coding theory. But the nature of burst errors differ from channel to channel depending upon the behavior of the channel and therefore different type of codes were developed to deal with different type of burst errors. Among many versions of burst errors are CT burst [3], closed loop burst, open loop burst, low density burst and high density burst etc. While studying different communication channels and type of burst errors, it was observed that among all the categories of errors, burst due to Fire [6] is the most common error that occurs during transmission. By a burst of length *b* or less, we mean a vector whose all the nonzero positions are confined to some *b* consecutive components the first and the last of which is non zero. In view of this burst error correcting codes have been developed. Some of such burst error correcting codes have found great applications in numerous areas of practical importance also and therefore have acquired important position in the literature in comparison to other variants of burst, and a good deal of research has gone into the development of bursts and multiple bursts error connecting codes. For references, see [2, 7, 9–12] and many others. Corresponding to various variants of the definition of burst, codes have been developed for correction of random burst or open loop burst errors, low and high density bursts, closed loop bursts and multiple bursts.

© Springer Nature Singapore Pte Ltd. 2018
D. Ghosh et al. (Eds.): ICMC 2018, CCIS 834, pp. 36–41, 2018.
https://doi.org/10.1007/978-981-13-0023-3_4

In very busy and fast communication channels which is the need of present time, it has been observed that errors repeat themselves more frequently during transmission. This phenomenon has shown that the normal burst error correcting codes cannot yield any positive result for repeated burst error detection and correction. As there is not any uniform terminology for multiple bursts, repeated bursts are also put in this category.

In view of this, it was desirable to develop codes detecting/correcting errors that were in the form of repeated bursts. Dass and Verma [5] took the initiative in this direction and developed codes for repeated burst error detection and correction in binary case.

Following the old technique of parity check matrix construction given by Varshamov-Gilbert-Sacks bound [8], Dass and Verma [5] obtained upper and lower bounds on the number of parity check digits required for correcting m repeated burst errors over GF(q) (see [4, Theorem 4]). Although the method was cumbersome, bounds were derived on the number of parity check digits. This complicated synthesis procedure involving unwieldy computations particularly in case of repeated bursts and to study repeated burst error correcting codes in detail for binary and non binary cases, it was desirable to simplify the parity check matrix construction procedure.

We will call such codes as $(5k, k; b, q, m)$ linear codes throughout this paper where n = 5k. For m repeated burst errors of any length with a specific value of m and length of the burst b, the matrix in binary case comes out to be

$$
H = \left(\begin{array}{cc} & \begin{array}{c} I_k \\ I_k \\ \cdot \\ \cdot \\ I_k \end{array} \\ I_{4k} & \end{array} \right) \tag{1}
$$

Such a matrix considered as parity check matrix shall give rise to a code that corrects m-repeated burst of length b or less. Here I_{4k} is an Identity matrix of order $(4k)$. I_k's are identity matrix of order k.

As an example, for $k = 3, m = 2, b = 3$, the parity check matrix for such a $(15, 3)$ binary code correcting 2 repeated bursts of length 3 or less may be written as

$$
\left(\begin{array}{ccccccccccccccc}
1 & 0 & 0 & 0 & 0 & 0 & 0 & 0 & 0 & 0 & 0 & 0 & 1 & 0 & 0 \\
0 & 1 & 0 & 0 & 0 & 0 & 0 & 0 & 0 & 0 & 0 & 0 & 0 & 1 & 0 \\
0 & 0 & 1 & 0 & 0 & 0 & 0 & 0 & 0 & 0 & 0 & 0 & 0 & 0 & 1 \\
\cdot & \cdot & \cdot & \cdot & \cdot & \cdot & \cdot & \cdot & \cdot & \cdot & \cdot & \cdot & \cdot & \cdot & \cdot \\
\cdot & \cdot & \cdot & \cdot & \cdot & \cdot & \cdot & \cdot & \cdot & \cdot & \cdot & \cdot & \cdot & \cdot & \cdot \\
0 & 0 & 0 & 0 & 0 & 0 & 0 & 0 & 0 & 0 & 0 & 1 & 0 & 0 & 1
\end{array} \right)
$$

In other words this the parity check matrix for $(15, 3)$ repeated burst correcting code in binary case. This code will always correct 2 repeated bursts of length 3. For detailed verification see [5].

We in this paper study non-binary repeated bursts error correcting codes over GF(3).

2   Non Binary Repeated Burst Error Correcting Linear Codes

Our purpose in this communication is to develop a generalized matrix method for all binary and non binary repeated burst error correcting codes as matrix formulation in (1) does not work in non binary cases. In view of this, we will have to first try to develop a different matrix over GF(3). Using hit and trial method, we came across a matrix that can be described as follows:

Let us define a diagonal matrix J whose diagonal elements are in a sequence 1, 2, 1, 2, 1, 2... etc. in such a way that a 2×2 J matrix may be written as

$$J_2 = \begin{pmatrix} 1 & 0 \\ 0 & 2 \end{pmatrix}$$

and $J_3 = \begin{pmatrix} 1 & 0 & 0 \\ 0 & 2 & 0 \\ 0 & 0 & 1 \end{pmatrix}$ is a 3×3 J matrix.

Using the result given in (1), we can now write a matrix H_J for (5k, k; b, q, m) non binary linear code over GF(3) as follows:

$$H = H_J = \begin{pmatrix} & J_k & \\ & \cdot & \\ J_{4k} & \cdot & \\ & \cdot & \\ & j_k & \end{pmatrix} \tag{2}$$

which is the parity check matrix of order $4k \times 5k$.

This formulation of general parity check matrix H in (2) shows that it is now easy to construct parity check matrices for (5k, k; b, q, m) non binary linear code also. Such a code will correct m repeated bursts of length b or less. Comparing this matrix with the usual procedure of constructing a parity check matrix H given by Varshamov-Gilbert-Sacks bound, it can be seen that a column h_j can be added to the matrix H_J provided that it is not a linear combination of immediately preceding $b-1$ columns together with any $(2m-1)b$ or less consecutive columns from the remaining $j-b$ columns. So we start with (1000...0), second column (0200...0) and keep on adding the columns in H to get a $2 mb \times (2m + 1)b$ matrix as defined above in (3). It can be easily verified that the condition given by Dass and Verma [5] in the proof of theorem 4 is satisfied. Thus the (5k, k; b, q, m) non binary code which is the null space the matrix H as constructed above will correct all m repeated bursts of length b or less.

Also it is clear that for any given feasible integer value of the parameters k, m and b, a matrix of the type H as given in (2) can always be constructed and can be used to correct m repeated bursts of length b or less.

3 Illustration of Burst Error Correcting Linear Code for k = 3, b = 2, q = 3, m = 2

Consider a parity check matrix for (15, 3) code as shown below. It can be verified from the Error Pattern - Syndrome table that the code so constructed corrects all repeated bursts of length 2 or less.

$$H = \begin{pmatrix}
1 & 0 & 0 & 0 & 0 & 0 & 0 & 0 & 0 & 0 & 0 & 0 & 1 & 0 & 0 \\
0 & 2 & 0 & 0 & 0 & 0 & 0 & 0 & 0 & 0 & 0 & 0 & 0 & 2 & 0 \\
0 & 0 & 1 & 0 & 0 & 0 & 0 & 0 & 0 & 0 & 0 & 0 & 0 & 0 & 1 \\
0 & 0 & 0 & 2 & 0 & 0 & 0 & 0 & 0 & 0 & 0 & 0 & 1 & 0 & 0 \\
0 & 0 & 0 & 0 & 1 & 0 & 0 & 0 & 0 & 0 & 0 & 0 & 0 & 2 & 0 \\
0 & 0 & 0 & 0 & 0 & 2 & 0 & 0 & 0 & 0 & 0 & 0 & 0 & 0 & 1 \\
0 & 0 & 0 & 0 & 0 & 0 & 1 & 0 & 0 & 0 & 0 & 0 & 1 & 0 & 0 \\
0 & 0 & 0 & 0 & 0 & 0 & 0 & 2 & 0 & 0 & 0 & 0 & 0 & 2 & 0 \\
0 & 0 & 0 & 0 & 0 & 0 & 0 & 0 & 1 & 0 & 0 & 0 & 0 & 0 & 1 \\
0 & 0 & 0 & 0 & 0 & 0 & 0 & 0 & 0 & 2 & 0 & 0 & 1 & 0 & 0 \\
0 & 0 & 0 & 0 & 0 & 0 & 0 & 0 & 0 & 0 & 1 & 0 & 0 & 2 & 0 \\
0 & 0 & 0 & 0 & 0 & 0 & 0 & 0 & 0 & 0 & 0 & 2 & 0 & 0 & 1
\end{pmatrix}$$

The Syndromes for the parity check matrix given above can be obtained easily with the help of MS-Excel. When we check these syndromes then we will see that these all syndromes are distinct. This verifies that the (15, 3) code can correct all repeated bursts of length 2 and less.

4 Generalization of the Parity Check Matrix for (5k, k; b, q, m) Linear Codes for All Values of q

Let us define a diagonal matrix A such that A_2 is a 2×2 matrix denoted as

$$A_2 = \begin{pmatrix} 1 & 0 \\ 0 & q-1 \end{pmatrix}$$

Similarly a 3×3 matrix is given as

$$A_3 = \begin{pmatrix} 1 & 0 & 0 \\ 0 & q-1 & 0 \\ 0 & 0 & 1 \end{pmatrix} \tag{3}$$

The diagonal elements of this matrix are in the sequence 1, q – 1, 1, q – 1, 1, q – 1, 1... etc. In general the parity check matrix for (5k, k; b, q, m) repeated burst error correcting codes for all feasible values of q may be written as

$$H = H_A = \begin{pmatrix} A_k \\ \cdot \\ A_k & \cdot \\ & \cdot \\ & A_k \end{pmatrix} \tag{4}$$

where A's are matrices of type (3).

5 Discussion

(a) Alternatively, matrix (4), comes out to be

$$H_A = \begin{pmatrix} 1 & 0 & 0\,0\,0\,0\,0\ldots & 0 & 1 & 0 & 0 \\ 0 & q-1 & 0\,0\,0\,0\,0\ldots & 0 & 0 & q-1 & 0 \\ 0 & 0 & 1\,0\,0\,0\,0\ldots & 0 & 0 & 0 & 1 \\ \cdot & \cdot & \cdots\cdots\cdots & \cdot & \cdot & \cdot & \cdot \\ \cdot & \cdot & \cdots\cdots\cdots & \cdot & \cdot & \cdot & \cdot \\ \cdot & \cdot & \cdots\cdots\cdots & \cdot & \cdot & \cdot & \cdot \\ 0 & 0 & 0\,0\,0\,0\,0\ldots & q-1 & 0 & 0 & 1 \end{pmatrix} \tag{5}$$

(b) Now substituting $q = 2$ in matrix (5) we get

$$H_A = H = \begin{pmatrix} I_k \\ I_k \\ I_{4k} & \cdot \\ & \cdot \\ & I_k \end{pmatrix}$$

Which is a parity check matrix given by Dass and Verma [4] for m repeated burst error correcting linear codes in binary case.

(c) Substituting $q = 3$ in (5), the resultant matrix comes out to be

$$H_A = H_J = \begin{pmatrix} J_k \\ J_k \\ J_{4k} & \cdot \\ & \cdot \\ & J_k \end{pmatrix}$$

Where H_A is the matrix given above resembles with parity check matrix given in (2) for non binary linear codes for $q = 3$.

6   Conclusion and Open Problem

We have shown in this paper that a non binary repeated burst error correcting code exists for $q = 3$, $m = 2$ and $b = 2$. We have also given in (4) a parity check matrix for all possible suitable integer values of k, m and burst length b. Although we have discussed in detail the formulation of a parity check matrix for repeated burst error correcting codes over GF(q), we could only verify the matrix for $q = 3$. It needs further verification for larger values of q.

Acknowledgement. The authors are thankful to Bharat Garg and Preeti for their technical assistance.

References

1. Abramson, N.M.: A class of systemic codes for non independent errors. IRE Trans. Inf. Theor. **IT-5**(4), 150–157 (1959)
2. Bridwell, J.D., Wolf, J.K.: Burst distance and multiple burst correction. Bell Syst. Tech. J. **99**, 889–909 (1970)
3. Chien, R.T., Tang, D.T.: Definition of a burst. IBM J. Res. Dev. **9**(4), 292–293 (1965)
4. Dass, B.K., Verma, R.: Construction of m-repeated bursts error correcting binary linear code. Discret. Math. Algorithms Appl. **4**(3), 1250043 (2012). (7 p.)
5. Dass, B.K., Verma, R.: Repeated burst error correcting linear codes. Asian Eur. J. Math. **1**(3), 303–335 (2008)
6. Fire, P.: A class of multiple error correcting binary Codes for non independent errors, Sylvania report RSL- E-2. Sylvania Reconnaissance Systems Laboratory, Mountain View (1959)
7. Hamming, R.W.: Error detecting and error correcting codes. Bell Syst. Tech. J. **29**, 147–160 (1950)
8. Peterson, W.W., Weldon Jr., E.J.: Error Correcting Codes, 2nd edn. The MIT Press, Mass (1972)
9. Posner, E.C.: Simultaneous error correction and burst error detecting using binary linear cyclic codes. J. Soc. Ind. Appl. Math. **13**(4), 1087–1095 (1965)
10. Srinivas, K.V., Jain, R., Saurav, S., Sikdar, S.K.: Small-world network topology of hippocampal neuronal network is lost in an in vitro glutamate injury model of epilosy. Eur. Neurosci. **25**, 3276–3280 (2007)
11. Wolf, J.K.: On codes derivable from the tensor product of matrices. IEEE Trans. Inf. Theor. **IT-11**(2), 281–284 (1965)
12. Wyner, A.D.: Low-density-burst-correcting codes. IEEE Trans. Inf. Theor. **9**, 124 (1963)

Amalgamations and Equitable Block-Colorings

E. B. Matson$^{(\boxtimes)}$ ⓘ and C. A. Rodger

Department of Mathematics and Statistics, Auburn University,
221 Parker Hall, Auburn, AL 36849, USA
{eab0052,rodgec1}@auburn.edu

Abstract. An H-decomposition of G is a partition P of $E(G)$ into blocks, each element of which induces a copy of H. Amalgamations of graphs have proved to be a valuable tool in the construction of H-decompositions. The method can force decompositions to satisfy fairness notions. Here the use of the method is further applied to (s, p)-equitable block-colorings of H-decompostions: a coloring of the blocks using exactly s colors such that each vertex v is incident with blocks colored with exactly p colors, the blocks containing v being shared out as evenly as possible among the p color classes. Recently interest has turned to the color vector $V(E) = (c_1(E), c_2(E), \ldots, c_s(E))$ of such colorings. Amalgamations are used to construct (s, p)-equitable block-colorings of C_4-decompositions of $K_n - F$ and K_2-decompositions of K_n, focusing on one unsolved case with each where c_1 is as small as possible and c_2 is as large as possible.

1 Introduction

An H-decomposition of a graph G is an ordered pair (V, B) where V is the vertex set of G and B is a partition of the edges of G into sets, each of which induces a copy of H. The elements of B are known as the blocks of the decomposition. An H-decomposition (V, B) is said to have an (s, p)-equitable block-coloring $E : B \mapsto C = \{1, 2, \ldots, s\}$ if:

(i) the blocks in B are colored with exactly s colors,
(ii) for each vertex $u \in V(G)$ the blocks containing u are colored using exactly p colors, and
(iii) for each vertex $u \in V(G)$ and for each $\{i, j\} \subset C(E, u)$,
    $|b(E, u, i) - b(E, u, j)| \leq 1$,

where $C(E, u) = \{i \mid \text{some block incident with } u \text{ is colored } i \text{ by } E\}$ and $b(E, u, i)$ is the number of blocks in B containing u that are colored i by E. Such colorings have been considered by several authors, including L. Gionfriddo, M. Gionfriddo, Hork, Li, Matson, Milazzo, Ragusa and Rodger (see [5–7,13,14]), the work focusing on cases where $H \in \{C_3, C_4\}$ and $G \in \{K_n, K_n - F\}$, where F is a 1-factor of K_n. The main focus in these papers was to find the smallest and largest possible values of s for each fixed value of p.

© Springer Nature Singapore Pte Ltd. 2018
D. Ghosh et al. (Eds.): ICMC 2018, CCIS 834, pp. 42–50, 2018.
https://doi.org/10.1007/978-981-13-0023-3_5

More recently, the research has turned to the structure of such colorings in the form of the color vector $V(E) = (c_1(E), c_2(E), \ldots, c_s(E))$ of an (s, p)-equitable block-coloring E of G, where $c_i(E)$ is the number of vertices in G incident with a block of color i arranged in non-decreasing order. Of most interest are the extreme values of $c_i(E)$, thus motivating the following definitions.

Definition 1. *For any graphs G and H and for $1 \le i \le s$, define*

(i) $\phi(H, G; s, p, i) = \{c_i(E) \mid E \text{ is an } (s, p)\text{-equitable block-coloring of an } H\text{-decomposition of } G\}$.
(ii) $\psi'(H, G; s, p, i) = \min \phi(H, G; s, p, i)$,
(iii) and $\overline{\psi'}(H, G; s, p, i) = \max \phi(H, G; s, p, i)$.

Considering the tightest cases where s is as small as possible for a given value of p is a naturally challenging problem. Often this means that the $s = p$ case is being addressed, and so it is natural to construct such colorings by using path interchange techniques that abound in graph theory. But in rarer cases it turns out that s is always greater than p, requiring new methods to make progress to construct the colorings, hence the motivation for proving Theorems 1 and 2 below. (Interchanging colors along paths can introduce new colors at the end blocks, potentially contravening the requirement that exactly p colors appear on blocks at each vertex.) Before stating Theorem 1, some notation needs to be introduced.

Throughout this paper the focus is on the case where $(H, G) = (C_4, K_{v'} - F)$, and the related case where $(H, G) = (K_2, K_v)$ described below. In this context it has been shown that the only situation where s is always greater than p is when $v' \equiv 4t + 2 \pmod{8t}$ (for example, see [14,15]), in which case if s is as small as possible then $(s, p) = (2t + 1, 2t)$ for some integer t. So for the rest of the paper we now assume that $(s, p) = (2t + 1, 2t)$ and that $v' = 8tx + 4t + 2$ for some integer x; so clearly $v' > 1$ and $t \ge 1$. It is also convenient to define $\psi'(H, G; 2t + 1, 2t, i) = \psi'_i(H, G)$ and $\overline{\psi'}(H, G; 2t + 1, 2t, i) = \overline{\psi'}_i(H, G)$. Since each vertex u in $K_{v'} - F$ obviously has degree $8tx + 4x$, which is divisible by $2p = 4t$, u is contained in exactly $b'(v') = \frac{v'-2}{4t} = 2x + 1$ blocks in each of the $p = 2t$ color classes appearing at u (each block, being a 4-cycle, contains 2 edges incident with u). We are now ready to state the following theorem.

Theorem 1 [15]. *Let $v' \equiv 4t + 2 \pmod{8t}$. Let $4t \le 2b'(v') + 2$. Then*

(1) $\psi'_1(C_4, K_{v'} - F) = 2b'(v') + 2$ *and*
(2) for $3 \le i \le 2t + 1$, $\overline{\psi'_i}(C_4, K_{v'} - F) = v' - 2$.

Notice that there is an unsolved case left in Theorem 1, namely finding $\overline{\psi'_2}(C_4, K_{v'} - F)$; this is the one value of i where $\overline{\psi'_i}(C_4, K_{v'} - F)$ is not always either the obvious lower or upper bound on the size of a color class, so it is particularly challenging to find. In this paper, $\overline{\psi'_2}(C_4, K_{v'} - F)$ is found (see Corollary 1) by solving a related edge-coloring problem in Theorem 4 which is proved using the method of amalgamations of graphs (graph homomorphisms). This construction is then modified in Sect. 4 to provide a new proof of Theorem 1.

Amalgamations provide a versatile proof technique that has been used in the study of factorizations of graphs and Steiner triple systems, but its use in block-colorings is relatively new.

Pursuing this approach in more detail, it is shown in [15] that Theorem 1 is a direct consequence of the existence of a $(2t+1, 2t)$-equitable edge-coloring of K_v, where $v = v'/2$ (or, more precisely, a $(2t+1, 2t)$-equitable block-coloring of the obvious K_2-decomposition of K_v), so $v = 4tx + 2t + 1$ for some integer x; clearly if $v > 1$, then $t \geq 1$. Each vertex u has degree $4tx + 2t$, which is clearly divisible by $p = 2t$, so u is contained in exactly $b(v) = \frac{v-1}{2t} = 2x + 1 = b'(v')$ blocks (edges) in each of the $2t$ color classes appearing at u. Note that $b(v)$ is odd. In Sect. 4, a new proof of the following result is presented (and by the discussion above, a new proof of Theorem 1 as well).

Theorem 2 [15]. *Let $v \equiv 2t + 1 \pmod{4t}$ with $v > 1$. Let $2t \leq b(v) + 1$. Then,*

(1) $\psi_1'(K_2, K_v) = b(v) + 1$ and
(2) for $3 \leq i \leq 2t + 1$, $\overline{\psi_i'}(K_2, K_v) = v - 1$.

It is worth noting that a more generalized result in [15] complements Theorems 1 and 2, addressing the cases where $4t \geq 2b'(v') + 2$ and $2t \geq b(v) + 1$, showing that then $\overline{\psi_2'}(C_4, K_{v'} - F) = v' - 1$ and $\overline{\psi_2'}(K_2, K_v) = v - 1$ respectively.

The following notation will be useful throughout the paper. Let $K[R]$ denote the complete graph defined on the vertex set R. Color i is said to appear at a vertex u if at least one block incident with u is colored i.

2   Some Preliminary Results

In order to find $\overline{\psi_2'}(C_4, K_{v'} - F)$ and $\overline{\psi_2'}(K_2, K_v)$, we begin by finding bounds on the value of c_2 in the following Lemmas, utilizing some results proved in [14,15]. For ease of notation define $\lfloor x \rfloor_e$ to be the largest even integer less than or equal to x.

Lemma 1. *For $v \equiv 2t + 1 \pmod{4t}$ and $v' = 2v$,*

$$\overline{\psi_i'}(K_2, K_v) \leq \left\lfloor \frac{2tv - \sum_{j=1}^{i-1} \psi_j'(K_2, K_v)}{2t + 2 - i} \right\rfloor_e \quad and$$

$$\overline{\psi_i'}(C_4, K_{v'} - F) \leq \left\lfloor \frac{2tv' - \sum_{j=1}^{i-1} \psi_j'(C_4, K_{v'} - F)}{2t + 2 - i} \right\rfloor_e .$$

Proof. Note the elements of the color vector are listed in non-decreasing order; and since in Lemma 2.5 of [14] it is shown that for any $(2t+1, 2t)$-equitable edge-coloring E of K_v and for any $(2t + 1, 2t)$-equitable C_4-coloring E' of $K_{v'} - F$, both $\sum_{i=1}^{2t+1} c_i(E) = 2tv$ and $\sum_{i=1}^{2t+1} c_i(E') = 2tv'$, the above holds.   □

Lemma 2. *Let* $v = v'/2 = 4tx + 2t + 1$ *for some integer* x *and* $b(v) + 1 = b'(v') + 1 \geq 2t$. *Then,*

$$\overline{\psi}_2'(K_2, K_v) \leq \left\lfloor v - \frac{x+1}{t} \right\rfloor_e \quad \text{and} \quad \overline{\psi}_2'(C_4, K_{v'} - F) \leq \left\lfloor v' - \frac{2x+2}{t} \right\rfloor_e.$$

Proof. Let $b(v)+1 \geq 2t$. By Theorem 3.5 of [15], $\psi_1'(K_2, K_v) = b(v)+1$. Therefore by Lemma 1:

$$\begin{aligned}
\overline{\psi}_2'(K_2, K_v) &\leq \lfloor \frac{2tv - (b(v) + 1)}{2t + 2 - 2} \rfloor_e \\
&= \lfloor v - \frac{(b(v) + 1)}{2t} \rfloor_e \\
&= \lfloor v - \frac{2x + 2}{2t} \rfloor_e \\
&= \lfloor v - \frac{x + 1}{t} \rfloor_e.
\end{aligned}$$

By Corollary 3.6 of [15], $\psi_1'(C_4, K_{v'} - F) = 2b'(v') + 2$. Therefore by Lemma 1:

$$\begin{aligned}
\overline{\psi}_2'(C_4, K_{v'} - F) &\leq \left\lfloor \frac{2tv' - (2b'(v') + 2)}{2t + 2 - 2} \right\rfloor_e \\
&= \left\lfloor v' - \frac{(2b'(v') + 2)}{2t} \right\rfloor_e \\
&= \left\lfloor v' - \frac{2x + 4}{2t} \right\rfloor_e \\
&= \left\lfloor v' - \frac{2x + 2}{t} \right\rfloor_e.
\end{aligned}$$

$\square$

3   Settling the Unsolved Cases in Theorems 1 and 2

Apart from completing the open case left in Theorems 1 and 2, in this paper the use of amalgamations in block-decompositions is further demonstrated. Hilton and Rodger [8,9] used this technique to find embeddings of edge-colorings into hamiltonian decompositions. Buchanan [2] used amalgamations to find hamiltonian decompositions of $K_n - E(U)$ for any 2-regular spanning subgraph U, and this was extended to various multipartite graphs by Leach and Rodger [10,12]. Leach and Rodger [11] went on to find hamilton decompositions of complete multipartite graphs where each hamilton cycle spreads its edges out as evenly as possible among the pairs of parts of the graph. This notion has recently been extended by Erzurumluoğlu and Rodger [3,4] to factorizations and holey factorizations of complete multipartite graphs and then to C_4-decompositions of $K_v - F$ and edge-decompositions of K_v by Matson and Rodger in [15].

Formally, a graph H is said to be an *amalgamation* of a graph G if there exists a function ψ from $V(G)$ onto $V(H)$ and a bijection $\psi' : E(G) \to E(H)$ such that $e = \{u, v\} \in E(G)$ if and only if $\psi'(e) = \{\psi(u), \psi(v)\} \in E(H)$. The function ψ is called an amalgamation function. We say that G is a *detachment* of H, where each vertex u of H splits into the vertices of $\psi^{-1}(\{u\})$. An η-*detachment* of H is a detachment in which each vertex u of H splits into $\eta(u)$ vertices.

To describe the amalgamation result used here more precisely, some notation will be needed. Let $x \approx y$ represent the fact that $\lfloor y \rfloor \leq x \leq \lceil y \rceil$. Furthermore, let $\ell(u)$ denote the number of loops incident with vertex u, where each loop contributes twice to the degree of u, let $G(j)$ denote the subgraph of G induced by the edges colored j, and let $m(u, v)$ denote the number of edges between the pair of vertices u and v in G.

The following is a special case of Theorem 3.1 in [1] (omitting the condition that ensures color classes are connected and a balanced property on the color classes for multigraphs since in our case G is simple).

Theorem 3 (Bahmanian and Rodger [1, Theorem 3.1]). *Let H be a k-edge-colored graph and let η be a function from $V(H)$ into $\mathbb{N}$ such that for each $v \in V(H)$, $\eta(v) = 1$ implies $\ell_H(v) = 0$. Then there exists a loopless η-detachment G of H in which each $v \in V(H)$ is detached into $v_1, \ldots, v_{\eta(v)}$, such that G satisfies the following conditions:*

1. $d_G(u_i) \approx d_H(u)/\eta(u)$ *for each $u \in V(H)$ and $1 \leq i \leq \eta(u)$;*
2. $d_{G(j)}(u_i) \approx d_{H(j)}(u)/\eta(u)$ *for each $u \in V(H)$, $1 \leq i \leq \eta(u)$, and $1 \leq j \leq k$;*
3. $m_G(u_i, u_{i'}) \approx \ell_H(u)/\binom{\eta(u)}{2}$ *for each $u \in V(H)$ with $\eta(u) \geq 2$ and $1 \leq i < i' \leq \eta(u)$; and*
4. $m_G(u_i, v_{i'}) \approx m_H(u, v)/(\eta(u)\eta(v))$ *for every pair of distinct vertices $u, v \in V(H)$, $1 \leq i \leq \eta(u)$, and $1 \leq i' \leq \eta(v)$.*

We now complete the open case left in Theorem 2 as stated here as Theorem 4. As explained in the introduction, as a result of Theorem 4, we also complete the open case left in Theorem 1, stated here as Corollary 1, using the method of amalgamations in both.

Theorem 4. *Let $v \equiv 2t + 1 \pmod{4t}$ with $v > 1$. Let $2t \leq b(v) + 1$. Then*

$$\overline{\psi'_2}(K_2, K_v) = \left\lfloor v - \frac{x+1}{t} \right\rfloor_e.$$

Proof. Let $v = 4tx + 2t + 1$ for some integer x. Form a complete graph $\mathcal{G}_0$ on the set of vertices $V_0 = \{u_1, \ldots, u_{2x+2}\}$ and color all the edges of $\mathcal{G}_0$ with color $2t + 1$. So each vertex in $\mathcal{G}_0$ is incident with $2x + 1 = b(v)$ edges colored $2t + 1$ as desired. Notice that in the final edge-coloring of K_v, each vertex is missing (i.e., is not incident with any edges of) exactly one color. We will arrange for $1 \leq i \leq 2t$, color $m(i) = i$ to be missing from vertex u_i, for $2t + 1 \leq i \leq 2x + 2$ color $m(i) = \lceil \frac{i-2t}{2} \rceil \pmod{2t} \in \{1, \ldots, 2t\}$ to be missing from u_i, and color $m(\alpha_i) = 2t + 1$ to be missing from the remaining $v - 2x - 2$ vertices (which will be named $\alpha_1, \ldots, \alpha_{\eta(\alpha)}$ below). For $1 \leq i \leq 2t$ let $M(i) = \{u_j \in V_0 \mid m(j) = i\}$. Note for $1 \leq i < j \leq 2t$, $\mid |M(i)| - |M(j)| \mid \in \{0, 2\}$ and $|M(i)|$ is odd for all i.

Next form a new edge-colored graph $\mathcal{G}_0^+$ from $\mathcal{G}_0$ as follows. Add a single vertex, α. The aim now is to complete the proof by using Theorem 3 with $\eta(u_i) = 1$ for $1 \leq i \leq 2x + 2$ and $\eta(\alpha) = v - 2x - 2$. For $1 \leq i \leq 2x + 2$ join u_i to α with $b(v)$ edges of each color in $\{1, 2, \ldots, 2t\} \setminus \{m(i)\}$. Thus for $1 \leq i \leq 2x + 2$ the number of edges joining u_i to α is $(2t-1)(2x+1) = 4tx + 2t + 1 - (2x+1) - 1 = v - 1 - (2x + 1) = \eta(\alpha)$, and $d_{\mathcal{G}_0^+}(u_i) = v - 1$.

Let $a(i)$ be the number of vertices in $\mathcal{G}_0^+$ where color i appears and let $\epsilon_i = 2$ for $1 \leq i \leq x + 1 - t \pmod{2t}$ and $\epsilon_i = 0$ otherwise. Therefore $a(2t + 1) = 2x + 2$ and for $1 \leq i \leq 2t$,

$$a(i) = 2x + 3 - |M(i)|$$
$$= 2x + 2 - 2 \left\lfloor \frac{2x + 2 - 2t}{4t} \right\rfloor - \epsilon_i.$$

Note since $x \geq 0$ and $t \geq 1$ for $1 \leq i \leq 2t$,

$$\eta(\alpha) - (a(i) - 1) = v - 2x - 2 - \left(2x + 2 - 2 \left\lfloor \frac{2x + 2 - 2t}{4t} \right\rfloor - \epsilon_i - 1 \right)$$
$$= v - 4x - 3 - 2 \left\lfloor \frac{2x + 2 - 2t}{4t} \right\rfloor + \epsilon_i$$
$$\geq 4tx + 2t - 2 - 4x - \left(\frac{2x + 2 - 2t}{2t} \right)$$
$$= 4x(t - 1) + 2t - 1 - \frac{x + 1}{t}$$
$$= (4x + 1)(t - 1) + t - \frac{x + 1}{t} \geq 0.$$

Thus for $1 \leq i \leq 2t$ add $(b(v)\eta(\alpha) - b(v)(a(i) - 1))/2$ loops colored i to α, thus resulting in $d_{\mathcal{G}_0^+(i)}(\alpha) = b(v)\eta(\alpha)$. By the above calculations we know we will be adding a non-negative number of loops for all colors $1, \ldots, 2t$.

Let $l(\alpha)$ be the number of loops incident with α and $E(V(G_0), \alpha)$ be the set of edges from a vertex in G_0 to α. Therefore,

$$l(\alpha) = \left(d_{G_0^+}(\alpha) - |E(V(G_0), \alpha)| \right) / 2$$
$$= (\eta(\alpha)b(v)2t - (2x + 2)[b(v)(2t - 1)]) / 2$$
$$= (\eta(\alpha)b(v)2t - (2x + 2)\eta(\alpha)) / 2$$
$$= \eta(\alpha) (b(v)2t - 2x - 2) / 2$$
$$= \eta(\alpha) (4tx + 2t + 1 - 2x - 3) / 2$$
$$= \eta(\alpha)(v - 2x - 2 - 1)/2$$
$$= \eta(\alpha)(\eta(\alpha) - 1)/2.$$

Now apply Theorem 3 to form the detachment $\mathcal{G}$ of $\mathcal{G}_0^+$ in which α is detached into the vertices $\alpha_1, \ldots, \alpha_{\eta(\alpha)}$. For $1 \leq i \leq 2x + 2$, since u_i is joined to α with $b(v)$ edges in $\mathcal{G}_0^+$, by condition (3) u_i is joined to each vertex α_j for $1 \leq j \leq \eta(\alpha)$ by exactly one edge in $\mathcal{G}$. Also, since α is incident with $\eta(\alpha)(\eta(\alpha)-1)/2$ loops in $\mathcal{G}_0^+$, by condition (4) α_i is joined to α_j by exactly one edge for $1 \leq i < j \leq \eta(\alpha)$ in $\mathcal{G}$. It follows that $\mathcal{G}$ is isomorphic to $K_{2x+2+\eta(\alpha)} = K_v$. By condition (2), for each vertex u in $\mathcal{G}$, each color which appears at u does so on $b(v)$ edges. Therefore the edge-coloring E of $\mathcal{G}$ is $(2t + 1, 2t)$-equitable. Furthermore, in $\mathcal{G}$, color $2t + 1$ appears at $b(v) + 1 \geq 2t$ vertices and for $1 \leq i \leq 2t$, the number of vertices where color i appears is

$$a(i) - 1 + \eta(\alpha) = (2x + 2) - 2\left\lfloor \frac{x + 1 - t}{2t} \right\rfloor - \epsilon_i - 1 + v - (2x + 2)$$

$$= v - 1 - 2\left\lfloor \frac{x + 1 - t}{2t} \right\rfloor - \epsilon_i.$$

Therefore, since $a(i)$ and $\eta(\alpha)$ are both odd integers, if $2t$ divides $(x + 1 - t)$, then $\epsilon_1 = 0$ and

$$a(1) - 1 + \eta(\alpha) = v - 1 - \frac{x + 1 - t}{t}$$

$$= v - \frac{x + 1}{t}$$

$$= \left\lfloor v - \frac{x + 1}{t} \right\rfloor_e ,$$

and if $2t$ does not divide $(x + 1 - t)$ then $\epsilon_1 = 2$ and

$$a(1) - 1 + \eta(\alpha) = v - 1 - \left(2\left\lfloor \frac{x + 1 - t}{2t} \right\rfloor + 2 \right)$$

$$= v - 1 - 2\left\lceil \frac{x + 1 - t}{2t} \right\rceil$$

$$= v - 1 + 2\left\lfloor \frac{-(x + 1 - t)}{2t} \right\rfloor$$

$$= 2\left\lfloor \frac{v - 1}{2} + \frac{1}{2} - \frac{x + 1}{2t} \right\rfloor$$

$$= 2\left\lfloor \frac{v}{2} - \frac{x + 1}{2t} \right\rfloor$$

$$= \left\lfloor v - \frac{x + 1}{t} \right\rfloor_e .$$

Therefore by Lemma 2, $\overline{\psi_2'}(K_2, K_v) = \lfloor v - \frac{x+1}{t} \rfloor_e$ and the proof is complete (after renaming color $2t + 1$ with 1 and renaming colors $1, 2, \ldots, 2t$ with $2, 3, \ldots, 2t + 1$ respectively). $\qquad \square$

Corollary 1. *Let $v' \equiv 4t + 2 \pmod{8t}$. Let $2t \leq b'(v') + 1$. Then*

$$\overline{\psi'_2}(C_4, K_{v'} - F) = \lfloor v - \frac{2x+2}{t} \rfloor_e.$$

Proof. By Theorem 4 fo $v = v'/2$ there exists a $(2t+1, 2t)$-equitable edge-coloring E of K_v such that $c_2(E) = \lfloor v - \frac{x+1}{t} \rfloor_e$. So as explained in [15] there exists a $(2t + 1, 2t)$-equitable C_4-coloring E' of $K_{v'} - F$ such that $c_2(E') = 2c_2(E) = 2\lfloor v - \frac{x+1}{t} \rfloor_e = \lfloor 2 \left(v - \frac{x+1}{t} \right) \rfloor_e = \lfloor 2v - \frac{2x+2}{t} \rfloor_e = \lfloor v' - \frac{2x+2}{t} \rfloor_e$. Therefore by Lemma 2, $\overline{\psi'_2}(C_4, K_{v'} - F) = \lfloor v - \frac{2x+2}{t} \rfloor_e$.   □

4   A New Proof of Theorems 1 and 2

By modifying the proof of Theorem 4 we obtain a new proof of Theorem 2, and as explained in the introduction, a new proof of Theorem 1 as well, using amalgamations.

Proof. Let $v = 4tx + 2t + 1$ for some integer x. Form $\mathcal{G}_0$ in the same way as in Theorem 4. Here color $m(i) = i$ will be missing from vertex u_i for $1 \leq i \leq 2t - 1$, color $m(i) = 2t$ will be missing from vertex u_i for $2t \leq i \leq 2x + 2$, and color $m(\alpha_i) = 2t + 1$ will be missing from the remaining $v - 2x - 2$ vertices (which will be named $\alpha_1, \ldots, \alpha_{\eta(\alpha)}$ below).

Next form a new edge-colored graph $\mathcal{G}_0^+$ as in Theorem 4 and again the aim now is to complete the proof using Theorem 3 with $\eta(u_i) = 1$ for $1 \leq i \leq 2x + 2$ and $\eta(\alpha) = v - 2x - 2$. For $1 \leq i \leq 2x+2$ join u_i to α with $b(v)$ edges of each color $\{1, 2, \ldots, 2t\} \setminus \{m(i)\}$ as in Theorem 4; again the number of edges joining u_i to α is $\eta(\alpha)$, and $d_{\mathcal{G}_0^+}(u_i) = v - 1$. For $1 \leq i \leq 2t - 1$ add $b(v)(\eta(v) - (2x+1))/2$ loops of color i to α; so α has degree $b(v)\eta(v)$ in color class i (where loops contribute 2 to the degree of the incident vertex). Also add $b(v)(\eta(v) - (2t - 1))/2$ loops of color $2t$ to α; so α has degree $b(v)\eta(v)$ in color class $2t$ as well. Notice that the number of loops incident with α is

$$\begin{aligned}
l(\alpha) &= (2t - 1)b(v)(\eta(\alpha) - (2x + 1)/2) + b(v)(\eta(\alpha) - (2t - 1))/2 \\
&= (2t(2x + 1)\eta(\alpha) - (2x + 1)(2t - 1)(2x + 2))/2 \\
&= (2x + 1)(2t\eta(\alpha) - (4xt - 2x - 4t - 2))/2 \\
&= (2x + 1)(2t\eta(\alpha) - (\eta(\alpha) + 2t - 1))/2 \\
&= (2x + 1)(\eta(\alpha) - 1)(2t - 1))/2 \\
&= \eta(\alpha)(\eta(\alpha) - 1)/2.
\end{aligned}$$

As in the proof of Theorem 4, Theorem 3 allows us to form $\mathcal{G}$ isomorphic to K_v from $\mathcal{G}_0^+$ so that the edge-coloring E of $\mathcal{G}$ is $(2t+1, 2t)$-equitable. Furthermore, in $\mathcal{G}$, color $2t+1$ appears at $b(v)+1$ vertices, color $2t$ appears at $v - 2t - 1$ vertices, and each other color appears at $v - 1$ vertices. Since in [15] it is shown in this case that $\psi'_i(K_2, K_v) \geq b(v) + 1$ and that $\overline{\psi'_i}(K_2, K_v) \leq v - 1$ for $1 \leq i \leq 2t + 1$, the proof is complete (after renaming the colors $1, 2, \ldots, 2t + 1$ with $2t + 1, 2t, \ldots, 1$ respectively).   □

References

1. Bahmanian, M.A., Rodger, C.A.: Multiply balanced edge-colorings of multigraphs. J. Graph Theory **70**, 297–317 (2012). https://doi.org/10.1002/jgt.20617
2. Buchanan, H.: Graph factors and hamiltonian decompositions. Ph.D. Dissertation. University of West Virginia (1998)
3. Erzurumluoğlu, A., Rodger, C.A.: Fair holey hamiltonian decompositions of complete multipartite graphs and long cycle frames. Discret. Math. **338**, 1173–1177 (2015). https://doi.org/10.1016/j.disc.2015.01.035
4. Erzurumluoğlu, A., Rodger, C.A.: Fair 1-factorizations and fair holey 1-factorizations of complete multipartite graphs. Graphs Comb. **32**, 1377–1388 (2016). https://doi.org/10.1007/s00373-015-1648-9
5. Gionfriddo, L., Gionfriddo, M., Ragusa, G.: Equitable specialized block-colourings for 4-cycle systems-I. Discret. Math. **310**, 3126–3131 (2010). https://doi.org/10.1016/j.disc.2009.06.032
6. Gionfriddo, M., Hork, P., Milazzo, L., Rosa, A.: Equitable specialized block-colourings for Steiner triple systems. Graphs Combin. **24**, 313–326 (2008). https://doi.org/10.1007/s00373-008-0794-8
7. Gionfriddo, M., Ragusa, G.: Equitable specialized block-colourings for 4-cycle systems-II. Discret. Math. **310**(13–14), 1986–1994 (2010). https://doi.org/10.1016/j.disc.2010.03.018
8. Hilton, A.J.W.: Hamilton decompositions of complete graphs. J. Comb. Theory Ser. B **36**, 125–134 (1984). https://doi.org/10.1016/0095-8956(84)90020-0
9. Hilton, A.J.W., Rodger, C.A.: Hamilton decompositions of complete regular s-partite graphs. Discret. Math. **58**, 63–78 (1986). https://doi.org/10.1016/0012-365X(86)90186-X
10. Leach, C.D., Rodger, C.A.: Non-disconnecting disentanglements of amalgamated 2-factorizations of complete multipartite graphs. J. Comb. Des. **9**, 460–467 (2001). https://doi.org/10.1002/jcd.1024
11. Leach, C.D., Rodger, C.A.: Fair hamilton decompositions of complete multipartite graphs. J. Comb. Theory Ser. B **85**, 290–296 (2002). https://doi.org/10.1006/jctb.2001.2104
12. Leach, C.D., Rodger, C.A.: Hamilton decompositions of complete multipartite graphs with any 2-factor leave. J. Graph Theory **44**, 208–214 (2003). https://doi.org/10.1002/jgt.10142
13. Li, S., Rodger, C.A.: Equitable block-colorings of C_4-deompostions of $K_v - F$. Discret. Math. **339**, 1519–1524 (2016). https://doi.org/10.1016/j.disc.2015.12.029
14. Li, S., Matson, E.B., Rodger, C.A.: Extreme equitable block-colorings of C_4-decompositions of $K_v - F$. Australas. J. Comb. **71**(1), 92–103 (2018)
15. Matson, E.B., Rodger, C.A.: More extreme equitable colorings of decompositions of K_v and $K_v - F$. Discret. Math. **341**, 1178–1184 (2018). https://doi.org/10.1016/j.disc.2017.10.018

Computing

Reduction in Execution Cost of k-Nearest Neighbor Based Outlier Detection Method

Sanjoli Poddar$^{(\boxtimes)}$ and Bidyut Kr. Patra$^{(\boxtimes)}$

National Institute of Technology Rourkela, Rourkela, Odisha, India
sanjoli0511@gmail.com, patrabk@nitrkl.ac.in

Abstract. Outlier detection is an important task as it leads to the discovery of critical information in a variety of the application domains. The variants of k-nearest neighbor based outlier detection method have been successfully applied over decades. However, these approaches have high execution time as they compute a score (known as outlier score) for each data point. In this paper, we propose a method to reduce the execution time of k-nearest neighbor based algorithms. Proposed method quickly identifies the data points which are normal and therefore outlier score for such points need not be computed in further processing. The proposed method is generic and can be applied to improve the execution efficiency of many density-based and distance-based outlier detection methods. Proposed work is compared with other existing methods and the result shows that the proposed work outperforms other methods.

Keywords: Density based outlier detection method
k-nearest neighbor · LOF · Execution time

1 Introduction

Outliers are the observations which deviates so much from the other observations as to arouse suspicions that it was generated from a different mechanism [1]. Efficient mining of the data is very important as it finds its application in various domains such as credit card fraud analysis, intrusion detection system, medical field, marketing etc. Both supervised and unsupervised learning methods are used to identify the outliers [8]. In unsupervised learning, no prior knowledge about the data set is known. This makes the unsupervised outlier detection methods very popular over supervised approach. Popular unsupervised algorithms include clustering techniques, distance-based outlier, density based-outlier and k-nearest neighbor based methods.

Among the unsupervised learning algorithms, density based outlier detection methods are very popular and efficient for identifying the outliers. The main idea behind density based methods is to compute outlier score for each data point and declare the points with high scores as outlier points. In order to compute the outlier score, k-nearest neighbors of each data point are computed and subsequently use their statistics according to the individual algorithm. Popular density based

© Springer Nature Singapore Pte Ltd. 2018
D. Ghosh et al. (Eds.): ICMC 2018, CCIS 834, pp. 53–60, 2018.
https://doi.org/10.1007/978-981-13-0023-3_6

outlier detection method includes LOF (local outlier factor) [5], COF (connectivity outlier factor) [11], INFLO. The distance based method is also found to be using k-nearest neighbor information [9].

The outlier score corresponding to normal data point is of limited use especially when the objective is to mine out the outliers. As density and distance based approaches calculates the outlier score for every the data point, it makes the method inefficient. The problem increases in many folds with increasing the size of the data set.

In this paper, we propose a method to improve the execution time of k-nearest neighbor based outlier detection methods. In proposed method, we introduce a novel measure termed as *devToMean* to identify the normal points for which outlier scores are not required to compute in further processing. The novel measure ensures that none of the outlier points is identified as normal points. Having filtered these normal points, we only compute outlier score of the remaining data points based on the individual algorithm. The experiments are carried out both on synthetic as well as real datasets and the results show a significant improvement in execution cost over the other methods.

The rest of the paper is organized as follows. Section 2 describes state-of-the-art works in this direction. We describe proposed work in Sect. 3. Experimental results and analysis are reported in Sect. 4. We conclude our paper in Sect. 5.

2 Related Work

The broad application of the outlier detection has made the literature very rich. Widely popular outlier detection techniques include statistical approach, distance-based, density-based, rule-based, neighborhood based, *etc.* [8].

Distance and density based outlier detection techniques are widely used when no prior knowledge about the dataset is known unlike statistical approach. Knorr and Ng [2] proposed first distance-based outlier detection technique. It uses the distance parameter to find the outliers. The notion of the distance based algorithm is further extended to k-nearest neighbor distance or statistics [9]. In [9], it uses the relative location of an object to its neighbor to determine the degree to which object deviates from its neighbors. Thus the objects with the higher LDOF score (Eq. 1) are regarded as outliers.

$$LDOF_k(x_p) = \frac{d_{x_p}}{D_{x_p}} \tag{1}$$

where, $D_{x_p} = \frac{1}{k(k-1)} \sum_{x_i,x_{i'} \in kNN(x_p)} dist(x_i, x_{i'})$, d_{x_p} is the average distance from point x_p to all its k nearest neighbors, kNN is the k-nearest neighbors of the point x_p excluding the point x_p itself.

Density based outlier detection techniques use local information/statistics of each data point for computing outlier score of the point. Some of the significant works done in this area includes LOF (local outlier factor) [5], COF (connectivity based outlier factor) [11], INFLO [7], *etc.* In the popular LOF method [5], local reachibility density of a point is computed using the statistics of k-nearest neighbors of the point. Finally, it computes a score called *lof* which is the average of the ratio of local reachibility density of a point p to the density of the point p's nearest neighbors. If the factor *lof* is close to 1 then the point is considered as normal. If the value of the lof $\gg 1$, then it is declared as an outlier. INFLO [7] outlier detection approach addresses the problem of LOF in a dataset with variable density over the feature space. It considers both k-neighbors and reverse k-neighbors statistics while computing outlier score [3,4,6,10].

All the methods discussed above are proven to be very powerful and efficient in term of finding outliers. However, these approaches have high execution time as they compute outlier score for each data point. This problem becomes severe with the increase of size of the dataset. Few methods are reported to improve the execution time but they are specific to particular density based outlier detection technique while compromising the accuracy (precision) of the method.

Some of the work done to improve upon the density based outlier detection methods include LOF' [6]. Authors argued that the MinPts-dist is sufficient to find the density of a point. Subsequently, the basic formula (Eq. 1) for computing the outlierness of a point is altered. In FastLOF [10], author argue that a good estimation is fair enough for normal data points but precise nearest neighbors are required for the outliers. To reduce the execution time of k-nearest neighbor search, data set is randomly divided into data chunks and search is performed only within a single chunk for each point in it. Subsequently, approach takes a decision that which data points can be further considered to find the outliers and other are safely pruned (removed). However, in this case all outlier points may not be detected. Other pruning strategies are also developed to reduce the execution time of density and distance based approaches [3,4]. Basic idea of these research is to identify the normal points and prune them for further processing. In [4], k-means clustering method is applied over the dataset and subsequently, pruning strategy is applied to individual cluster. The points within a cluster are pruned (removed) if they locate close to the centriod (within the radius of the cluster). Finally, LDOF is applied to remaining points in the dataset. However, the genuine outlier point located close to the centriod of a cluster can be considered as normal point and it can be pruned in this approach. Another major drawback of the pruning approach is that pruning a normal point may change the characteristics of its neighbors (*i.e., normal to outlier point*). Therefore, it increases the false positive rate. Reduction of execution time of these approaches is achieved at the cost of accuracy of outlier detection mechanism.

3 Proposed Work

We address the problem of computation overhead involved in the methods discussed in previous section in a novel way. In those methods, outlier score is calculated for each data point. Intuitively, outlier scores corresponding to the normal points are not of significant use and hence this calculation can be avoided. In this proposed approach, we quickly identify most of the normal points and computation of the outlier score for these points are not performed in subsequent step. It can be noted that we do not prune these points unlike pruning strategy. Therefore, accuracy of the individual method is not compromised in this approach and it leads to reduce the execution cost of the density and distance based outlier detection methods.

Our proposed approach works in two phases. In the first phase, a linear clustering method (k-means) is applied to partition the data into a number of chunks (cluster) and centroid point of each of the cluster is computed. We aim to mark all normal data points from each of these cluster. It can be intuitively said that the normal points lie in dense region, hence its deviation (density deviation) from its neighbors is small, whereas outliers lie in the sparse region and its deviation from its neighbors would be more as compared to normal points. We use this assumption in order to identify the normal points within each cluster. We introduce a metric termed as *devToMean* for each point within a cluster. Let C_i be a cluster and m_{C_i} be the mean of C_i. The *devToMean* of a point $x \in C_i$ is the ratio of deviation of the point x from mean point to the average deviation of its neighbors from the mean of the cluster. This is computed using the following Eq. 2.

$$devToMean(x) = \frac{||x - m_{C_i}||}{\frac{1}{k} \sum_{x_i \in k-NN(x)} ||x_i - m_{C_i}||} \tag{2}$$

The *devToMean* determines how much an object deviates from the mean of the cluster to which it belongs with respect to its neighbors. If this deviation of a point is similar to that of its neighbors, then the value of *devToMean* is close to 1 and the point is considered as a normal point. The value of *devToMean* for outlier point can be high ($>> 1$) (outlier point far away from mean point) or close to 0 (outlier point close to the mean).

We mark the normal points (*devToMean* ≈ 1) and avoid computation of outlier score for these points in next phase. The marked points are not pruned (removed) from the dataset. Remaining unmarked points are sorted based on their *devToMean* values. We apply 10 percentile rule to find probable outliers in the dataset. We select the *top ten (10) percentile* and *bottom ten (10) percentile* of these sorted unmarked points as probable outliers and investigate them in the next phase. The idea of this rule is that there are few outlier points compared to the normal points and the value of *devToMean* for genuine outlier is either close to 0 or very high ($>> 1$).

The second stage involves the calculation of the outlier score of the points obtained using 10 percent rule. For computing the outlier score, one can use any popular distance or density based approach discussed in Sect. 2. While computing the k-nearest neighbor statistics of these selected points, all data points are used including the marked normal points. Therefore, accuracy of the outlier detection method is not compromised and results obtained by our approach will be the same as that of the original approach applied on the same dataset. The proposed approach is depicted in Algorithm 1. Finally, a ranking list is made for all the outlier score and top-n outliers are selected.

Algorithm 1. DevToMean outlier detection

Input: D, *num-clust*, k
Output: *outlier-score*
1: $Clusters \leftarrow k\text{-}meansClustering(D, num\text{-}clust)$
2: **for** each $c_i \in Clusters$ **do**
3:     **for** each $p_j \in c_i$ **do**
4:         Compute $devToMean(p_j)$
5:         Mark the point p_j if $devToMean(p_j) = 1$.
6:     **end for**
7: **end for**
8: $POutlier \leftarrow$ Filter the unmarked points by applying 10 percentile rule.
9: **for** each $p_i \in POutlier$ **do**
10:     Compute outlier score of p_i using entire dataset D.
11: **end for**

As the outlier score corresponding to a very less number of the data point is calculated, the execution time of the proposed algorithm is quite less as compared to the time taken by the original algorithm. The number of clusters to be provided in the first phase is very important factor. The number of clusters should be such that the size of the cluster is neither too big nor too small. If the size of a cluster is too small, then the genuine outlier point might get overlooked as it would contain the value of *devToMean* factor close to 1. Also, if the size of the cluster is too large, then the reduction in time complexity would be quite small as large volume of data set would be examined to calculate the value of *devToMean* factor. Thus, specifying the appropriate number of clusters according to the size of the data set is very important. For testing purpose, we consider that the minimum size of the cluster to be 100.

4   Experimental Results

The proposed method is tested on synthetic as well as real data set. The synthetic data set was uniformly distribution within a region. We also injected few outliers to the dataset. We took one classification dataset named *Cover Type Data* from UCI machine learning repository and converted into One class data with few injection of outlier points from other classes.

We introduce a metric termed as *speed up factor S* which measures the percentage of decrease in the execution time of the proposed method from the execution time of the original approach. The metric is normalized by the maximum decrease in execution time over various input sizes of a dataset. Let t_{dev}^m and t_o^m be the execution time of proposed approach and original approach while both of them applied on a subset of size m of a dataset, respectively. The speed up factor S_m is computed as follows.

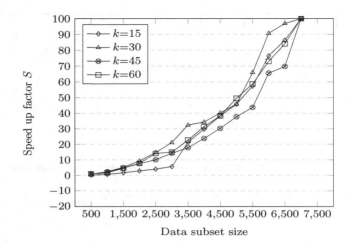

Fig. 1. Speedup factor S with varying data set and parameter k for LOF

$$S_m = \frac{t_o^m - t_{dev}^m}{\max_l\{t_o^l - t_{dev}^l\}} \times 100 \tag{3}$$

where, l is the size of a data subset. The minimum value of $S_m = 0$, when proposed approach takes exactly the same time ($t_o^m = t_{dev}^m$) as that of the original approach.

We speed LOF (density based outlier detection algorithm) up using our approach and speed up factor of the proposed approach is plotted in varying data size of the synthetic dataset in Fig. 1. It can be easily inferred from the plot that the proposed method's efficiency increases with increase in size of the data set in terms of execution cost. For the considered size of the data set, the reduction in time is more than 50% for higher values of k. We achieved a significant reduction in execution time of the LOF method over increasing the size of the dataset. This shows that proposed approach is effective in large size data. The popular distance based outlier detection method LDOF [9] speed-ed up using proposed approach and reported in Fig. 2. Similar trends are observed.

Few works are reported to reduce the execution time of the density based outlier detection techniques using pruning strategies while some of them modified the method for finding outlier detection method [4,6,10]. We compare our proposed method with FastLOF which belongs to first category on real Cover Type dataset. Results are recorded in Fig. 3. The results clearly show that our proposed method outperforms FastLOF. It can be noted that comparison with other pruning based approaches discussed in Sect. 2 are not reported here as these methods cannot produce exactly the same detection accuracy as that of the original approaches.

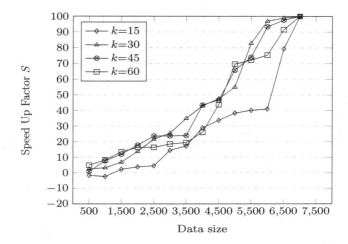

Fig. 2. Efficiency factor S with varying data set and parameter k for LDOF

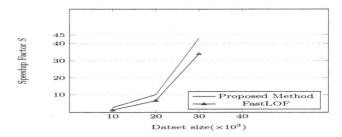

Fig. 3. Comparison of the proposed approach with FastLOF [10] with $k = 70$.

5   Conclusion

In this paper, we proposed a framework to speed up a set of popular outlier detection methods which compute outlier score for each data point using k-nearest neighbor statistics. We introduced a metric $devToMean$ which quickly identifies normal points and computation of outlier scores for these points are avoided in decision making. It is observed from experimental results that the proposed framework is very effective for large dataset and for grater value of the parameter k. In future, we can further reduce the execution time in speeding up the identification process of normal points ($devToMean$).

References

1. Hawkins, D.M.: Identification of Outliers, vol. 11. Chapman and Hall, London (1980)
2. Knorr, E.M., Ng, R.T.: A unified notion of outliers: properties and computation. In: Proceedings of the Third International Conference on Knowledge Discovery and Data Mining 1997 (KDD 1997), pp. 219–222 (1997)

3. Pamula, R.: Data Pruning Based Outlier Detection (Doctoral Dissertation) (2015). http://gyan.iitg.ernet.in/handle/123456789/631
4. Pamula, R., Deka, J.K., Nandi, S.: An outlier detection method based on clustering. In: Proceeding of International Conference on Emerging Applications of Information Technology, Kolkata, India (2011)
5. Breunig, M.M., Kriegel, H.-P., Ng, R.T, Sander, J.: LOF: identifying density-based local outliers. In Proceedings of the 2000 ACM SIGMOD International Conference on Management of Data 2000 (SIGMOD 2000), pp. 93–104. ACM (2000)
6. Chiu, A.L., Fu, A.W.: Enhancements on local outlier detection. In: Proceedings of Seventh International Conference on Database Engineering and Applications Symposium, pp. 298–307. IEEE (2003)
7. Jin, W., Tung, A.K.H., Han, J., Wang, W.: Ranking outliers using symmetric neighborhood relationship. In: Ng, W.-K., Kitsuregawa, M., Li, J., Chang, K. (eds.) PAKDD 2006. LNCS (LNAI), vol. 3918, pp. 577–593. Springer, Heidelberg (2006). https://doi.org/10.1007/11731139_68
8. Chandola, V., Banerjee, A., Kumar, V.: Anomaly detection: a survey. ACM Comput. Surv. (CSUR) 41(3), 15 (2009)
9. Zhang, K., Hutter, M., Jin, H.: A new local distance-based outlier detection approach for scattered real-world data. In: Theeramunkong, T., Kijsirikul, B., Cercone, N., Ho, T.-B. (eds.) PAKDD 2009. LNCS (LNAI), vol. 5476, pp. 813–822. Springer, Heidelberg (2009). https://doi.org/10.1007/978-3-642-01307-2_84
10. Goldstein, M.: FastLOF: an expectation-maximization based local outlier detection algorithm. In: Proceeding of 21st International Conference on Pattern Recognition 2012 (ICPR 2012), pp. 2282–2285 (2012)
11. Tang, J., Chen, Z., Fu, A.W., Cheung, D.W.: Enhancing effectiveness of outlier detections for low density patterns. In: Chen, M.-S., Yu, P.S., Liu, B. (eds.) PAKDD 2002. LNCS (LNAI), vol. 2336, pp. 535–548. Springer, Heidelberg (2002). https://doi.org/10.1007/3-540-47887-6_53

ECG Biometric Recognition

Anita Pal$^{(\boxtimes)}$ and Yogendra Narain Singh

Department of Computer Science and Engineering,
Institute of Engineering and Technology,
Dr. A. P. J. Abdul Kalam Technical University, Lucknow, Uttar Pradesh, India
anitapal13@gmail.com, singhyn@gmail.com

Abstract. This paper presents a human recognition system using single lead electrocardiogram (ECG). The method corrects the ECG signal from noise as well as other artifacts to it and extracts major features from P-QRS-T waveforms. Finite Impulse Response (FIR) equiripple high pass filter is used for denoising ECG signal. Haar wavelet transform is used to detect the R peaks. By using this novel approach, different extensive information like heart rates, interval features, amplitude features, angle features area features are received among dominant fiducials of ECG waveform. The feasibility of ECG as a new biometric is tested on selected features that report the recognition accuracy to 97.12% on the data size of 100 recordings of PTB database. The results obtained from the proposed approach surpasses the other conventional methods for biometric applications.

Keywords: ECG · Heartbeats · Biometric recognition

1 Introduction

In this modern digital era, an unique and accurate identity is essential need of the society. Traditional strategies for recognition include PIN numbers, tokens, passwords and ID cards raise serious security concerns of identity theft. The major benefit of security systems based on biometrics is that they work on an individual physiological or behavioral characteristics. One of the flaws of commonly used biometrics are the ease of falsification of credentials. For example, a photo can be counterfeited a face, the iris can be falsified by contact lenses and even the fingerprint can be fooled from a gel or latex finger.

In order to overcome the issues of conventional biometrics the bioelectric signals are one of the better choice. They are subjective to an individual and therefore, harder to mimic them. They are highly secure and are prevent from any fear of imitation. The electrocardiogram (ECG) is one of the known bioelectrical signal used to monitor the health of an individual heart. An ECG records changes in the electric potential of cardiac cells and possesses unique characteristics. The ECG records the electrophysiologic pattern of depolarizing and repolarizing during each heartbeat as shown in Fig. 1. Studies show that ECG exhibits discriminatory patterns among individuals [5–16].

© Springer Nature Singapore Pte Ltd. 2018
D. Ghosh et al. (Eds.): ICMC 2018, CCIS 834, pp. 61–73, 2018.
https://doi.org/10.1007/978-981-13-0023-3_7

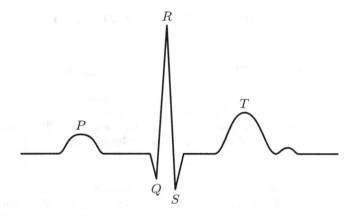

Fig. 1. ECG waveform

Outmoded medicine through efforts to universalize ECG signal to produce a common diagnostic method applicable to most individual [1], but the uniqueness of ECG among individuals is an advantage in biometrics as well as challenge in medicine [17]. Several studies have demonstrated ECG-based recognition is a robust biometric method. To ascertain that it is possible to identify individual using ECG, Biel *et al.* extracted the features from P, QRS and T waveforms and evaluate the feasibility of ECG signal for human recognition [5]. They performed multivariate analysis for classification and achieved 100% recognition rate. Israel *et al.* have demonstrated the Wilks Lambda technique for feature selection and linear discriminant analysis for classification [10]. This framework was tried on a database of 29 subjects with 100% human recognition rate was achieved. Shen *et al.* presented one lead ECG based on identity verification with seven fiducial based features that are related to QRS complex [6]. The consequence of identity verification has discovered to be 95% using template matching, 80% using decision based neural network and 100% for consolidating the two methods from a gathering of 20 people. Singh and Gupta have proposed P and T wave delineators along with QRS complex to extract different features from dominant fiducials of the electrocardiogram on each heartbeat [16]. The proposed system is tested on 50 subjects and achieved the classification accuracy to 98%.

In this paper, a robust and an efficient method of ECG biometric recognition is proposed. For denoising ECG signal, FIR equiripple high pass filter is used that removes baseline noise. The FIR equiripple low pass filter removes the power interference noise. Haar wavelet transform is used for accurately detection of the R peaks (R_{peak}). All other dominant features of the ECG waveform are detected with respect to the R peaks by setting of the windows whose sizes depend on the length of the corresponding wave duration and location. Features of the ECG signal including interval features, amplitude features, angle features and area features where successfully despite. The algorithm has been applied on 100 ECG signals of PTB database from physionet bank and could detect 39 features from every ECG signals. By applying PCA and Kernel PCA reduction methods on 39 features. Finally the similarities within the components of feature set are calculated on the basis of Euclidean distance.

The rest of the paper is organized as follows. Section 2 presents the methodology for the recognition system based on ECG. The delineation techniques of P and T wave are demonstrated with detailed description of ECG data. The experiment results of recognition system presented in Sect. 3. Finally, conclusions are drawn in Sect. 4.

2   Methodology

The framework of ECG biometric recognition system is shown in Fig. 2. The method is implemented in a series of steps: (1) ECG data preprocessing: includes correction of signal from noise artifacts. (2) Data representation: includes delineation of dominant waveform and recognition of dominant features between the diagnostic points. (3) Recognition: that matches test template with the template stored in the database using a suitable technique.

2.1   ECG Data Preprocessing

An electrocardiogram exhibits the electric potential actually electrical voltages are higher in the heart, it can be characterize as P, Q, R, S, and T waves. When an ECG is recorded, it contaminates several kind of noises. The contamination of different artifacts such as baseline wander noise and power line interference may change the levels, values of amplitudes and time periods of the ECG waveform, respectively.

Equiripple highpass filter is capable of removing baseline wander noise without affecting the dominant fiducials of the ECG. The equiripple highpass filter uses a filter order of 2746, cutoff frequency of 1 Hz, stop frequency of 2 Hz, and stop attenuation of 80 dB. The power interference noise appears as spike in frequency components analysis at 50 Hz. This frequency component can be removed by using notch filter. The FIR equiripple lowpass filter is used with filter order of 508 and cutoff frequency is set to 40 Hz. This filter is followed by an IIR filter to reach sharp frequency notch and avoid phase distortion.

2.2   Data Representation

The ECG signal is now ready to process for features extraction. In this stage, a systematic analysis of ECG is done using different techniques. The Haar wavelet transform method is used to extract the ECG features. Haar wavelet gives promising performance to delineate P-QRS-T wave fiducials.

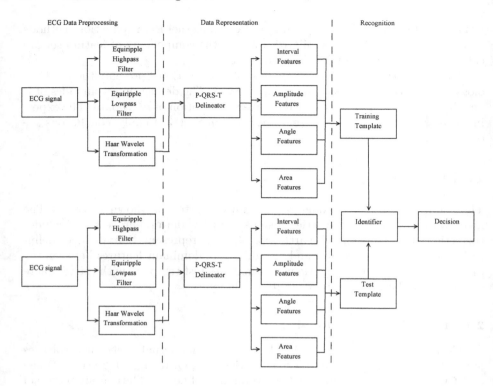

Fig. 2. ECG biometric recognition system [16].

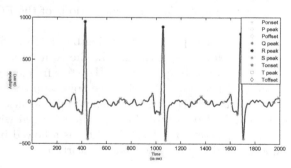

Fig. 3. Detection of ECG waveforms, Ponset, P, Poffset, Q, R, S, Tonset, T and Toffset.

Peak Detection. Using Harr wavelet transform R peaks are easily detected due to the multiresolution analysis of the ECG signal. In reference to the R peak location, the P, Q, S and T waveforms are detected. The rhythm of heartbeat is calculated using the following formula:

Number of heartbeats = R peaks * Length of signal/(Frequency * 60 s) per minute.

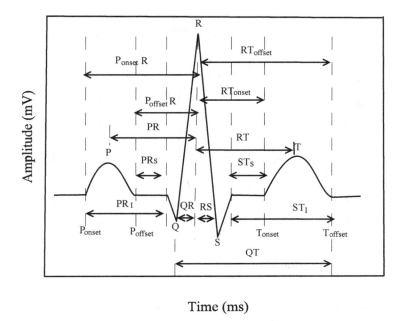

Time (ms)

Fig. 4. Interval features of ECG waveform [16].

The procedure to detect P-peak is shown as follows: To detect P_{peak} location, window of 160 samples is set. This window extends from 200 to 60 samples to the left of R_{peak}. Within the window, P_{peak} is located at the samples that have the maximum amplitude value. Another window of 90 samples is set. The window boundary from 100 to 10 samples to the left of R_{peak} location. Q_{peak} is located where the minimum amplitude value is found within the window. For S_{peak}, the window of size 95 samples is set and window extends from 5 to 100 samples to the right side of R_{peak} location. The minimum amplitude value within the window is the S_{peak} location. T_{peak} are the farthest waves from R_{peak}. T_{peak} are detected using window of 300 samples of width. These windows start at 100 samples on the right of R_{peak} and end at 400 samples away from R_{peak}. T_{peak} is located at the maximum amplitude value from right side of R_{peak} within the window. A window of size 300 samples is set. Within this window the minimum amplitude value at 150 samples from the left of T_{peak} is T_{onset} location and minimum amplitude value at 150 samples to the right of T_{peak} location is T_{offset} location. Thus all peaks are successfully detected. Figure 3 shows detected P, Q, R, S, T, P_{offset}, P_{onset}, T_{offset} and T_{onset} waves.

2.3 Feature Extraction

Once the ECG is delineated, peak and limits of QRS complex, P wave and T wave are known. From known fiducials 39 features which are extracted from each heartbeat where each derives from one of the classes:

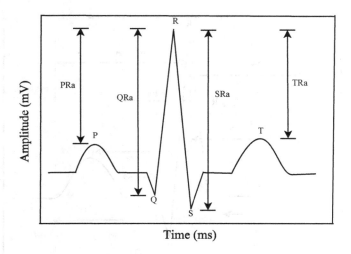

Fig. 5. Amplitude features of ECG waveform [16].

Interval Features. Following features related to heartbeat intervals are computed. The PR_I is the time interval between P_{peak} and R_{peak} fiducials. PR_S is the time interval between P_{offset} to QRS_{onset} fiducials. The QT is the corrected time interval between QRS_{onset} to T_{offset} fiducials, according to Bazett's formula. The ST_S is the time interval from QRS_{offset} to T_{onset} fiducials and ST_I is the time intervals from QRS_{offset} to T_{offset} fiducials. Other interval features are computed relative to R_{peak} fiducial. The time interval from R_{peak} to P wave fiducials, P_{offset}, P_{peak} and P_{offset} are defined as $P_{offset}R$, PR and $PonR$, respectively. The time interval from R_{peak} to Q_{peak} is defined as QR and time interval from R_{peak} to S_{peak} is defined as RS. Similarly, time interval from R_{peak} to T wave fiducials, T_{onset} and T_{offset} are defined as RT, RT_{onset} and RT_{offset} respectively.

The computed time interval features are shown in Fig. 4. Along to these interval features within a beat three interbeat interval features set RR, PP and TT are also extracted. RR is defined as the time interval between two successive R-peaks, similarly PP and TT are also detected. The RR feature is also used to correct the QT interval from the effects of change in heartrate [16].

Amplitude Features. Following amplitude features are computed relative to the amplitude of R peak. This class of features are dependent to QRS complex which is usually invariant to change in the heart rate. The QRa feature is defined as the difference in amplitude of R and Q waves. The SRa feature is defined as the difference in amplitude between R and S waves. Similarly, the difference in amplitude of P wave and T wave to R wave are defined as PRa and TRa, respectively [16]. These amplitude features are shown in Fig. 5.

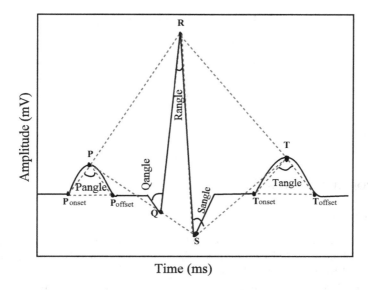

Fig. 6. Angle features of ECG waveform [16].

Angle Features. Following features related to angular displacement between different peak fiducials of P, Q, R, S and T waves are extracted from each heartbeat. Hence the aim is to extract a class of features which are stable and prone to the change in heart rate. The $\angle Q$ is defined as the angular displacement between directed lines joining from Q_{peak} to P_{peak} and Q_{peak} to R_{peak} fiducials [43]. Using Cosine rule $\angle Q$ can be computed as follows:

$$\cos Q = \frac{PQ^2 + QR^2 - PR^2}{2 * PQ * QR} \tag{1}$$

$\angle R$ is defined as the angular displacement between directed lines joining from R_{peak} to Q_{peak} and from R_{peak} to S_{peak} fiducials. Similarly, $\angle S$ is defined as the angular displacement between directed lines joining from S_{peak} to R_{peak} and from S_{peak} to T_{peak} fiducials. $\angle P$ is defined as the angular displacement between directed lines joining from P_{onset} to P_{peak} and from P_{peak} to P_{offset} fiducials. $\angle T$ is defined as the angular displacement between directed lines joining from T_{onset} to T_{peak} and from T_{peak} to T_{offset} fiducials. These angle features are shown in Fig. 6.

Area Features. We compute another set of feature called area features formed among ECG wave fiducials as follows (Table 1):

The procedure used to compute the area of a triangle having known vertices $(A_x, A_y), (B_x, B_y)$ and (C_x, C_y) in a 2D space is given as follows [44]:

$$\textbf{Area of Triangle ABC} = \frac{A_x(B_y - C_y) + B_x(C_y - A_y) + C_x(A_y - B_y)}{2} \tag{2}$$

Table 1. Area features of a heartbeat

Area features	Representation
Area P	Area of $\triangle P$
Area Q	Area of $\triangle Q$
Area R	Area of $\triangle R$
Area S	Area of $\triangle S$
Area T	Area of $\triangle T$

3   Recognition Results

3.1   Database

Physikalisch-Technische Bundesanstalt (PTB), the National Metrology Institute of Germany, has provided the digitized ECG for research [41]. The ECG signal were collected from healthy volunteers and patients with different heart diseases by Professor Michael, M.D., at the Department of Cardiology of University Clinic Benjamin Franklin in Berlin, Germany. The PTB database contain total records 549 from 290 subjects with the conventional 12 leads is represented as i, ii, iii, avr, avl, avf, v1, v2, v3, v4, v5, v6 together with three Frank ECG leads that is vx, vy, vz. Each signal is digitized at 1000 samples per second, with 16 bit resolution over a range of ±16.384 mV. The performance of the ECG biometric recognition system is evaluated on the ECG recordings of 100 subjects from the class Physikalisch Technishe Bundesanstalt (PTB) database. The proposed methodology is tested on 100 subject of PTB database from each of these subject 6 windows of 30 s is created. A feature vector of 600×39 (PTB) dimension. 6 windows from each subject is used as training template from which distance was calculated for each subject.

3.2   Feature Selection

Feature selection is the process of selecting a subset of relevant features from the feature vector collected from ECG identification model. In this paper two dimensionality reduction methods are used that is principal components analysis (PCA) and kernel principal components analysis (KPCA) [3]. PCA is a very popular technique for dimensionality reduction. Suppose a data set is of n-dimensions, the aim of the PCA is to find a linear subspace of d-dimension which is less than n than this data points lies on the linear subspace. Such a reduced subspace attempts to maintain the inconsistency of the data. The PCA approach can be described in five steps: (1) Calculate the covariance matrix of the given d-dimensional data set. (2) After that calculate the eigenvalues and eigenvector of the given data set and sort the eigenvalues in a decreasing order. (3) Select the k eigenvectors that belong to k largest eigenvalues and k is the dimension of the new feature space. (4) Compute the W projection matrix of the k selected eigenvectors. (5) Finally, transform the given data set X to obtain the k-dimensional feature subspace Y

$$Y = W^T . X \tag{3}$$

PCA is designed for linear capabilities in high-dimensional data set. However, high dimensional data sets are nonlinear [3]. In some cases the high-dimensional data lay on boundary or near the boundary of a nonlinear manifold, so in this case PCA cannot variability of the data correctly. In kernel PCA, the kernel is used in PCA to calculate the high-dimensional feature vector efficiently in nonlinear mapping on the given input data set. The formulation of kernel PCA as follows:

$$\sum_i^t \Theta(x_i) - Z_q Z_q^T \Theta(x_i) \tag{4}$$

where Z_q consist of eigenvectors and (x_i) is data set.

3.3   Recognition Performance

For recognition we generate the genuine and imposter matching scores. A matching score is a similarity measures between features derived from the test and training template. For different individuals, the test template is compared to the template stored in the gallery set using Euclidean distance as the similarity measure to generate matching scores (Table 2).

Table 2. Evaluation of recognition performance using different method

GAR (%)	FAR (%)		
	Euclidean distance	PCA	Kernel PCA
100	24.57	25.06	19.67
90	7.7	3.02	1.63
80	4.94	1.02	0.41
70	1.8	0.2	0.29
60	0.4	0.04	0.08
50	0.16	0.00	0.04

The receiver operating characteristic (ROC) curve plot is a function of the decision threshold which plots the rate of false acceptance against the false rejection. The equal error rate (EER) is defined as the rate at which the false acceptance rate equals the false rejection rate. The accuracy of the recognition system is determined from subtracting the EER value to 100.

The equal error rate (EER) of the identification system is found to be 2.88% and accuracy is 97.12% by applying kernel PCA for dimensionality reduction. By

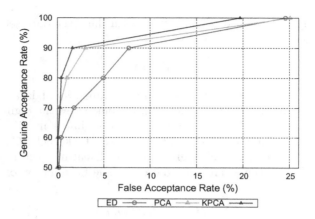

Fig. 7. ROC curves representing equal error rate.

Table 3. Comparision of proposed method with other known methods

Methods	ECG database/ Population size	Features used	Classification method	Recognition results
Tantawi et al. [34]	PTB/38	QRS- complex	Radial basis neural network	97%
Safie et al. [33]	PTB/112	QRS- complex	Matching score	91.01%
This paper	PTB/100	Amplitude & Interval	Matching score	97.12%
		Angle & Area		

using PCA EER is 8.86% and accuracy is 91.14% and Euclidean distance having an EER is 8.98% and accuracy is 91.01%. The performance of the ECG biometric recognition system is represented using receiver operator characteristic (ROC) curve as shown in Fig. 7. It shows that the system has genuine acceptance rate (GAR) for kernel PCA is 100% at 19.67% false acceptance rate (FAR), GAR for PCA is 100% at 25.06% FAR and GAR for euclidean distance is 100% at 24.57%. The recognition performance is found better for kernel PCA reduction method. In comparision to other methods, the proposed ECG biometric recognition system give outstanding performance on PTB database and this is shown in Table 3.

4   Conclusion

This study has proposed a method of biometric recognition of individuals using their heartbeats. The method has delineated the dominant fiducials of ECG waveform and then interval, amplitude, angle and area features are computed. The recognition results are shown that the proposed method of ECG biometric recognition is and useful to distinguish the heartbeats of normal as well as the inpatient subjects.

Universally, individuals have a heart and the nature of the way it beats and once used as a biometric proves the life of the user in a natural way. Therefore, no test of liveness is required. Finally, each individual has a unique set of heartbeat features. Thus, the proposed techniques can be used as a potential biometric for human recognition which is very secure and robust from falsification.

References

1. Draper, H.W., Peffer, C.J., Stallmann, F.W., Littmann, D., Pipberger, H.V.: The corrected orthogonal electrocardiogram and vector cardiogram in 510 normal men (Frank lead system). Circulation **30**(6), 853–864 (1964)
2. Webster, J.G.: Medical Instrumentation: Application and Design. Wiley, Philadelphia (1997)
3. Duda, R.O., Hart, P.E., Stork, D.G.: Pattern Classification, 2nd edn. Wiley-Interscience, New York (2000)
4. Wayman, J.L.: Fundamentals of biometric authentication technologies. Int. J. Image Graph. **1**(1), 93–113 (2001)
5. Biel, L., Pettersson, O., Philipson, L., Wide, P.: ECG analysis: a new approach in human identification. IEEE Trans. Instrum. Meas. **50**(3), 808–812 (2001)
6. Shen, T.W., Tompkins, W.J., Hu, Y.H.: One-lead ECG for identity verification. In: Proceedings 2nd Joint Conference on IEEE Engineering in Medicine and Biology Society and the Biomedical Engineering Society, Houston, pp. 280–305 (2002)
7. Stankovic, R.S., Falkowski, B.J.: The Haar wavelet transform: its status and achievements. Comput. Electr. Eng. **29**, 25–44 (2003)
8. Palaniappan, R., Krishnun, S.M.: Identifying individuals using ECG beats. In: Proceedings International Conference on Signal Processing and Communications, pp. 569–572 (2004)
9. Saechia, S., Koseeyaporn, J., Wardkein, P.: Human identification system based ECG signal. In: Proceedings TENCON 2005 IEEE Region 10, pp. 1–4 (2005)
10. Israel, S.A., Irvine, J.M., Cheng, A., Wiederhold, M.D., Wiederhold, B.K.: ECG to identify individuals. Pattern Recognit. **38**, 133–142 (2005)
11. Kim, K.-S., Yoon, T.-H., Lee, J.-W., Kim, D.-J., Koo, H.-S.: A robust human identification by normalized time-domain features of electrocardiogram. In: Proceedings 2005 IEEE Engineering in Medicine and Biology 27th Annual Conference, Shanghai, China, pp. 1114–1117 (2005)
12. Molina, G.G., Bruekers, F., Presura, C., Damstra, M., Veen, M.V.: Morphological synthesis of ECG signals for person authentication. In: Proceedings of Signal Processing Conference, Poznan, Poland (2007)
13. Watson, A.: Biometrics: easy to steal, hard to regain identity. Nature **449**(7162), 535 (2007)
14. Singh, Y.N., Gupta, P.: Quantitative evaluation of normalization techniques of matching scores in multimodal biometric systems. In: Lee, S.-W., Li, S.Z. (eds.) ICB 2007. LNCS, vol. 4642, pp. 574–583. Springer, Heidelberg (2007). https://doi.org/10.1007/978-3-540-74549-5_61
15. Wubbeler, G., Stavridis, M., Kreiseler, D., Bousseljot, R., Elster, C.: Verification of humans using the Electrocardiogram. Pattern Recognit. Lett. **28**(10), 1172–1175 (2007)
16. Singh, Y.N., Gupta, P.: ECG to individual identification. In: Proceedings 2nd IEEE International Conference on Biometrics Theory, Applications and Systems, pp. 1–8 (2008)

17. Chan, A.D.C., Hamdy, M.M., Badre, A., Badee, V.: Wavelet distance measure for person identification using electrocardiograms. IEEE Trans. Instrum. Meas. **57**(2), 248–253 (2008)
18. Mane, V.M., Jadhav, D.V.: Review of multimodal biometrics: applications, challenges and research areas. Int. J. Biom. Bioinform. **3**(5), 90–95 (2009)
19. Boumbarov, O., Velchev, Y., Sokolov, S.: ECG personal identification in subspaces using radial basis neural networks. In: Proceedings IEEE International Workshop on Intelligent Data Acquisition and Advanced Computing Systems: Technology and Applications (IDAACS 2009), pp. 446–451 (2009)
20. Singh, Y.N., Gupta, P.: A robust delineation approach of electrocardiographic P Waves. In: Proceedings of the 2009 IEEE Symposium on Industrial Electronics and Applications (ISIEA), vol. 2, pp. 846–849 (2009)
21. Singh, Y.N., Gupta, P.: A robust and efficient technique of T wave delineation from electrocardiogram. In: Proceedings of Second International Conference on Bioinspired Systems and Signal Processing (BIOSIGNALS), IEEE-EMB, pp. 146–154 (2009)
22. Singh, Y.N., Gupta, P.: Biometrics method for human identification using electrocardiogram. In: Tistarelli, M., Nixon, M.S. (eds.) ICB 2009. LNCS, vol. 5558, pp. 1270–1279. Springer, Heidelberg (2009). https://doi.org/10.1007/978-3-642-01793-3_128
23. Fatemian, S.Z., Hatzinakos, D.: A new ECG feature extractor for biometric recognition. In: Proceedings 16th International Conference on Digital Signal Processing, pp. 1–6 (2009)
24. Odinaka, I., Lai, P.-H., Kaplan, A.D., Sullivan, J.A.O., Sirevaag, E.J., Kristjansson, S.D., Sheffield, A.K., Rohrbaugh, J.W.: ECG biometrics: a robust short-time frequency analysis. In: Proceedings 2010 IEEE International Workshop on Information Forensics and Security (WIFS), pp. 1–6 (2010)
25. Ye, C., Coimbra, M., Kumar, B.: Investigation of human identification using two-lead electrocardiogram ECG signals. In: Proceedings 4th IEEE International Conference Biometrics: Theory Applications and Systems (BTAS), pp. 1–8 (2010)
26. Coutinho, D.P., Fred, A.L.N., Figueiredo, M.A.T.: One-lead ECG based personal identification using Ziv-Merhav cross parsing. In: Proceedings of 20th International Conference on Pattern Recognition (ICPR), pp. 3858–3861 (2010)
27. Venkatesh, N., Jayaraman, S.: Human electrocardiogram for biometrics using DTW and FLDA. In: Proceedings of 20th International Conference on Pattern Recognition (ICPR), pp. 3838–3841 (2010)
28. Li, M., Narayanan, S.: Robust ECG biometrics by fusing temporal and cepstral information. In: Proceedings of 20th International Conference on Pattern Recognition (ICPR), pp. 1326–1329 (2010)
29. Tawfik, M.M., Kamal, H.S.T.: Human identification using QT signal and QRS complex of the ECG. Online J. Electron. Electr. Eng. **3**(1), 383–387 (2011)
30. Singh, Y.N.: Challenges of UID environment. In: Proceedings of the UID National Conference on Impact of Aadhaar in Governance, pp. 37–45, December 2011
31. Singh, Y.N., Singh, S.K.: The state of information security. In: Proceedings of the Artificial Intelligence and Agents: Theory and Applications, pp. 363–367, December 2011
32. Singh, Y.N., Gupta, P.: Correlation based classification of heartbeats for individual identification. J. Soft Comput. **15**, 449–460 (2011)
33. Sae, S.I., Soraghan, J.J., Petropoulakis, L.: Electrocardiogram (ECG) biometric authentication using pulse active ratio (PAR). IEEE Trans. Inf. Forensics Secur. **6**, 1315–1322 (2011)

34. Tantawi, M., Revett, K., Tolba, M., Salem, A.: A novel feature set for deployment in ECG based biometrics. In: 2012 Seventh International Conference on International Conference of Computer Engineering Systems (ICCES), pp. 186–191 (2012)
35. Israel, S.A., Irvine, J.M.: Heartbeat biometrics: a sensing system perspective. Int. J. Cogn. Biom. **1**(1), 39–65 (2012)
36. Luo, Y., Hargraves, R.H., Bai, O., Ward, K.R.: A hierarchical method for removal of baseline drift from biomedical signals application in ECG analysis. Sci. World J. **2013**, 1–2 (2013)
37. Singh, Y.N., Singh, S.K.: Identifying individuals using eigenbeat features of electrocardiogram. J. Eng. **2013**, 1–8 (2013)
38. Singh, Y.N., Singh, S.K.: A taxonomy of biometric system vulnerabilities and defences. J. Biom. **5**, 137–159 (2013)
39. Wang, Z., Zhang, Y.: Research on ECG biometric in cardiac irregularity condition. In: Proceedings of IEEE International Conference on Medical Biometrics (ICMB), pp. 157–163 (2014)
40. Singh, Y.N.: Human recognition using fishers discriminant analysis of heartbeat interval features and ECG morphology. Neurocomputing **167**, 322–335 (2015)
41. Physionet, "Physiobank Archives", Physikalisch-Technische Bundesanstalt, Abbestrasse 2–12, 10587 Berlin, Germany. https://www.physionet.org/physiobank/database/ptbdb. Accessed 2016
42. Maths open Reference. http://in.mathworks.com/help/wavelet/ref/wavedec.html. Accessed June 2017
43. Maths open Reference. http://www.mathopenref.com/lawofcosines.html. Accessed June 2017
44. Maths open Reference. http://www.mathopenref.com/coordtrianglearea.html. Accessed June 2017

A Survey on Automatic Image Captioning

Gargi Srivastava^(✉)⬤ and Rajeev Srivastava

Department of Computer Science and Engineering,
Indian Institute of Technology (BHU) Varanasi,
Varanasi 221005, U.P., India
{gargis.rs.cse16,rajeev.cse}@iitbhu.ac.in

Abstract. Automatic image captioning is the process of providing natural language captions for images automatically. Considering the huge number of images available in recent time, automatic image captioning is very beneficial in managing huge image datasets by providing appropriate captions. It also finds application in content based image retrieval. This field includes other image processing areas such as segmentation, feature extraction, template matching and image classification. It also includes the field of natural language processing. Scene analysis is a prominent step in automatic image captioning which is garnering the attention of many researchers. The better the scene analysis the better is the image understanding which further leads to generate better image captions. The survey presents various techniques used by researchers for scene analysis performed on different image datasets.

Keywords: Image captioning · Scene analysis · Computer vision

1 Introduction

Automatic image captioning is the process of providing natural language captions for images automatically. The area is garnering attention from researchers because of the huge unorganized multimedia data pouring in every second. Automatic image captioning is a step ahead of automatic image tagging where images are tagged with relevant keywords related to the contents in the image. Various researchers have come up with the definition of automatic image captioning. In [1], authors in their work define automatic image captioning as the process by which a computer system automatically assigns metadata in the form of captioning or keywords to a digital image. Mathews et al. [12] in their paper define it as automatically describing the objects, people and scene in an image. Wang et al. [21] in their paper give the definition as recognition of visual objects in an image and the semantic interactions between objects and translate the visual understanding to sensible sentence descriptions. Liu et al. [22] mention that the grammar must be error-free and fluent. For summing up image captioning can be defined as generating short descriptions representing contents (object, scene and their interaction) of an image in human-like language automatically.

© Springer Nature Singapore Pte Ltd. 2018
D. Ghosh et al. (Eds.): ICMC 2018, CCIS 834, pp. 74–83, 2018.
https://doi.org/10.1007/978-981-13-0023-3_8

Automatic image captioning is viewed as an amalgamation of computer vision and natural language processing. The computer vision part is about recognizing the contents of an image and the natural language processing part is about converting the recognition into sentences. Research has flourished in both the fields. Computer vision researchers try to better understand the image and natural language processing research try to better express the image. Because of this integration, automatic image captioning has come out as an emerging field in artificial intelligence.

1.1 Applications

Automatic image captioning is an interesting area because of its application in various fields. It can be used in image retrieval system to organize and locate images of interest from a database. It is also useful for video retrieval. It can be used for the development of tools that aid visually impaired individuals to access pictorial information. It finds application in query-response interfaces. Journalists find the application useful in finding and captioning images related to their articles. Human-machine interaction systems can also employ the results of automatic image captioning. Such systems are also helpful in locating images verbally. It can also be used for military intelligence generation, surveillance systems, goods annotation in warehouse and self-aware systems (Fig. 1).

Fig. 1. Example of automatically captioned images [3].

1.2 Scene Analysis

Scene analysis is a module in automatic image captioning and has gained importance recently. In image captioning, generally, the output is the main object in the image without caring about what the background of the image is. This negligence makes the description of the image very vague and unclear.

Consider an image where a person is standing and in the background there is a river and another image where the background is a desert. If the focus is only on the object, both the images will be captioned as a person. If the background scene is taken into consideration, the first image may be captioned as a person standing in front of a river and the second image may be captioned as a person in desert. Suppose a journalist wants a sample image for her article and sends a query in the image database as keywords person, river. In first case of image annotation both the images will be retrieved whereas in second case only the first image will be retrieved. Thus, scene analysis is very important for proper image captioning which leads to better image retrieval results (Figs. 2 and 3).

Fig. 2. Without scene analysis: a person. With scene analysis: a person in front of river (Sample image taken from internet)

Fig. 3. Without scene analysis: a person. With scene analysis: a person in a desert (Sample image taken from internet)

For scene analysis, the image needs to be broken down into segments for understanding. This leads to the inclusion of another image processing field - image segmentation. Various segmentation techniques exist and several are coming up as to segment the images in a way that the machine understands the image better and can generate better captions. Another field included in scene analysis is object recognition which in itself is a very broad research area.

Object detection can be enhanced by adding contextual information. Scene analysis provides the required contextual information. As the number of scenes is finite, scene analysis is also considered as scene classification problem. Since objects are related to the scenes, the probability distribution of each object over different scenes is different. Convolutional neural networks have been trained over 25 million images of Places dataset to predict approximately 200 different scene-types.

In a nutshell, scene analysis of an image is very important. Without this there is no scope for meaningful captioning of images.

2   Related Works

A lot of research has been done in the field of automatic image captioning. The whole procedure of generating image captions by machines follow a common framework which is discussed below.

2.1   Framework

On the whole, the entire procedure can be subdivided into 2 parts: image processing and language processing. Image processing part includes: image segmentation, feature extraction and classification. Feature extraction and classification can be together referred to as object recognition.

After the object recognition, we obtain the keywords corresponding to the identified objects in the images. These keywords are then fed to language processing unit which results in forming meaningful captions for images.

Each of the three modules are independent and can be researched upon individually. Techniques applied for one of them does not affect the one used for the other module. It is beneficial as each module can be studied and analyzed in isolation (Fig. 4).

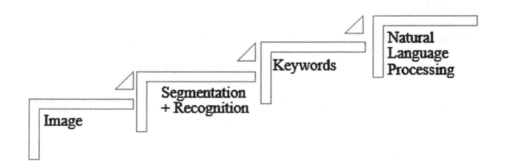

Fig. 4. Steps in automatic image captioning

2.2   Approaches

For segmentation and recognition various techniques can be used: supervised learning, unsupervised learning, neural networks or probabilistic methods (Fig. 5).

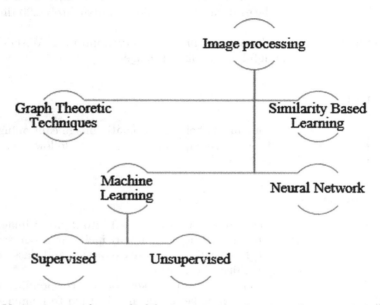

Fig. 5. Various approaches applied for image processing part of automatic image captioning

3   Comparative Study

See Table 1.

Table 1. A comparative study of different works in automatic image captioning

S. No.	Year and Author	Abstract	Result	Datasets	Limitations	Merits
1	Sumathi and Hemalatha [1]	Approach: treat annotation as a retrieval problem. Features: low level image features Simple combination of basic distances using JEC to find the nearest neighbor. Classification: SVM	Precision: 77% Recall: 35% F1 score: 51%	Flickr	Only annotations no captions	Simple distance-based technique

(continued)

Table 1. (*continued*)

S. No.	Year and Author	Abstract	Result	Datasets	Limitations	Merits
2	Yu and Sein [2]	Approach: intensity invariant approach. Preprocessing steps: gray scale converting, noise filtering, image enhancing. Segmentation: based on color intensity	No comparative results mentioned	Not mentioned	Only annotation tags are generated	Simple graph technique
3	Ushiku et al. [3]	Approach: generate a sentential caption for the input image by summarizing captions	BLEU: 0.63 NIST: 0.82	PASCAL sentence	Caption accuracy is sensitive to the retrieval precision of training samples	Instead of generating captions from scratch, they have used summarization technique
4	Federico and Furini [4]	Approach: automatic speech recognition with caption alignment mechanism	No comparative results mentioned	Video lectures of university professors	Since the behavior of ASR depends on the acoustic and language models, the audio markup insertion is likely to affect the performance of the speech analyzer	Cost effective solution
5	Feng and Lapata [5]	Probabilistic image annotation model for content selection. Extractive and abstractive surface realization model	Translation edit rate: 1.77. Grammaticality: 6.42. Relevance: 4.10	BBC dataset	1. Finite topics 2. Very little linguistic knowledge 3. Local features	Extremely helpful to journalists
6	Xi and Im Cho [6]	Features: weighted feature clustering based on statistical distribution. Annotation: maximal conditional probability	Precision: 40–60% Recall 40–60%	Corel	Dominated by weakly relevant features	Gives an insight to information gain theory
7	Horiuchi et al. [7]	Approach: collect general phrases for generating image descriptions	17/20 of image descriptions are scored higher than the image descriptions selected by 1 million. 6/7 of image descriptions are scored higher than the image descriptions of Integer Linear Programming	PASCAL visual object classes challenge	1. Image descriptions too concise 2. Image descriptions affected by quality of image retrieval system	Selecting phrases based on frequency results in fewer errors and more relevant descriptions

(continued)

Table 1. (continued)

S. No.	Year and Author	Abstract	Result	Datasets	Limitations	Merits
8	Ramnath, et al. [8]	Approach: uploading photo to a cloud service and running parallel modules to generate captions	Of 2385 ratings, 49.6% were very good	Personal photos	1. Few recognition capabilities	Keyword based search for personal photos
9	Sivakrishna Reddy, et al. [9]	Features: SIFT. Annotations: clustering	No comparative results mentioned	Not mentioned	Not generating new captions	Reusing existing caption
10	Shivdikar et al. [10]	Approach: combination of feature detection algorithms, context-free grammar	F1 Score: 94.33% BLEU: 0.75	Flickr8K Flickr30K COCO SBU	1. Changing nature of output by single layer learning 2. Smaller n-grams	Forms the base for increasingly complex and accurate neural network algorithms
11	Mathews [11]	Predicting human-like names for visual objects Sentences expressing a strong positive/negative sentiment	Among the 2633 visual concepts the method improves upon the Most frequent name baseline for 1222 concepts	ImageNet	Accuracy not good for ambiguous names	Introducing style is good for generating customized caption
12	Mathews et al. [12]	Predicting basic-level names using a series of classification and ranking tasks	Precision: 34.7%	ImageNet-Flickr	Only picture-to-word, no captions	Naming visual concepts is important part of automatic image captioning
13	Plummer et al. [13]	Coreference chains. Manually annotated bounding boxes	Recall: 76.4%	Flickr30k	1. No attempt to match regions in a query image and phrases in a candidate matching sentence 2. Does not care about the nature of the auxiliary data	Technique helps to localize entities in an image which is helpful for continued progress in image captioning
14	Vijay and Ramya [14]	Create captions for news images	No comparative results mentioned	News articles	No auxiliary information used	Easy understanding of news articles
15	Shahaf et al. [15]	Influence of the language of cartoon captions on the perceived humorousness of the cartoons	The classifier picked funnier of the two captions 64% of the time	Crowdsourced cartoon captions	1. Brief context and anomaly analysis 2. Ignored visual concept of cartoons 3. Cannot identify weaknesses of captions	Opens scope for caption understanding
16	Li et al. [16]	Generate Chinese sentence descriptions for unlabeled images	Machine translated neural image captioning is more suited for Chinese captioning	Flickr8K	Not much improvement observed	Expanding the scope to languages can help build the system for people from different linguistic backgrounds

(continued)

Table 1. (*continued*)

S. No.	Year and Author	Abstract	Result	Datasets	Limitations	Merits
17	Jin and Nakayama [17]	Approach: forms image annotation task as a sequence generation problem predict proper length of tags	Precision: 36% Recall: 37% F1: 34% N+: 267	Corel 5K, ESP Game, IAPR TC12	Nave approach to decide order	Order of tags in training phase has a great impact on the final annotation performance
18	Shi and Zou [18]	Fully convolutional networks	Precision: 95.3% Recall: 94.1%	Google Earth, GaoFen-2	1. Different geographical levels are not considered 2. Lesser ground features	Expansion of remote sensing images
19	Shetty et al. [19]	Approach: augment CNN features with scene context features	CIDEr Score: 0.954	MSCOCO	1. Fails to learn relationship between object and image 2. Cannot count objects properly 3. Vocabulary size is small	Employing scene analysis gives better information about the image
20	Li et al. [20]	Scene oriented CNN	BLEU: 0.68 METEOR: 22.8	MSCOCO	Scope for sentiment addition to captions	Including scene information
21	Wang et al. [21]	Deep CNN Approach: use of history and future context information Data augmentation techniques: multi-crop, multi-scale, vertical mirror	BLEU: 0.67 METEOR: 19.5 CIDEr: 66.0	Flickr8K Flickr30K MSCOCO	Less focus on language representation	Focus on language representation to generate better caption
22	Liu et al. [22]	Approach: formulate image captioning as a multimodal translation task, Represent the input image as a sequence of detected objects	No comparative results mentioned	Not mentioned	No detailed methods mentioned	High-level features enrich the visual part
23	Blandfort et al. [23]	Approach: deep convolutional neural network for detecting adjective noun pairs graphical network. Architecture: concept and Syntax Transition Network	31.5% of the captions generated were reported as more human-like in comparison to the original caption. In 62.5% of images atleast one subject chose their caption over the original one	YFCC100M	1. Grammar is given less weightage. 2. Considering concept scores only for thresholding not for ranking 3. Generation of similar sentences 4. Scope for network optimization	Includes sentiment factor
24	Tariq and Foroosh [24]	Approach: extract contextual cues from available sources of different data modalities and transforms them into a probability space	METEOR: 0.053 TER: 1.75	TIME magazine	Only annotations no captions	Importance of weighted auxiliary information

4   Issues and Challenges

A number of open research issues and challenges have been identified in this field. A few of them are listed below:

1. Large collections of digital images exist without annotations.
2. The quality and quantity of training set becomes an important factor in determining the quality of captions that are generated.
3. Images with low resolution, low contrast complex background and texts with multiple orientation, style, color and alignment increase the complexity of image understanding.
4. The training set must include as much variety as possible.
5. Searching the optimal method for each of them is very expensive and it has a major effect on the performance of the overall system.
6. Capturing sentiments in the captions is a major challenge as not many datasets are available that include sentiment based annotations.
7. Few datasets are available that provide captions in different languages and moreover machine translation results are not always relevant.

5   Conclusion and Future Work

Automatic image captioning is an emerging area in the field of artificial intelligence and computer vision. The area has real life applications in various fields. It is an ensemble of various modules which opens a lot of area for exploration. Better captions can be generated with proper segmentation. Enhanced descriptions can be made using sentiment addition, activity recognition, background identification and scene analysis. Moreover the areas of deep learning for faster and accurate results can also be explored further. If the hardware resource cost is a limitation, traditional machine learning algorithms can also be investigated for the purpose.

References

1. Sumathi, T., Hemalatha, M.: A combined hierarchical model for automatic image annotation and retrieval. In: International Conference on Advanced Computing (2011)
2. Yu, M.T., Sein, M.M.: Automatic image captioning system using integration of N-cut and color-based segmentation method. In: Society of Instrument and Control Engineers Annual Conference (2011)
3. Ushiku, Y., Harada, T., Kuniyoshi, Y.: Automatic sentence generation from images. In: ACM Multimedia (2011)
4. Federico, M., Furini, M.: Enhancing learning accessibility through fully automatic captioning. In: International Cross-Disciplinary Conference on Web Accessibility (2011)
5. Feng, Y., Lapata, M.: Automatic caption generation for news images. IEEE Trans. Pattern Anal. Mach. Intell. **35**(4), 797–811 (2013)

6. Xi, S.M., Im Cho, Y.: Image caption automatic generation method based on weighted feature. In: International Conference on Control, Automation and Systems (2013)
7. Horiuchi, S., Moriguchi, H., Shengbo, X., Honiden, S.: Automatic image description by using word-level features. In: International Conference on Internet Multimedia Computing and Service (2013)
8. Ramnath, K., Vanderwende, L., El-Saban, M., Sinha, S.N., Kannan, A., Hassan, N., Galley, M.: AutoCaption: automatic caption generation for personal photos. In: IEEE Winter Conference on Applications of Computer Vision (2014)
9. Sivakrishna Reddy, A., Monolisa, N., Nathiya, M., Anjugam, D.: A combined hierarchical model for automatic image annotation and retrieval. In: International Conference on Innovations in Information Embedded and Communication Systems (2015)
10. Shivdikar, K., Kak, A., Marwah, K.: Automatic image annotation using a hybrid engine. In: IEEE India Conference (2015)
11. Mathews, A.: Captioning images using different styles. In: ACM Multimedia Conference (2015)
12. Mathews, A., Xie, L., He, X.: Choosing basic-level concept names using visual and language context. In: IEEE Winter Conference on Applications of Computer Vision (2015)
13. Plummer, B.A., Wang, L., Cervantes, C.M., Caicedo, J.C., Hockenmaier, J., Lazebnik, S.: Flickr30k entities: collecting region-to-phrase correspondences for richer image-to-sentence models. In: International Conference on Computer Vision (2015)
14. Vijay, K., Ramya, D.: Generation of caption selection for news images using stemming algorithm. In: International Conference on Computation of Power, Energy, Information and Communication (2015)
15. Shahaf, D., Horvitz, E., Mankoff, R.: Inside jokes: identifying humorous cartoon captions. In: International Conference on Knowledge Discovery and Data Mining (2015)
16. Li, X., Lan, W., Dong, J., Liu, H.: Adding Chinese captions to images. In: International Conference in Multimedia Retrieval (2016)
17. Jin, J., Nakayama, H.: Annotation order matters: recurrent image annotator for arbitrary length image tagging. In: International Conference on Pattern Recognition (2016)
18. Shi, Z., Zou, Z.: Can a machine generate humanlike language descriptions for a remote sensing image? IEEE Trans. Geosci. Remote Sens. **55**(6), 3623–3634 (2016)
19. Shetty, R., Tavakoli, H.R., Laaksonen, J.: Exploiting scene context for image captioning. In: Vision and Language Integration Meets Multimedia Fusion (2016)
20. Li, X., Song, X., Herranz, L., Zhu, Y., Jiang, S.: Image captioning with both object and scene information. In: ACM Multimedia (2016)
21. Wang, C., Yang, H., Bartz, C., Meinel, C.: Image captioning with deep bidirectional LSTMs. In: ACM Multimedia (2016)
22. Liu, C., Wang, C., Sun, F., Rui, Y.: Image2Text: a multimodal caption generator. In: ACM Multimedia (2016)
23. Blandfort, P., Karayil, T., Borth, D., Dengel, A.: Introducing concept and syntax transition networks for image captioning. In: International Conference on Multimedia Retrieval (2016)
24. Tariq, A., Foroosh, H.: A context-driven extractive framework for generating realistic image descriptions. IEEE Trans. Image Process. **26**(2), 619–631 (2017)

Texture and Color Visual Features Based CBIR Using 2D DT-CWT and Histograms

Jitesh Pradhan$^{(\boxtimes)}$ (iD), Sumit Kumar, Arup Kumar Pal, and Haider Banka

Department of Computer Science and Engineering,
Indian Institute of Technology (ISM), Dhanbad 826004, India
jiteshpradhan@cse.ism.ac.in

Abstract. In content based image retrieval (CBIR) process, every image has been represented in a compact set of local visual features i.e. color, texture, and/or shape of images. This set of local visual features is known as feature vector. In the CBIR process, feature vectors of images have been used to represent or to identify similar images in adequate way. As a result, feature vector construction has always been considered as an important issue since it must reflect proper image semantics using minimal amount of data. The proposed CBIR scheme is based on the combination of color and texture features. In this work initially, we have converted the given RGB image into HSV color image. Subsequently, we have considered H (hue), S (saturation), and V (intensity) components for extraction of visual image features. The texture features have been extracted from the V component of the image using 2D dual-tree complex wavelet transform (2D DT-CWT) where it analyzes the textural patterns in six different directions i.e. $\pm 15°$, $\pm 45°$, and $\pm 75°$. At the same time, we have computed the probability histograms of H and S components of the image respectively and subsequently those are divided into non-uniform bins based on cumulative probability for extraction of color based features. So, in this work both the color and texture features have been extracted simultaneously. Finally, the obtained features have been concatenated to attain the final feature vector and same is considered in image retrieval process. We have tested the novelty and performance of the proposed work in two Corel, two objects, and, a texture image datasets. The experimental results reveal the acceptable retrieval performances for different types of datasets.

Keywords: Content-based image retrieval
Dual-tree complex wavelet transform · Color and texture features
Probability histogram

1 Introduction

1.1 Background

In the present digital era, a radical expansion has been observed in the field of digital and Internet technology. As a result, the uses of digital communication

© Springer Nature Singapore Pte Ltd. 2018
D. Ghosh et al. (Eds.): ICMC 2018, CCIS 834, pp. 84–96, 2018.
https://doi.org/10.1007/978-981-13-0023-3_9

and Internet applications have been exponentially increasing steadily. People are exceedingly getting addicted with the Internet applications and spending more time on the web. Consequently, Internet traffic and digital data on web repositories are also escalating exponentially. In these web repositories, a huge chunk of digital data is in form of image. Handling of these gigantic web repositories by human annotation process is considered as an impractical task and at the same time, retrieval of images from these web repositories has become even more difficult. Since, these particular images are not significantly described by human annotation process. Hence, an effective image retrieval scheme is directly associated with construction of salient visual features. In literature, three types of image retrieval schemes have been introduced by contemporary researchers. These three image retrieval schemes are text-based image retrieval (TBIR) [1], content-based image retrieval (CBIR) [2], and semantic-based image retrieval (SBIR) [3]. In TBIR, meta-data (i.e. location name, file name, keywords, tags, etc.) associated with the images are used for the retrieval purpose. Since these meta-data does not represent any actual image information, it is impractical to use them to represent any image. As a result, image retrieval by TBIR systems usually includes lots of junk images. To overcome limitations of TBIR, researchers have introduced a CBIR system [4–6] which works on salient visual contents of the image i.e. shape, texture, and/or color features which are also referred as low-level image features. These actual image features have been represented in a single feature set known as feature vector in image retrieval. Researchers have used shape, texture, and color features alone as well as in different combinations for CBIR applications.

1.2   Literature Review

In past two decades, many CBIR systems have already been introduced based on primitive low-level or combination of low-level image features. In 2008, Chun et al. [4] have proposed a CBIR system in which they have combined the multi-resolution texture and color features together. They have used color autocorrelograms for color features and BDIP and BVLC techniques for texture feature extraction. They have applied wavelet to achieve multi-resolution images and extracted color and texture features form each resolution. Later, they have merged all features for retrieving images. In 2009, Lin et al. [6] have introduced a smart CBIR system which works on three different image features. They have used difference between pixels of scan pattern (DBPSP) and color co-occurrence matrix (CCM) to achieve texture and color features simultaneously. Further, they have also used color histogram for K-mean (CHKM) to extract third image feature based on the distribution of color values. Finally, they have used the different combinations of these three image features for CBIR process. In 2010, Feng et al. [5] have combined the visual attention model with CBIR process to approximate the users perception. They have also used relevance feedback approach to estimate the high-level semantics of the image. They have used visual attention model to extract prominent edges from the image for shape feature extraction. Further, they have used salient region adjacency graphs along with edge histogram descriptors based feature extraction approach for CBIR application. In 2011, Yue et al. [7] have used color and texture image features for image retrieval. They have used texture co-occurrence matrix along with the color histogram to create a feature vector for CBIR application. Yue et al. have shown that the combination of different features works better for image retrieval. In 2012, Youssef [8] have used integrated discrete curvelets for CBIR application. They have proposed a new sub-band clustering technique based on region vector codebook for color feature extraction from color histogram. Subsequently, they have used the most similar highest priority (MSHP) principal based image matching approach for CBIR process. In 2013, Subrahmanyam et al. [9] have introduced modified color motif co-occurrence matrix (MCMCM) for CBIR process. Here, MCMCM extracts the pixel wise inert-correlation of different color components of an RGB image. They have integrated the DBPSP approach with the proposed MCMCM technique for better feature extraction for CBIR process. In 2014, ElAlami [10] has introduced a new image matching technique for CBIR application. He has used DBPSP along with CCM to extract texture and color features simultaneously. Subsequently, he has reduced the feature set by a dimension reduction approach. Later, he has applied artificial neural network for classification and calculated the minimum area between two vectors to compute the distance for CBIR process. In 2015, Guo and Prasetyo [11] have introduced halftoning and truncation coding based CBIR approach. They have used ordered-dither block truncation coding (ODBTC) technique for image compression and extracted bit pattern features (BPF) and color co-occurrence feature (CCF) for CBIR. In 2016, Varish et al. [12] have used color and texture features in hierarchical way to filter out the irrelevant images. They have used different visual features in each level of the hierarchy for image filtration and retrieval from the database. In 2017, Cui et al. [13] have introduced a hybrid learning technique based on textual and visual information. They have used this hybrid learning approach to extract the textual meta-data of the image and combined it with the visual information for CBIR application.

Another image retrieval approach is SBIR which works on high level image information. SBIR does not use the low-level image features as like CBIR, it works on the overall semantic perception of the image. Object detection and recognition, image classification, semantic templates, bag-of-visual-words, image semantic tag assignment [3,14] is some fundamental techniques used in SBIR. SBIR approach needs high pre-processing cost, storage space and CPU time. These are the main limitations of SBIR approaches.

1.3   Major Contribution

In this paper, we have proposed a novel CBIR scheme which works on color and texture visual image features. In this work, we have converted the RGB image into HSV image because in RGB image, color components are highly correlated as a result color chromatic information gets lost. We have used 2D dual-tree complex wavelet transform (2D DT-CWT) for texture analysis in V (intensity) component of the image. Simultaneously, we have created normalized histograms of H (hue) and S (saturation) components of the image. Later, we have computed the probability histograms of the H and S components. Further, we have divided it into 10 non-uniform bins according to the cumulative probability model. Finally, we have combined all visual image features for CBIR process.

1.4 Paper Organization

The Sect. 1 gives the brief introduction about CBIR and current state-of-arts in CBIR. In Sect. 2, we have explained 2D dual-tree complex wavelet transform. The proposed CBIR scheme has been explained in Sect. 3. In Sect. 4, we have presented the retrieval results and furnished the performance comparisons with related CBIR schemes. At last, Sect. 5 shows the conclusions and future works.

2   Dual-Tree Complex Wavelet Transforms (DT-CWT)

Gabor filters and discrete wavelet transform are the most commonly used techniques for texture analysis but these techniques have some serious disadvantages. Gabor filters takes more time to extract texture features and also these are not orthogonal. Similarly, DWT analyzes textural patterns with less number of directions (i.e. $0°$, $45°$, and $90°$) and also it is not shift invariant. DT-CWT [12] overcomes all the above problems of Gabor filter and DWT. The DT-CWT works in similar way like DWT but it generates two different trees of DWT. Both DWT trees are real with a low-pass and high-pass filters. Both DWT trees analyze the textural patterns parallelly where, first one is considered as real part whereas second one is considered as complex part of DT-CWT. In this manner DT-CWT uses two real DWT with 4 filters to produce real and imaginary parts of the transform. The 2D DT-CWT analyzes the textural patterns on six different directions by producing wavelets in $\pm15°$, $\pm45°$, and $\pm75°$ for real and imaginary parts.

Let $I(x, y)$ is an image and 2D DT-CWT decomposes this image by applying six complex wavelets along with a complex scale function. Let, h is high-pass and low-pass filter set of real parts and g is the high-pass and low-pass filter set of imaginary parts of 2D DT-CWT. Here h_1 and g_1 are high-pass filter sets. Similarly, h_2 and g_2 are low-pass filter sets. Based on these assumptions, the complex wavelet function of 2D DT-CWT is defined as:

$$f(x, y) = f(x) \times f(y) \tag{1}$$

$$f(t) = f_h(t) + j \times f_g(t), \ Such \ that \ t = x \ or \ y \tag{2}$$

where, $f_h(t)$ and $f_g(t)$ are real and imaginary parts of 2D DT-CWT. So, the complex wavelet function can be expanded as follows:

$$f(x, y) = \{f_h(x) + j \times f_g(x)\}\{f_h(y) + j \times f_g(y)\} \tag{3}$$

$$f(x, y) = \{f_h(x)f_h(y) - f_g(x)f_g(y)\} + j \times \{f_h(x)f_g(y) + f_g(x)f_h(y)\} \tag{4}$$

where, $Real(f(x, y)) = \{f_h(x)f_h(y) - f_g(x)f_g(y)\}$ and $Imaginary(f(x, y)) = \{f_h(x)f_g(y) + f_g(x)f_h(y)\}$.

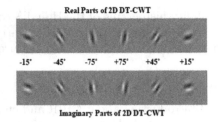

Fig. 1. Six impulse responses of real and imaginary parts of 2D DT-CWT

Figure 1 shows the six directional (i.e. ±15°, ±45°, and ±75°) impulse responses produced by the real and imaginary parts of the dual-tree complex wavelet function. Figure 2 shows the filter arrangement diagram of 2D DT-CWT up to 3 levels of decompositions.

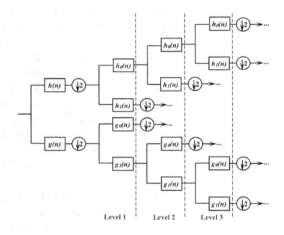

Fig. 2. Filter arrangement in 2D DT-CWT for 3 levels of decompositions

3   Proposed CBIR Scheme

In this section, proposed CBIR scheme has been explained in detail. Our proposed scheme works in two different stages where in first stage we have extracted the texture and color visual image features simultaneously. In second stage, image retrieval has been performed. In feature extraction stage, first we have converted the RGB image into HSV image. Subsequently, we have applied 2D DT-CWT up to n levels of decomposition on value (V) components. Here, we have used $n = 3$ for the retrieval experiments and each level of 2D DT-CWT produces 4 approximated coefficients, 2 form real parts and 2 from imaginary parts.

Along with approximated images, it also generates six directional wavelets (i.e. ±15°, ±45°, and ±75°) from real parts and six directional wavelets (i.e. ±15°, ±45°, and ±75°) from imaginary parts. Further, we have calculated statistical parameters from all wavelet coefficients of each level of decomposition and approximated coefficients of final level of decomposition. Simultaneously, we have calculated normalized histograms of hue (H) and saturation (S) components of the image. Further, we have calculated probability histograms and divided it into m non-uniform bins and in experiments we have used $m = 10$. Later, we have extracted statistical parameters from each bin of the histogram. Finally, all features have been combined together for CBIR process. Figure 3 shows the schematic block diagram of the proposed CBIR scheme. Further, Algorithms 1, 2, and 3 explains the detailed steps of CBIR process.

Algorithm 1 takes an RGB image and converts it into HSV image. Further, 2D DT-CWT has been employed on the V component of the image since it contains most of the textural visual features. The 2D DT-CWT will extract the

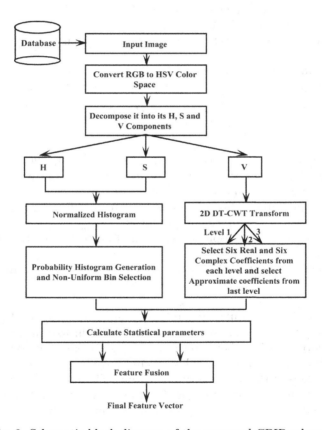

Fig. 3. Schematic block diagram of the proposed CBIR scheme

Algorithm 1. Texture Feature Extraction Algorithm (TFEA).

Input: An RGB color image.
Output: Final set of texture visual image features.
1: Take an RGB colored image as an input from user.
2: Convert RGB image to HSV color image.
3: Select value (V) components from the HSV color image.
4: Apply 2D DT-CWT on V component for 3 levels of decomposition.
5: Calculate mean and standard deviation statistical parameters of all detailed coefficients of all decomposition levels and approximate coefficients from last decomposition level.
6: Store all calculated statistical features in a form of texture feature vector.

Algorithm 2. Color Feature Extraction Algorithm (CFEA).

Input: An RGB colored image.
Output: Final set of color visual image features.
1: Take an RGB colored image as an input from user.
2: Convert RGB image to HSV color-space image.
3: Select hue (H) and saturation (S) components from the HSV color-space image.
4: Create normalized histograms of H and S components.
5: Calculated probability histograms of both H ans S histograms as follows:

$$Ph_k(i) = \frac{Hg_k(i)}{\sum_{j=1}^{n} Hg_k(j)} \tag{5}$$

where, i represents the ith component of a histogram, Ph_k represents the probability histogram of kth image, Hg_k represents the histogram of kth image and n is the total count of coefficients.
6: Divide the probability histograms into m non-uniform groups where the cumulative probability of each group must be $\leq \frac{1}{m}$.
7: Calculate standard deviation, skewness, and kurtosis statistical parameters from all bins.
8: Store all calculated statistical features in a form of color feature vector.

---

textural features from six different directions. Simultaneously, in Algorithm 2 H and S components have been used to generated color histograms. Further, both these color histograms have been converted into probability histogram to reduce the feature dimension. Statistical color features have been extracted from the m non-uniform beans of the probability histograms. Final feature vector combines both the texture ans color features for CBIR process. Finally, Algorithm 3, have been used to extract the similar images from the image dataset.

Algorithm 3. Content-Based Image Retrieval Algorithm (CBIRA).

Input: An RGB color image.
Output: Retrieved set of similar images.
1: Take an RGB color image as an input from user.
2: Extract texture and color visual features using algorithms TFEA and CFEA.
3: Combine the normalized texture and color features in a form of single feature vector.
4: Select an image dataset with n numbers of images and extract texture and color features for all images using algorithms TFEA and CFEA.
5: Combine the normalized texture and color features of each dataset image in a form of single feature vector.
6: Create a feature space which congregate the feature vectors of all dataset images with same index value as in dataset.
7: Compute the Euclidean distance in between query feature vector and feature space feature vectors.
8: Retrieve first k (where $k \leq n$) images from dataset having minimum distance values.

4 Results and Performance Analysis

In this section, we have presented the retrieval results of the proposed CBIR scheme in terms of precision, recall, and f-score which are defined as follows:

$$\mu_P(\%) = \frac{RI}{RI + NI} \times 100 \qquad (6)$$

$$\mu_R(\%) = \frac{RI}{RI + DI} \times 100 \qquad (7)$$

$$\mu_{Fs}(\%) = \frac{2 \times \mu_P \times \mu_R}{\mu_P + \mu_R} \times 100 \qquad (8)$$

Here, μ_P is the precision, μ_R is the recall, and μ_{Fs} is the f-score. RI is similar image in retrieved image set and NI is the dissimilar image in the retrieved image set. DI is the similar image present in dataset other than RI. Later, we have used five different image datasets to check the retrieval performance of the proposed CBIR scheme. In these five datasets, first two are natural image datasets i.e. Corel-1000 [12] and GHIM-10K [12]. The next two image datasets are object image datasets i.e. COIL-100 [15] and Produce-1400 [16]. The fifth dataset is a texture image dataset which is Outex [17]. Table 1 gives the brief description about all five datasets. Later, sub-section explains the feature vector length calculation. Sub-section shows the time performance of the proposed CBIR. Sub-section shows the retrieval performance. Finally, sub-section shows the performance comparisons.

Table 1. Brief description of all five image datasets used in retrieval experiment

Image dataset	Nature of dataset	No. of classes	Total no. of images	No. of images in each class
Corel-1000	Corel	10	1000	100
GHIM-10K	Corel	20	10000	500
COIL-100	Object	100	7200	72
Produce-1400	Object	14	1400	100
Outex	Texture	24	4320	180

4.1 Feature Vector Analysis

In this work, we have used 2D DT-CWT with 3 levels of decompositions for texture analysis where we have used it in value (V) components. In 2D DT-CWT, each level produces two approximate images with real and imaginary coefficients for further decomposition. So, in total it generates four approximate coefficients in each level of decomposition. Along with approximate coefficients, it also produces six real and six imaginary wavelet coefficients. Here, we have selected all wavelets coefficients of all 3 levels with 4 approximated coefficients of third level. Further, we have calculated mean and standard deviation from each coefficient. Hence, texture feature vector is having 80 feature elements (i.e. 3 levels × 12 wavelet coefficients × 2 parameters + 4 approximated coefficients × 2 parameters = 80 elements). At the same time, we have calculated 3 statistical parameters from each bin of hue and saturation probability histograms. Hence, number of elements in color feature vector is 60 (i.e. 2 histograms × 10 bins × 3 parameters = 60 elements). As a result, final feature is having 140 feature elements (i.e. 80 texture elements + 60 color elements = 140 elements).

4.2 Time Performance Analysis

Here, we have presented the time performance analysis in terms of CPU time required by any process. In this proposed CBIR scheme, there are 4 different types of processes which are texture feature extraction, color feature extraction,

Table 2. CPU time (in seconds) analysis for different process of proposed CBIR scheme

Image dataset	Texture feature extraction	Color feature extraction	Feature fusion	Image retrieval
Corel-1000	0.099 s	0.035 s	0.010 s	0.092 s
GHIM-10K	0.106 s	0.037 s	0.010 s	0.412 s
COIL-100	0.065 s	0.026 s	0.010 s	0.305 s
Produce-1400	0.512 s	0.157 s	0.010 s	0.104 s
Outex	0.064 s	0.025 s	0.010 s	0.215 s

feature fusion, and image retrieval. Table 2 shows the average CPU time taken by each process of CBIR while performing image retrieval from all 5 datasets.

4.3 Retrieval Performance Analysis

In this sub-section, we have presented the retrieval performance of the proposed CBIR scheme for top 10 and top 20 retrieved images in terms of precision (μ_P), recall (μ_R), and f-score (μ_{Fs}). We have retrieved these images from five different image datasets. Table 3 gives the retrieval performance in terms of average precision, recall and f-score for all five datasets for top 10 and top 20 retrieved images. In this table, we can see that all retrieval precisions are above 75%. This shows that, the proposed method is performing well for different types of images. Later, Fig. 4(i) to (iv) shows the precision graph of all five image datasets for top

Table 3. Overall average performance of proposed CBIR for top 10 and top 20 retrieved images from all five image datasets

Image dataset	$\mu_P(\%)$		$\mu_R(\%)$		$\mu_{Fs}(\%)$	
	Top 10	Top 20	Top 10	Top 20	Top 10	Top 20
Corel-1000	87.67	81.50	8.77	16.30	15.94	27.17
GHIM-10K	87.00	76.10	1.74	3.05	3.41	5.85
COIL-100	93.18	81.08	12.94	22.52	22.73	35.25
Produce-1400	91.19	78.69	9.12	15.74	16.58	26.23
Outex	95.94	87.79	5.33	9.75	10.10	17.56

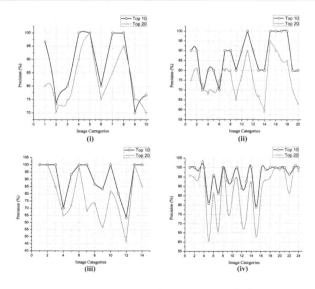

Fig. 4. Precision graph of (i) Corel-1000 (ii) GHIM-10K (iii) Produce-1400 (iv) Outex datasets for top 10 and top 20 retrieved images

10 and top 20 retrieved images. These all precision graphs show the group wise average precisions values. In all these graphs, we can see that most of the nodes are showing above 75% precision value so these are acceptable results. Since Coil-100 has 100 different image groups so we have not presented the precision graph for Coil-100.

4.4   Performance Comparison

In this paper, we have used 3 standard state-of-arts CBIR scheme to compare with our proposed CBIR scheme. These schemes have been introduced by ElAlami [10], Guo and Prasetyo [11], and Zeng et al. [18]. In Table 4, we have demonstrated the comparison of our anticipated CBIR scheme with these three other standard schemes. Table 4 shows the comparison for top 20 retrieved images in terms of average precision, recall, and f-score from Corel-1000 dataset. In this table we can see that, our anticipated CBIR scheme has shown better performance with respect to other three schemes. Later, Fig. 5 shows the category

Table 4. Comparison in terms of average precision, recall, and f-score for top 20 retrieved images from Corel-1000 dataset

Image retrieval methods	Corel-1000		
	$\mu_P(\%)$	$\mu_R(\%)$	$\mu_{Fs}(\%)$
ElAlami	76.10	16.10	25.90
Guo and Prasetyo	77.90	15.58	26.85
Zeng et al.	80.57	16.11	25.96
Proposed method	81.50	16.30	27.17

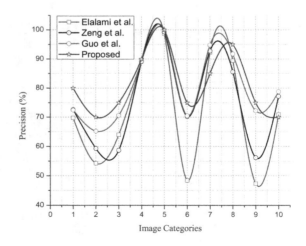

Fig. 5. Performance comparison in terms of precision for top 20 retrieved images from Corel-1000 dataset (Color figure online)

wise comparisons for top 20 retrieved images from Corel-1000 dataset. In Fig. 5, we can see that the blue line represents the proposed CBIR scheme and it has shown maximum high peaks. From Fig. 5, it has been clear that our proposed CBIR scheme is performing better for most of the cases as compare to other three schemes.

5 Conclusions

In RGB color space all color components shows high inter-correlations and due to which chromatic information of the image gets distorted. Hence, HSV color-space is better option for feature visual extraction. So in this work, the authors have presented a novel CBIR scheme which extracts texture and color visual image features simultaneously from HSV color image rather than considering RGB color-space. Here, we have applied 2D DT-CWT on value (V) components with 3 levels of decompositions because, it is shift invariant and it analyzes textures in six directions. Further, we have picked all six directional wavelet coefficients from real and imaginary part from each level of decomposition. We have also picked 4 approximated coefficients from third level of decompositions. Later, statistical parameters have been computed to preserve the texture features. Simultaneously, we have computed the probability histograms of hue (H) and saturation (S) components and we have divided these histograms into m non-uniform bins. Here, the bin division is based on the cumulative probability approach such that each bin will have approximately same number of pixel. As a result, color property of the image will get evenly distributed among all bins. Finally, the resultant feature vector contains better low-level visual color and texture features. We have also performed comparisons between our anticipated CBIR and other schemes in which our scheme has shown better performance. We have tested the robustness of our anticipated CBIR scheme by performing image retrieval form 5 different image datasets. The retrieval results validate the novelty and robustness of the proposed CBIR scheme with respect to different standard image datasets.

References

1. Gudivada, V.N., Raghavan, V.V.: Design and evaluation of algorithms for image retrieval by spatial similarity. ACM Trans. Inf. Syst. (TOIS) **13**(2), 115–144 (1995)
2. Liapis, S., Tziritas, G.: Color and texture image retrieval using chromaticity histograms and wavelet frames. IEEE Trans. Multimed. **6**(5), 676–686 (2004)
3. Guérin, C., Rigaud, C., Bertet, K., Revel, A.: An ontology-based framework for the automated analysis and interpretation of comic books images. Inf. Sci. **378**, 109–130 (2017)
4. Chun, Y.D., Kim, N.C., Jang, I.H.: Content-based image retrieval using multiresolution color and texture features. IEEE Trans. Multimed. **10**(6), 1073–1084 (2008)
5. Feng, S., Xu, D., Yang, X.: Attention-driven salient edge(s) and region(s) extraction with application to CBIR. Signal Process. **90**(1), 1–15 (2010)

6. Lin, C.-H., Chen, R.-T., Chan, Y.-K.: A smart content-based image retrieval system based on color and texture feature. Image Vis. Comput. **27**(6), 658–665 (2009)
7. Yue, J., Li, Z., Liu, L., Fu, Z.: Content-based image retrieval using color and texture fused features. Math. Comput. Model. **54**(3), 1121–1127 (2011)
8. Youssef, S.M.: ICTEDCT-CBIR: integrating curvelet transform with enhanced dominant colors extraction and texture analysis for efficient content-based image retrieval. Comput. Electr. Eng. **38**(5), 1358–1376 (2012)
9. Subrahmanyam, M., Wu, Q.M.J., Maheshwari, R.P., Balasubramanian, R.: Modified color motif co-occurrence matrix for image indexing and retrieval. Comput. Electr. Eng. **39**(3), 762–774 (2013)
10. ElAlami, M.E.: A new matching strategy for content based image retrieval system. Appl. Soft Comput. **14**, 407–418 (2014)
11. Guo, J.-M., Prasetyo, H.: Content-based image retrieval using features extracted from halftoning-based block truncation coding. IEEE Trans. Image Process. **24**(3), 1010–1024 (2015)
12. Varish, N., Pradhan, J., Pal, A.K.: Image retrieval based on non-uniform bins of color histogram and dual tree complex wavelet transform. Multimed. Tools Appl. **76**(14), 15885–15921 (2017)
13. Cui, C., Lin, P., Nie, X., Yin, Y., Zhu, Q.: Hybrid textual-visual relevance learning for content-based image retrieval. J. Vis. Commun. Image Represent. **48**, 367–374 (2017)
14. Pradhan, J., Pal, A.K., Banka, H.: A prominent object region detection based approach for CBIR application. In: 2016 Fourth International Conference on Parallel, Distributed and Grid Computing (PDGC), pp. 447–452. IEEE (2016)
15. Nene, S.A., Nayar, S.K., Murase, H., et al.: Columbia object image library (COIL-20) (1996)
16. tropical-fruits-db-1024x768.tar.gz. http://www.ic.unicamp.br/~rocha/pub/downloads/tropical-fruits-DB-1024x768.tar.gz/. Accessed 18 Aug 2017
17. site www, vision & image: lagis-vi.univ-lille1.fr (2017). http://lagis-vi.univlille1.fr/datasets/outex.html. Accessed 18 Aug 2017
18. Zeng, S., Huang, R., Wang, H., Kang, Z.: Image retrieval using spatiograms of colors quantized by Gaussian mixture models. Neurocomputing **171**, 673–684 (2016)

A Filtering Technique for All Pairs Approximate Parameterized String Matching

Shibsankar Das$^{(\boxtimes)}$

Department of Mathematics, Institute of Science,
Banaras Hindu University, Varanasi 221005, Uttar Pradesh, India
shibsankar@bhu.ac.in, shib.iitm@gmail.com

Abstract. The paper deals with all pairs approximate parameterized string matching problem with error threshold k, among two sets of equal length strings. Let $P = \{p_1, p_2, \ldots, p_{n_P}\} \subseteq \Sigma_P^m$ and $T = \{t_1, t_2, \ldots, t_{n_T}\} \subseteq \Sigma_T^m$ be two sets of strings where $|\Sigma_P| = |\Sigma_T|$. For each $p_i \in P$, the problem is to find $t_j \in T$ which is approximately parameterized closest to p_i under the threshold. The solution has complexity $O(n_P\, n_T\, m)$. We introduce Parikh vector filtering technique in order to preprocess the given strings and avoid the unwanted paired comparisons. The PV-filtering does not change the asymptotic time complexity but rapidly improves running time for small error threshold as shown by experiments.

Keywords: Approximate parameterized string matching
Hamming distance $\cdot$ $\gamma(k)$-match of vectors $\cdot$ Parikh vector
PV-filtering technique

1 Introduction

The problem of searching a given string in a text has a wide range of applications such as in text-editing programs, search engines and searching for patterns in a DNA sequence. There are non-indexed and indexed versions of this problem. In the indexed version, it is allowed to preprocess the string (pattern or text) before searching for the pattern in the text. The motivation of preprocessing is to improve the efficiency of the search. The standard variations of string matching problems are exact string matching [7,14], parameterized string matching [3–5], approximate string matching [23] and approximate parameterized string matching [8,26].

A preliminary version of this work is submitted as a technical report in Department of Theoretical Computer Science, Faculty of Information Technology, Czech Technical University in Prague [11]. Shibsankar Das was supported by the fellowship of HERITAGE Erasmus Mundus Partnership project for Ph.D. exchange mobility.

D. Ghosh et al. (Eds.): ICMC 2018, CCIS 834, pp. 97–109, 2018.
https://doi.org/10.1007/978-981-13-0023-3_10

The Approximate Parameterized String Matching (APSM) problem is a well studied problem [1,2,13,19,21,25]. In [19], Hazay et al. have given reduction between the maximum weighted bipartite matching problem [12,15–17,20] and APSM problem for two equal length strings. They have used the maximum weighted bipartite (decomposition) algorithm, originally proposed by Kao et al. [20], to solve the APSM problem between two equal length strings $p \in \Sigma_P^*$ and $t \in \Sigma_T^*$ in time $O(m^{1.5})$, where $|p| = |t| = m$.

In this paper, we investigate All Pairs (best) Approximate Parameterized String Matching (APAPSM) problem with error threshold k (with respect to Hamming distance error model) among two sets of equal length strings. Let $P = \{p_1, p_2, \ldots, p_{n_P}\} \subseteq \Sigma_P^m$ and $T = \{t_1, t_2, \ldots, t_{n_T}\} \subseteq \Sigma_T^m$ be two sets of strings where $1 \leq i \leq n_P$, $1 \leq j \leq n_T$ and $|\Sigma_P| = |\Sigma_T| = \sigma$. The APAPSM problem is to find: for each $p_i \in P$, a string $t_j \in T$ which is approximately parameterized closest to p_i under k threshold.

Section 2 describes the required preliminaries to understand the APAPSM problem which is explained in detail in the next section. In Sect. 3, we discuss a solution to the APAPSM problem with worst-case complexity $O(n_P\, n_T\, m)$, assuming a constant size alphabet. Next, we design a filtering technique by using Parikh vector [24] in order to preprocess the given strings and reduce the number of pair comparisons for solving APSM between the pair of strings with k error threshold. We call it PV-filter. Even though the filter does not improve the asymptotic bound theoretically, practical results in Sect. 4 show that it performs well for small error threshold. Finally, Sect. 5 summarizes the results.

2 Preliminaries and Related Results

We use some basic notions throughout the paper. An alphabet is a non-empty finite set of symbols. A *string* over a given alphabet is a finite sequence of symbols. We denote Σ^* as the set of all finite-length strings over alphabet Σ. The empty string is denoted by ε. The length of any string w is the total number of symbols in w and is denoted by $|w|$; so $|\varepsilon| = 0$. Let $\Sigma^+ = \Sigma^* \setminus \{\varepsilon\}$ and for a given $m \in \mathbb{N}_0$, Σ^m is the set of all strings of length m over the alphabet Σ [26].

Let $w = xyz$ be a string where $x, y, z \in \Sigma^*$. We call y as a *substring* of string w. If $x = \varepsilon$ then y is a *prefix* of w. If $z = \varepsilon$ then y is a *suffix* of w. The i-th symbol of a string w is denoted by $w[i]$ for $1 \leq i \leq |w|$. We denote substring y of string w as $w[i..j]$ if y starts at position i and ends at position j for $1 \leq i \leq j \leq |w|$, and string $w[i..j] = \varepsilon$ if $i > j$ [26]. Let $\mathbb{N}_0$ be the set of non-negative integers.

Approximate String Matching (ASM): *ASM* problem considers the string matching problem with errors. It is an important problem in many branches of computer science, with several applications to text searching, computational biology, pattern recognition, signal processing etc. [9,23,26].

Let $d: \Sigma^* \times \Sigma^* \to \mathbb{N}_0$ be the *distance function*. The *distance* $d(x, y)$ between two strings $x = x[1..n] \in \Sigma^*$ and $y = y[1..m] \in \Sigma^*$ is the minimal cost of a sequence of operations that transform x into y (and ∞ if no such sequence exists).

The cost of a sequence of operations is the sum of the costs of the individual operations. In general, the set of possible operations are insertion, deletion, substitution or replacement and transposition [23]. Therefore under the distance measure, the *ASM problem* becomes minimizing the total cost to transform the pattern and its occurrence in a text to make them equal and find the text positions where this cost is low enough. Some of the most classical distance metrics are Levenshtein distance [22], Damerau distance [10] and Hamming distance [18]. Hamming Distance (HD), denoted as d_H, allows only replacements. It is restricted to equal length strings. In the literature, the search problem in many cases is called "string matching with mismatches" [23,26].

Since in this paper, the APAPSM problem is considered under HD, the following definitions are considered using HD applied to equal length strings. From now onwards we assume that $d = d_H$, for notational simplicity. Given an error threshold $k \in \mathbb{N}_0$, a pair of strings $u \in \Sigma_u^*$ and $v \in \Sigma_v^*$ where $m = |u| = |v|$, consider the following definitions. Without loss of generality, we presume both the alphabet sizes are equal when dealing with a bijection between the alphabets.

Parameterized String Matching (PSM): String $u = u[1..m]$ is said to be a *parameterized match* or *p-match* with v (denoted as $u \mathrel{\widehat{=}} v$) if there exists a bijection $\pi \colon \Sigma_u \to \Sigma_v$ such that $\pi(u) = \pi(u[1])\pi(u[2]) \ldots \pi(u[m]) = v$ [3].

Approximate Parameterized String Matching (Without Error Threshold): Given a bijection $\pi \colon \Sigma_u \to \Sigma_v$, the π-mismatch between u and v is the HD between the image of u under π and v, i.e., $d(\pi(u), v)$ [19]. We denote this by π-mismatch(u, v). Note that, there is an exponential number of possible bijections from Σ_u to Σ_v. Also, such π for which $d(\pi(u), v)$ is minimum, may not be unique.

The Approximate Parameterized String Matching (APSM) between u and v is to find a π such that over all bijections π-mismatch(u, v) is minimized. We denote this by $APSM(u, v)$. Formally, $APSM(u, v) = \{\pi \mid d(\pi(u), v)$ is minimum over all $\pi\}$. We define the *cost of* $APSM(u, v)$ as $cost(APSM(u, v)) = d(\pi(u), v)$ where $\pi \in APSM(u, v)$.

Parameterized String Matching (PSM) with k Mismatches: PSM with k mismatch seeks to find a bijection $\pi \colon \Sigma_u \to \Sigma_v$ such that the π-mismatch$(u, v) \leq k$. We then say that u parameterized matches v with k threshold. In literature, this problem is also known as *string comparison problem with threshold k* [19]. However, any π with π-mismatch$(u, v) \leq k$ will be satisfactory in this case (i.e., π-mismatch(u, v) need not be the minimum one over all $\pi \colon \Sigma_u \to \Sigma_v$).

Both the above problems were solved in $O(m^{1.5})$ time [19] by reducing them to maximum weight bipartite matching problem and using Kao et al.'s algorithm [20]. Let us define APSM problem with k error threshold as follows.

Approximate Parameterized String Matching with k Error Threshold:
APSM with k error threshold, denoted as $APSM(u, v, k)$, seeks to find a π: $\Sigma_u \to$
Σ_v (over all bijections) such that $d(\pi(u), v)$ is minimum but not greater than k
[19]. More formally, $APSM(u, v, k) = \{\pi \mid \pi \in APSM(u, v) \ \wedge \ d(\pi(u), v) \leq k\}$.
We define *cost of* $APSM(u, v, k)$ as $cost(APSM(u, v, k)) = d(\pi(u), v)$, where
$\pi \in APSM(u, v, k)$. In case, $APSM(u, v, k) = \emptyset$, then $cost(APSM(u, v, k)) = \infty$.

Example 1 in page 4 shows the difference between the above definitions.

3   All Pairs Approximate Parameterized String Matching

In this section, we investigate all pairs (best) approximate parameterized string
matching (APAPSM) problem with k error threshold (with respect to Hamming
distance error model) among the two sets P and T of equal length strings. The
problem definition is the following along with the other required definitions[1].

**Definition 1 (Pair Approximate Parameterized String Matching
(PAPSM) with Error Threshold k).** *Given a string $p \in \Sigma_P^m$ and $T =$
$\{t_1, t_2, \ldots, t_{n_T}\} \subseteq \Sigma_T^m$, where $|\Sigma_P| = |\Sigma_T| = \sigma$ and $0 \leq k \leq m$. The PAPSM
problem with k error threshold is to find j such that $APSM(p, t_j, k)$ gives π_j over
all bijections and $d(\pi_j(p), t_j)$ is minimum over all j where $1 \leq j \leq n_T$.*

Denote this problem as $PAPSM(p, T, k)$. In more formal notation, $PAPSM$
$(p, T, k) = \{j \mid \pi_j \in APSM(p, t_j, k) \ \wedge \ d(\pi_j(p), t_j) = \min_{1 \leq i \leq n_T}\{cost(APSM$
$(p, t_i, k))\}\}$. In other words, the problem is to find $t_j \in T$ which is approximately
parameterized closest to p with k error threshold. We call $d(\pi_j(p), t_j)$ as the
cost of $PAPSM(p, T, k)$ and let us denote this by $cost(PAPSM(p, T, k))$. In case,
$PAPSM(p, T, k) = \emptyset$, then $cost(PAPSM(p, T, k)) = \infty$.

Example 1. Given $p = abab \in \Sigma_P^4 = \{a, b\}^4$, $T = \{t_1 = cdcd, \ t_2 = dcdc, \ t_3 =$
$ccdd, \ t_4 = cccd\} \subseteq \Sigma_T^4 = \{c, d\}^4$ and $k = 1$. Now,

$$APSM(p, t_1, k) = \{\pi_1 = \{a \to c, b \to d\}\}, \quad d(\pi_1(p), t_1)) = 0;$$
$$APSM(p, t_2, k) = \{\pi_2 = \{a \to d, b \to c\}\}, \quad d(\pi_2(p), t_2)) = 0;$$
$$APSM(p, t_3, k) = \emptyset;$$
$$APSM(p, t_4, k) = \{\pi_4 = \{a \to c, b \to d\}\}, \quad d(\pi_4(p), t_4)) = 1.$$

Observe that, $\pi_3 = \{a \to c, \ b \to d\} \in APSM(p, t_3)$ but $d(\pi_3(p), t_3)) = 2 > k$.
So, $APSM(p, t_3, 1) = \emptyset$. Hence, $PAPSM(p, T, 1) = \{1, 2\}$. Also note that, if $k = 3$,
then for $\pi_4' = \{a \to d, b \to c\}$, π_4'-mismatch$(p, t_4) \leq k$. Hence just finding π_4'
is also satisfactory to say that p is parameterized matched with t_4 under $k = 3$
error threshold; whereas $APSM(p, t_4, 3) = \{\pi_4\}$ and $\pi_4' \notin APSM(p, t_4, 3)$.   □

---

[1] These definitions can also be extended with respect to other error models.

Note that it is sufficient to report a string from T which is closest to p under a given error threshold k. Also, it is possible to enumerate all $t_i \in T$ which are closest to p. Observe that, if $PAPSM(p, T, k) = \{i, j\}$ corresponding to the strings t_i and t_j, then $cost(PAPSM(p, T, k)) \le k$ and more importantly, $cost(PAPSM(p, T, k)) = d(\pi_i(p), t_i) = d(\pi_j(p), t_j)$.

Theorem 1. *Given $p \in \Sigma_P^m$ and $T = \{t_1, t_2\} \subseteq \Sigma_T^m$. If p is an approximate parameterized matched with t_1 and $t_1 \cong t_2$, then p is also approximate parameterized matched with t_2 and its cost equal to $cost(APSM(p, t_1))$.*

Proof. The proof consists of two phases. Since p is approximate parameterized matched with t_1 (without any error threshold), then say $\pi_1 \in APSM(p, t_1)$. As a consequence, $cost(APSM(p, t_1)) = d(\pi_1(p), t_1)$ and moreover it is minimum over all bijections from Σ_P to Σ_T. Also, since $t_1 \cong t_2$, there exist a bijection, say $\pi \colon \Sigma_T \to \Sigma_T$ such that $\pi(t_1) = t_2$ and so $cost(APSM(t_1, t_2)) = d(\pi(t_1), t_2) = 0$.

Let $\pi_2 = \pi \circ \pi_1 \colon \Sigma_P \to \Sigma_T$ and is defined as $\pi \circ \pi_1(u) = \pi(\pi_1(u))$ where $u \in \Sigma_P^m$. It can be easily proved by contradiction that $d(\pi_2(p), t_2)$ is minimum over all bijections. So we skip it.

Now, $cost(APSM(p, t_2)) = d(\pi_2(p), t_2) = d(\pi(\pi_1(p)), t_2) = d(\pi(\pi_1(p)), \pi(t_1))$ $= d(\pi_1(p), t_1) = cost(APSM(p, t_1))$. Therefore, $\pi_2 = \pi \circ \pi_1 \in APSM(p, t_2)$ and its cost equal to $cost(APSM(p, t_1))$ unit. □

The above theorem is extended for APSM problem with k error threshold.

Theorem 2. *Given $p \in \Sigma_P^m$ and $T = \{t_1, t_2\} \subseteq \Sigma_T^m$ and $0 \le k \le m$. If p is an approximate parameterized matched with t_1 under the k error threshold and $t_1 \cong t_2$, then p is also approximate parameterized matched with t_2 under the k error threshold and with the cost equal to $cost(APSM(p, t_1, k))$.*

Definition 2 (All Pairs Approximate Parameterized String Matching (APAPSM) with k Threshold). *Let $P = \{p_1, p_2, \ldots, p_{n_P}\} \subseteq \Sigma_P^m$ and $T = \{t_1, t_2, \ldots, t_{n_T}\} \subseteq \Sigma_T^m$. The problem is to find a mapping $\eta \colon [1, n_P] \to [1, n_T]$ such that sum of the $cost(APSM(p_i, t_{\eta(i)}, k))$ over all i $(1 \le i \le n_P)$ is minimum.*

Let us denote this problem as $APAPSM(P, T, k)$. The problem is to search: for each $p_i \in P$ $(1 \le i \le n_P)$, a $t_j \in T$ $(1 \le j \le n_T)$ which is approximately parameterized closest to p_i under k error threshold. More formally, $APAPSM(P, T, k) = \{(PAPSM(p_1, T, k), PAPSM(p_2, T, k), \ldots, PAPSM(p_{n_P}, T, k))\}$.

Theorem 3. *The above problem can be solved in $O(n_P \, n_T \, m^{1.5})$ time.*

Proof. It is direct from the solution of APSM problem proposed by Hazay et al. [19] by considering all possible pairs between P and T.

Definition 3 ($\gamma(k)$ -match of strings). *Let $k \in \mathbb{N}_0$. For two given strings $u = u[1..m]$, $v = v[1..m] \in \Sigma^*$ and the alphabet set $\Sigma = \{a_1, a_2, \ldots, a_\sigma\}$ where each $a_i \in \mathbb{N}_0$, u is said to be $\gamma(k)$-matched with v if and only if $\sum_{i=1}^m |u_i - v_i| \le k$.*

The term $\gamma(k) - match$ is a suitably renamed version of the terminology $\gamma - approximate$ which was prescribed in [6] and defined on strings. Similar as above, we define $\gamma(k)$-match on two equal cardinality vectors of numbers.

Definition 4 (γ-distance, $\gamma(k)$-match of vectors). *Given two vectors* $u = (u_1, u_2, \ldots, u_m)$, $v = (v_1, v_2, \ldots, v_m)$ *where* $u_i, v_j \in \mathbb{N}_0, 1 \leq i, j \leq m$ *and* $l, k \in \mathbb{N}_0$. γ-*distance between* u *and* v *is* l *(denoted as* $\gamma(u, v) = l$*) if and only if* $l = \sum_{i=1}^{m} |u_i - v_i|$. *We say that* u, $\gamma(k)$-*matches with* v, *if and only if* $\gamma(u, v) = \sum_{i=1}^{m} |u_i - v_i| \leq k$.

The notion of Parikh mapping or vector was introduced by R.J. Parikh in [24]. It provides numerical properties of a string in terms of a vector by counting the number of occurrences of the symbols in the string. Parikh vector of a string w is denoted as $\psi(w)$.

Definition 5 (Parikh Vector (PV)). *Let* $\Sigma = \{a_1, a_2, \ldots, a_\sigma\}$. *Given* $w \in \Sigma^*$, $\psi(w) = (f(a_1, w), f(a_2, w), \ldots, f(a_\sigma, w))$ *where* $f(a_i, w)$ *gives the frequency of the symbol* $a_i \in \Sigma$ *($1 \leq i \leq \sigma$) in the string* w.

For example, if $\Sigma = \{c, d\}$ then $\psi(cddcc) = (3, 2)$. However, much information is lost in the transition from a string to its PV. Note that Parikh mapping is not injective as many strings over an alphabet may have the same PV and so the information of a string is reduced while changing the string to a PV. For example, the strings $cccdddd$ and $dcdcdcd$ have the same Parikh vector $(3, 4)$.

Definition 6 (Normalized Parikh Vector (NPV)). *NPV of a string* $w \in \Sigma^*$ *is* $\widehat{\psi}(w) = (g_1, g_2, \ldots, g_\sigma)$ *such that* $\forall i, 1 \leq i < \sigma, g_i \geq g_{i+1}$ *and there exists a bijective mapping* $\rho: \{1..\sigma\} \to \{1..\sigma\}$ *such that* $g_i = f(a_{\rho(i)}, w)$.

In other words, we sort the elements of $\psi(w)$ in non-increasing order to get the $\widehat{\psi}(w)$ of string w. For example, $\psi(dcdcdcd) = (3, 4)$ and $\widehat{\psi}(dcdcdcd) = (4, 3)$.

Theorem 4. *Given a pair of equal length strings* $u \in \Sigma_P^*$ *and* $v \in \Sigma_T^*$, *if* $u \cong v$ *then* $\gamma(\widehat{\psi}(u), \widehat{\psi}(v)) = 0$.

Proof. Since $u \cong v$, then by definition there exists a bijection $\pi: \Sigma_P \to \Sigma_T$ such that $\pi(u) = v$, i.e. $\pi(u)$ is obtained by renaming each character of u using π. Though symbols of Σ_P are renamed by π, the frequency of each symbol $a \in \Sigma_P$ in u will be same as the frequency of $\pi(a) \in \Sigma_T$ in $v = \pi(u)$. As a consequence, $\widehat{\psi}(v) = \widehat{\psi}(u)$, even though there may be the case $\psi(u) \neq \psi(v)$. $\square$

However, the converse is not always true. To show that, we shall give the following example.

Example 2. Given $p = ababa, \in \Sigma_P^* = \{a, b\}^*$ and $T = \{t_1 = cdcdd, t_2 = dcdcd\} \subseteq \Sigma_T^* = \{c, d\}^*$. Now,

$$\psi(p) = (3, 2) \text{ and } \widehat{\psi}(p) = (3, 2);$$

$$\psi(t_1) = (2, 3) \text{ and } \widehat{\psi}(t_1) = (3, 2);$$

$$\psi(t_2) = (2, 3) \text{ and } \widehat{\psi}(t_2) = (3, 2).$$

As mentioned in Theorem 4, $p \mathrel{\widehat{=}} t_2$ and so $\gamma(\widehat{\psi}(p), \widehat{\psi}(t_2)) = 0$, even though $\psi(p) \neq \psi(t_2)$. Conversely, $\widehat{\psi}(t_1) = \widehat{\psi}(p) = \widehat{\psi}(t_2) = (3, 2)$, but $p \mathrel{\widehat{\neq}} t_1$ and $p \mathrel{\widehat{=}} t_2$. Hence, in case $\gamma(\widehat{\psi}(u), \widehat{\psi}(v)) = 0$, it is required to check if $u \mathrel{\widehat{=}} v$ or not. □

In general, a *filter* is a device or subroutine that processes the feasible inputs and tries to remove some undesirable component. We design an interesting filtering technique by using Parikh vector in order to preprocess the given strings of P and T and to reduce the number of pair comparisons for solving approximate parameterized string matching between the pair of strings under k error threshold. We name the filter which is mentioned in Theorem 7 as *PV-filter* and the process of filtering the input data by PV-filter as *PV-filtering*.

The following theorems are useful in minimizing the number of pairs comparisons for APAPSM problem to improve the solution from the practical aspect. Theorem 5 is applicable for ASM problem. It is extended in Theorems 6 and 7 in the context of APSM problem without and with k error threshold, respectively.

Theorem 5. *Let $u, v \in \Sigma^*$ be a pair of equal length strings and $k = d(u, v)$, is the Hamming distance. Then $\gamma(\widehat{\psi}(u), \widehat{\psi}(v)) \leq 2k$ and $\gamma(\psi(u), \psi(v)) \leq 2k$.*

Proof. We prove it by the principle of mathematical induction on k.

Base case: For $u = v$, $k = d(u, v) = 0$ and $\gamma(\widehat{\psi}(u), \widehat{\psi}(v)) = 0$.
Hypothesis: Assume that for any k with $0 \leq k = d(u, v) \leq i$, $\gamma(\widehat{\psi}(u), \widehat{\psi}(v)) \leq 2k$.
Inductive step: Let, after introducing one more error by replacement (symbol $a \in \Sigma$ is replaced by $b \in \Sigma$ in any position of u) operation in u we get u' such that $d(u', u) = 1$ and $k = d(u', v) = i + 1$. However, while changing u to u' with $d(u', u) = 1$, there may be only other case that $k = d(u', v) = i - 1$ for which also the inequality is true (by the induction hypothesis). So we have to argue for the former case: $k = i + 1$. While introducing one error by replacement, $\gamma(\widehat{\psi}(u'), \widehat{\psi}(u))$ will be increased by at most 2 as the frequency of symbol a is decreased by one and the frequency of b is increased by one. Hence, $\gamma(\widehat{\psi}(u'), \widehat{\psi}(v)) \leq 2i + 2 = 2(i + 1)$ while $k = d(u', v) = i + 1$.

Hence the proof of the first inequality, by the principle of mathematical induction.

For the other one also, the proof justification is similar. □

Theorem 6. *Given a pair of strings $u \in \Sigma_P^m$, $v \in \Sigma_T^m$, let $k = cost(APSM(u, v))$. Then $\gamma(\widehat{\psi}(u), \widehat{\psi}(v)) \leq 2k$.*

Proof. Let $\pi \in APSM(u, v)$. Therefore by definition, $k = cost(APSM(u, v)) = d(\pi(u), v)$ is minimum over all bijections. Let $\pi(u) = u' \in \Sigma_T^m$. Since $u \mathrel{\widehat{=}} u'$ under π, $\widehat{\psi}(u) = \widehat{\psi}(u')$, by Theorem 4. Hence $\gamma(\widehat{\psi}(u), \widehat{\psi}(v)) = \gamma(\widehat{\psi}(u'), \widehat{\psi}(v))$. By using Theorem 5, we have $\gamma(\widehat{\psi}(u), \widehat{\psi}(v)) = \gamma(\widehat{\psi}(u'), \widehat{\psi}(v)) \leq 2k$. □

Theorem 7. *Given $u \in \Sigma_P^m$ and $v \in \Sigma_T^m$. Let $\widehat{k} = cost(APSM(u, v, k))$. Then $\gamma(\widehat{\psi}(u), \widehat{\psi}(v)) \leq 2\widehat{k}$ (which we call as PV-filter).*

Proof. The proof is very similar as Theorem 6. Let $\pi \in APSM(u, v, k)$. Accordingly, there exists a bijection $\pi \colon \Sigma_P \to \Sigma_T$ such that $\widehat{k} = cost(APSM(u, v, k)) = d(\pi(u), v)$ is minimum but not greater than k. Let $u' = \pi(u) \in \Sigma_T^m$. With similar argument as above, we have $\gamma(\widehat{\psi}(u), \widehat{\psi}(v)) = \gamma(\widehat{\psi}(u'), \widehat{\psi}(v)) \leq 2\widehat{k}$. □

We use this PV-filter as a subroutine during the design of a simple algorithm to solve the APAPSM problem with error threshold k between two sets P and T of equal length strings. In worst-case (i.e. none of the pairs are filtered out by PV-filter), it takes $O(n_P\, n_T\, m)$.

Computing APAPSM Under Error Threshold: Let $P = \{p_1,\, p_2, \ldots, p_{n_P}\} \subseteq \Sigma_P^m$ and $T = \{t_1,\, t_2, \ldots, t_{n_T}\} \subseteq \Sigma_T^m$ be two sets of strings where $|\Sigma_P| = |\Sigma_T| = \sigma$. In Algorithm 1, we compute APAPSM problem with error threshold $k \in \mathbb{N}_0$ among two sets P and T of equal length strings. In Step 3, clustering is precisely recommended, in case in advance it is known that there are many exact and parameterized repetition of strings in P and T. To create the equivalence classes in P and T separately, with respect to parameterization, clustering is done based on the converse of Theorem 4, i.e., in case for any two strings $u, v \in P$ (and T, respectively) if $\gamma(\widehat{\psi}(u), \widehat{\psi}(v)) = 0$, then and only then check for $u \cong v$. If $u \cong v$ holds, them put u and v into the same cluster.

Algorithm 1. Compute $APAPSM(P, T, k)$ after using the PV-filter

Input: The sets P, T of equal length strings and an error threshold k.
Output: $APAPSM(P, T, k)$ with respect to Hamming distance error model.

APAPSM(P, T, k)
   1: **for** $i \leftarrow 1 : n_P$ **do** compute NPV of p_i.
   2: **for** $i \leftarrow 1 : n_T$ **do** compute NPV of t_i.
   3: **do** parameterized clustering of P and T, i.e., for any $(p_1, p_2) \in P \times P$ or $T \times T$
      of a cluster, $p_1 \cong p_2$. To speed up the clustering, if $\gamma(\widehat{\psi}(p_1), \widehat{\psi}(p_2)) = 0$ **then**
      only check for $p_1 \cong p_2$, or otherwise, $p_1 \not\cong p_2$ (Negation of Theorem 4).
   4: **for** each parameterized cluster of P, pick a representative, say p_i
         for each parameterized cluster of T, pick a representative, say t_j
            if $\gamma(\widehat{\psi}(p_i), \widehat{\psi}(t_j)) \leq 2k$ (which is the *PV-filtering*)
               then compute $APSM(p_i, t_j, k)$.
            end if
         end for
      end for

Complexity Analysis: Steps 1–3 of Algorithm 1 are the preprocessing steps for computing $APAPSM(P, T, k)$; Steps 1–2 takes $O(m(n_P + n_T))$ and Step 3 takes $O(m(n_P^2 + n_T^2))$ time, assuming a constant size alphabets. But as mentioned earlier, clustering is optional, it might be skipped depending on the circumstances.

In Step 4, for any pair $(p_i, t_j) \in P \times T$, computation of $APSM(p_i, t_j, k)$ can be done by reducing the problem to maximum weight bipartite matching (MWBM) problem [19]. Let $G = (V, E, W)$ be an undirected, weighted (non-negative integer weight) bipartite graph where V, E and W are the vertex set, edge set and total weight of G, respectively. MWBM problem can be solved in $O(\sqrt{|V|}W')$ time, where $|E| \leq W' \leq W$ [12]. It is a fine-tuned version of the existing decomposition solution [20]. Using the fine-tuned decomposition solution for MWBM, $APSM(p_i, t_j, k)$ can be solved in $O(m\sqrt{\sigma})$ where $W' = O(W) = O(m)$ and $V = O(\sigma)$ [12,13]. In the worst-case scenario: each of the clusters will have just a single string either from P or T and PV-filter in Step 4 does not filter out any pair $(p_i, t_j) \in P \times T$. Therefore worst-case running time of the Algorithm 1 is $O(n_P\, n_T\, m\sqrt{\sigma})$, which is $O(n_P\, n_T\, m)$, if we assume a constant alphabet.

4   Experimental Results

To test the efficiency of the PV-filter, we performed several experimental studies, but only a few are reported in this section because of page limitation. Algorithm 1 which solves $APAPSM(P, T, k)$, is implemented in *MATLAB Version 7.8.0.347 (R2009a)*. All the experiments are conducted on a PC Laptop with an *Intel$^{\circledR}$ Core^{TM} 2 Duo (T6570 @ 2.10 GHz) Processor, 3.00 GB RAM and 500 GB Hard Disk*, running the *Microsoft Windows 7 Ultimate (32-bit Operating System)*.

Data Description: We generate the input data sets P and T by using the pre-defined `randi` function. It helps to generate uniformly distributed pseudorandom integers. The function `randi(imax,m,n)` returns an `m`-by-`n` matrix containing pseudorandom integer values drawn from the discrete uniform distribution on `1:imax`.

Efficiency of PV-Filter: The experimental results show that the PV-filter is efficient, essentially for small error threshold k, to avoid unwanted pairs (u, v) comparison for $APSM(u, v, k)$, where $u \in P$ and $v \in T$. According to the random experiment, if the error threshold $k \leq \frac{m}{3}$, then almost more than one-third of the total pairs comparison can be skipped. Moreover, very smaller threshold gives much better filtering. Please see the experiments.

Experiment 1. *Consider, alphabet sets* $\Sigma_P = \{\mathsf{a, b, c, d, e, f, g, h, i, j}\}$, $\Sigma_T = \{\mathsf{a', b', c', d', e', f', g', h', i', j'}\}$; $P \in \Sigma_P^*$, $T \in \Sigma_T^*$; *cardinality of each of the sets* P *and* T *is* $|P| = |T| = 100$; *and* $|p_i| = |t_j| = 6$ *for* $1 \leq i, j \leq |P| = |T|$.

According to the data set generated in Experiment 1, a total of 10,000 (u, v) pairs of comparisons for $APSM(u, v, k)$, where $u \in P$ and $v \in T$, are required without PV-filtering. Figure 1 shows the efficiency graph of the filter on the input data set. Each blue "$*$" point in the graph indicates the number of elimination of pairs comparison for a given error threshold, after using the PV-filter.

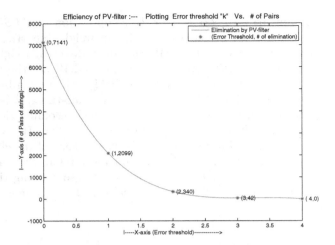

Fig. 1. Elimination graph of pairs of strings after using PV-filter for the input data set with $|\Sigma_P| = |\Sigma_T| = 10$; $|P| = |T| = 100$; $|p_i| = |t_j| = 6$ for $1 \le i, j \le |P| = |T|$, as mentioned in Experiment 1.

Table 1. PV-filtering for the data set in Experiment 1.

Number of pairs ↓	$k = 0$	$k = 1$	$k = 2$	$k = 3$	$k = 4$
Before using PV-filter	10,000	10,000	10,000	10,000	10,000
Eliminated by PV-filter	7,141	2,099	340	42	0
Passed through PV-filter	2,859	7,901	9,660	9,958	10,000
Whose APSM cost $\le k$	570	3,986	8,699	9,869	10,000

For example, each point (i, j) in Fig. 1 represents that for $k = i$ error threshold, j number of (u, v) pairs of strings have skipped the comparison for $APSM(u, v, i)$.

Table 1 gives more light to the Experiment 1. The second row represents that for a given k, a total number of (u, v) pairs are to be checked for $APSM(u, v, k)$, initially before using PV-filter; the third row says, for respective k the number of pairs of strings are eliminated by PV-filter; simultaneously, the fourth row describes that how many string pairs are passed by the filter; and finally, the last row mentions, for how many (u, v) pairs, actually $cost(APSM(u, v)) \le k$ among the passed pairs.

Experiment 2. *Consider the alphabet sets $\Sigma_P = \{a, b, c, \ldots, x, y, z\}$, $\Sigma_T = \{a', b', c', \ldots, x', y', z'\}$ with $|\Sigma_P| = |\Sigma_T| = 26$; $P \in \Sigma_P^*$, $T \in \Sigma_T^*$; cardinality of the sets P and T is $|P| = |T| = 100$; and $|p_i| = |t_j| = 2000$ for $1 \le i, j \le |P| = |T|$. Figure 2 gives the elimination graph. The corresponding table is skipped due to space limitation.*

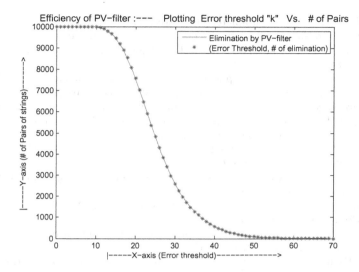

Fig. 2. Elimination graph of pairs of strings after using PV-filter for the input data set with $|\Sigma_P| = |\Sigma_T| = 26$; $|P| = |T| = 100$; $|p_i| = |t_j| = 2000$ for $1 \leq i, j \leq |P| = |T|$, as mentioned in Experiment 2.

5   Conclusions

In this paper, we have explored all pairs approximate parameterized string matching problem with k Hamming distance error threshold between two sets of equal length stings. We have presented a solution with worst-case complexity $O(n_P\, n_T\, m)$, assuming constant alphabet size. In order to minimize number of paired comparisons for solving APSM between pair of strings with error threshold, we have proposed a PV-filtering technique by using Parikh vector. Although the filter does not improve the worst-case asymptotic bound, but the using it as a subroutine, we can avoid some of the unwanted paired comparisons for APSM. Experimental results show that the PV-filter is efficient for small error threshold.

Acknowledgement. The author is grateful to Dr. Jan Holub for his helpful comments and suggestions.

References

1. Apostolico, A., Erdős, P.L., Jüttner, A.: Parameterized searching with mismatches for run-length encoded strings. Theor. Comput. Sci. **454**, 23–29 (2012). Formal and Natural Computing Honoring the 80th Birthday of Andrzej Ehrenfeucht
2. Apostolico, A., Erdős, P.L., Lewenstein, M.: Parameterized matching with mismatches. J. Discret. Algorithms **5**(1), 135–140 (2007)

3. Baker, B.S.: A theory of parameterized pattern matching: algorithms and applications. In: Symposium on Theory of Computing, pp. 71–80. ACM (1993)
4. Baker, B.S.: Parameterized pattern matching: algorithms and applications. J. Comput. Syst. Sci. **52**(1), 28–42 (1996)
5. Baker, B.S.: Parameterized duplication in strings: algorithms and an application to software maintenance. SIAM J. Comput. **26**(5), 1343–1362 (1997)
6. Cambouropoulos, E., Crochemore, M., Iliopoulos, C.S., Mouchard, L., Pinzon, Y.J.: Algorithms for computing approximate repetitions in musical sequences. Int. J. Comput. Math. **79**(11), 1135–1148 (2002)
7. Charras, C., Lecroq, T.: Handbook of Exact String Matching Algorithms. King's College Publications, London (2004)
8. Crochemore, M., Hancart, C., Lecroq, T.: Algorithms on Strings. Cambridge University Press, New York (2007)
9. Crochemore, M., Rytter, W.: Jewels of Stringology: Text Algorithms. World Scientific Press, Singapore (2002)
10. Damerau, F.J.: A technique for computer detection and correction of spelling errors. Commun. ACM **7**(3), 171–176 (1964)
11. Das, S., Holub, J., Kapoor, K.: All pairs approximate parameterized string matching. Technical report FIT-2014-01, Department of Theoretical Computer Science, Faculty of Information Technology, Czech Technical University in Prague, Thăkurova 2700/9, 160 00 Praha 6, Czech Republic, March 2014
12. Das, S., Kapoor, K.: Fine-tuning decomposition theorem for maximum weight bipartite matching. In: Gopal, T.V., Agrawal, M., Li, A., Cooper, S.B. (eds.) TAMC 2014. LNCS, vol. 8402, pp. 312–322. Springer, Cham (2014). https://doi.org/10.1007/978-3-319-06089-7_22
13. Das, S., Kapoor, K.: Weighted approximate parameterized string matching. AKCE Int. J. Graphs Comb. **14**(1), 1–12 (2017)
14. Faro, S., Lecroq, T.: The exact online string matching problem: a review of the most recent results. ACM Comput. Surv. **45**(2), 13:1–13:42 (2013)
15. Fredman, M.L., Tarjan, R.E.: Fibonacci heaps and their uses in improved network optimization algorithms. J. ACM **34**(3), 596–615 (1987)
16. Gabow, H.N.: Scaling algorithms for network problems. J. Comput. Syst. Sci. **31**(2), 148–168 (1985)
17. Gabow, H.N., Tarjan, R.E.: Faster scaling algorithms for network problems. SIAM J. Comput. **18**(5), 1013–1036 (1989)
18. Hamming, R.W.: Error detecting and error correcting codes. Bell Syst. Tech. J. **29**(2), 147–160 (1950)
19. Hazay, C., Lewenstein, M., Sokol, D.: Approximate parameterized matching. ACM Trans. Algorithms **3**(3) (2007). https://doi.org/10.1145/1273340.1273345
20. Kao, M.Y., Lam, T.W., Sung, W.K., Ting, H.F.: A decomposition theorem for maximum weight bipartite matchings. SIAM J. Comput. **31**(1), 18–26 (2001)
21. Lee, I., Mendivelso, J., Pinzón, Y.J.: $\delta\gamma$ – parameterized matching. In: Amir, A., Turpin, A., Moffat, A. (eds.) SPIRE 2008. LNCS, vol. 5280, pp. 236–248. Springer, Heidelberg (2008). https://doi.org/10.1007/978-3-540-89097-3_23
22. Levenshtein, V.: Binary codes capable of correcting deletions, insertions and reversals. Sov. Phys. Dokl. **10**(8), 707–710 (1966)
23. Navarro, G.: A guided tour to approximate string matching. ACM Comput. Surv. **33**, 31–88 (2001)
24. Parikh, R.J.: On context-free languages. J. ACM **13**(4), 570–581 (1966)

25. Prasad, R., Agarwal, S.: Study of bit-parallel approximate parameterized string matching algorithms. In: Ranka, S., Aluru, S., Buyya, R., Chung, Y.-C., Dua, S., Grama, A., Gupta, S.K.S., Kumar, R., Phoha, V.V. (eds.) IC3 2009. CCIS, vol. 40, pp. 26–36. Springer, Heidelberg (2009). https://doi.org/10.1007/978-3-642-03547-0_4

26. Smyth, B.: Computing Patterns in Strings. Pearson Addison-Wesley, New York (2003)

On Leaf Node Edge Switchings
in Spanning Trees of De Bruijn Graphs

Suman Roy[1(✉)] [iD], Srinivasan Krishnaswamy[1], and P. Vinod Kumar[2]

[1] Department of Electronics and Electrical Engineering,
Indian Institute of Technology Guwahati, Guwahati 781039, Assam, India
{suman.roy,srinikris}@iitg.ernet.in
[2] Bharat Broadband Network Limited, Trivandrum, India
saivinod.potnuru@gmail.com

Abstract. An n-th order k-ary de Bruijn sequence is a cyclic sequence of length k^n which contains every possible k-ary subsequence of length n exactly once during each period. In this paper, we show that, if we fix the initial n bits, any n-th order de Bruijn sequence can be transformed to another using a sequence of transformations.

Keywords: De Bruijn sequences · De Bruijn graph
Pseudorandom sequence generator · Shift register

1 Introduction

An n-th order k-ary de Bruijn sequence, $DB_n(k)$, is a periodic sequence of length k^n having every possible k-ary subsequence of length n exactly once in each period. An example of $DB_2(3)$ is 001021122. In [3], it has been shown that there exist $((k-1)!)^{k^{n-1}} k^{k^{(n-1)}-n}$ k-ary de Bruijn sequences of order n. De Bruijn sequences satisfy many statistical properties associated with randomness such as balance property, span-n property, etc. Thus, they find many applications ranging from cryptography and coding theory to communication systems [11,12]. This paper deals with binary de Bruijn sequences although the results can be easily extended for k-ary de Bruijn sequences. Feedback Shift Registers (FSRs) have been used to generate such sequences for many decades [7]. There are a number of algorithms available in literature to generate de Bruijn sequences using shift registers [5]. One way of generating de Bruijn sequences is by joining various FSRs of shorter cycles [4,8]. Given a k-ary de Bruijn sequence of order n, other such sequences can be obtained by using cross-join pairs [6,10]. Recursive algorithms to produce higher order de Bruijn sequences from the lower order de Bruijn sequences have been discussed in [1,9]. In this paper, we use the enumerative construction given in [2] to show that any n-th order de Bruijn sequence can be generated from another by using a set of transformations. Here all de Bruijn sequences that are cyclic shifts of each other shall be considered equivalent.

© Springer Nature Singapore Pte Ltd. 2018
D. Ghosh et al. (Eds.): ICMC 2018, CCIS 834, pp. 110–117, 2018.
https://doi.org/10.1007/978-981-13-0023-3_11

The remainder of this paper is organized as follows. Section 2 introduces de Bruijn graphs and contains a brief description of the algorithm given in [2]. Section 3 contains the main results of the paper. In Sect. 4, we summarize the results and conclude the paper.

2   Preliminaries

Let $G = (V, E)$ be a directed graph where V and E denote the vertex (or node) set and the edge set in G respectively. Every edge $e \in E$ is directed from the source vertex $s(e)$ to the target vertex $t(e)$. For all $v \in V$, $indeg(v)$ and $outdeg(v)$ are the number of incoming and outgoing edges respectively. An Eulerian cycle (or Eulerian circuit) in a directed graph is a directed cycle which uses every edge $e \in E$ exactly once. A graph that contains an Eulerian cycle is called an Eulerian graph. For a connected directed balanced graph G, there exists at least one Eulerian circuit. De Bruijn sequences are closely associated with special directed Eulerian graphs known as de Bruijn graphs [3]. A binary de Bruijn graph of order n, denoted as G_n, is a directed graph with 2^n vertices, each labeled with a unique n bit string. Each edge of the graph is labeled with a binary string of length $(n + 1)$. The edge labeled as $s_0 s_1 \ldots s_n$ connects the source vertex labeled $s_0 s_1 \ldots s_{n-1}$ with the target vertex labeled $s_1 s_2 \ldots s_n$.

Example 1. Figure 1 represents a second order binary de Bruijn graph G_2.

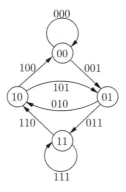

Fig. 1. G_2: De Bruijn graph of order 2.

In G_n, each vertex $v = (s_0 s_1 \ldots s_{n-1})$ has two out-edges $(s_0 s_1 \ldots s_{n-1}1)$ and $(s_0 s_1 \ldots s_{n-1}0)$; these edges are known as the one-edge and the zero-edge of v respectively. Since de Bruijn graphs are connected and balanced they always contain an Eulerian cycle. Observe that we can obtain an $(n + 1)$-th order de Bruijn sequence by considering the sequence of most significant bits of edges in an Eulerian cycle of G_n. Clearly, there exists a one-to-one correspondence between Eulerian cycles of G_n and $(n+1)$-th order de Bruijn sequences. We now proceed to briefly describe the process of generating Eulerian cycles of G_n given in [2].

An oriented spanning tree is an acyclic subgraph of a directed graph $G = (V, E)$. It has a vertex $r \in V$ known as root vertex such that $outdeg(r) = 0$ and there exists a path from every vertex $v \in V \backslash \{r\}$ to r. For a directed graph shown in Fig. 1, an oriented spanning tree T rooted at 00 is given in Fig. 2.

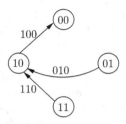

Fig. 2. T: Spanning tree of G_2 rooted at 00.

Let T be a spanning tree of $G_n = (V, E)$ where V is the set of vertices and E is the edge set in G_n. Now, for all $v \in V$, construct the lists l_v having $indeg(v)$ number of edges. This array of lists is known as *tree array*. For all $v \in V \backslash \{r\}$, the last element of l_v will be the unique outgoing edge of v which occurs in the spanning tree T. In case of the root vertex r, last element of l_r is the symbol Ω and the first entry of l_r can be any of its outgoing edges in G_n.

Example 2. Consider the de Bruijn graph G_2 shown in Fig. 1. A spanning tree of G_2 is shown in Fig. 2. A tree array l_v corresponding to T is given as follows:

$$l_{00} = \{(000), \quad \Omega\}$$
$$l_{01} = \{(011), (010)\}$$
$$l_{10} = \{(101), (100)\}$$
$$l_{11} = \{(111), (110)\}$$

Now, let T be a spanning tree of G_n rooted at the vertex r and consider its corresponding tree array. One can obtain the Eulerian cycle of G_n as follows. Starting with the unique edge of G_n which does not lie in the tree array, each edge x is followed by the first unused outgoing edge of $t(x)$ in the tree array. This process stops when the only unused entry of the tree array is Ω. Thus, given a spanning tree and a tree array we can generate an Eulerian cycle in G_n. Further, it has been shown in [2] that the correspondence between a spanning tree - tree array pair and Eulerian cycles is one-to-one and one can easily obtain one from the other.

Example 3. Consider the G_2 given in Fig. 1 and its spanning tree T rooted at 00 shown in Fig. 2. A tree array for the spanning tree is given in Example 2.

We start with the edge 001 followed by the edge 011 which is the first unused outgoing edge of the vertex $01 = t(001)$. By repeating this process, we get the Eulerian cycle of G_2 shown in Fig. 3. The corresponding 3-rd order de Bruijn sequence is 00111010.

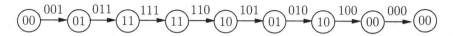

Fig. 3. Eulerian cycle of the de Bruijn graph G_2.

3 Construction of Any Spanning Tree of G_n from a Spanning Tree of G_n

Recall that, a binary de Bruijn graph G_n having a vertex $s = (s_0 s_1 \dots s_{n-1})$ has two out-edges, namely zero-edge and one-edge, which connect s to its successor vertex $(s_1 s_2 \dots s_{n-1} 0)$ and $(s_1 s_2 \dots s_{n-1} 1)$ respectively. These vertices are called conjugates of each other and corresponding edges are called conjugate edges. In a spanning tree of G_n, any vertex s is connected to one of its successor vertices. The all-zero and all-one vertices have only one possible successor in the spanning tree. Therefore, we consider these vertices merged with their respective successor vertices in the spanning tree as a single vertex. For example, consider the spanning tree of G_3 shown in Fig. 4. Here 000 and 001 (similarly, 111 and 110) are jointly treated as a single vertex $000 - 001$ $(111 - 110)$.

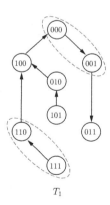

T_1

Fig. 4. A spanning tree of G_3 rooted at 011.

Note that fixing the root vertex and its unique outgoing edge that does not lie in the tree array essentially fixes the starting edge of the Eulerian cycle (therefore, the first n-bits of the de Bruijn sequence). Therefore, if we fix the entry corresponding to the root vertex in the tree array, we have a one-to-one correspondence between de Bruijn sequences with a given initial state and oriented spanning trees of G_n. In the remainder of this section we will show that from a given spanning tree of G_n rooted at a particular vertex we can generate any other spanning tree of G_n having the same root by a sequence of transformations.

The process of replacing the outgoing edge of one vertex in the spanning tree by its conjugate edge is known as edge switching. In a spanning tree T, a leaf node is a node that does not have any incoming edge. For example, consider one of the binary spanning trees of G_2, T, rooted at 00 given in Fig. 2. Here, 01 and 11 are the leaf nodes.

Lemma 1. *Switching the outgoing edge of a leaf node in a spanning tree of G_n generates another spanning tree.*

Proof. Let T be a spanning tree of G_n. Suppose an outgoing edge e of a vertex $v \in T$ is replaced by its conjugate edge e', then this switching results in a cycle only if $t(e)$ is a vertex from which there exists a path to v. Otherwise the resulting graph will be an another spanning tree. Now, if the node $v \in T$ is a leaf node then there exists no path to v from any other vertex. Therefore, when the outgoing edge of a leaf node is switched the resulting graph is another spanning tree of G_n.

Example 4. Consider the spanning tree, T_1, of G_3 as given in Fig. 4. T_1 has two leaf nodes viz. 101 and the merged pair $111 - 110$. Switching of these edges, 101 and $111 - 110$, produce two different spanning trees (see Fig. 5).

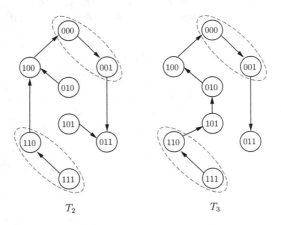

$$T_2 \qquad\qquad T_3$$

Fig. 5. Spanning trees of G_3 rooted at 011.

In a directed spanning tree T a node x is known as an ancestor of a node v if there exists a path from x to v in T. For example, in Fig. 4 the ancestors of the node 100 are 101, 010 and the merged vertex pair $111 - 110$. Now, we proceed to prove our main result.

Theorem 1. *Given a spanning tree T of G_n rooted at r, any other spanning tree T' of G_n having the same root can be obtained from T by a sequence of leaf nodes edge switching.*

Proof. Let the number of nodes in T whose outgoing edges are same as that in T' be κ. Now, consider a node $v \neq r$ in T whose outgoing edge is different from that in T' and all of whose ancestor nodes have the same out-edges as in T'. Now, we apply post-order depth first traversal on the sub-tree of T rooted at v and switch the nodes in the order in which they occur in this traversal. Clearly, this is a sequence of leaf node edge switchings culminating in the switching of the outgoing edge of v. Once the node v is switched, we switch all the other nodes that we had switched before v in the reverse order. We now have a new graph wherein the number of nodes in T whose outgoing edges are same as that in T' is $\kappa + 1$. We keep repeating this process till the outgoing edges of all vertices in T become same as T'. Thus, we transform the spanning tree T into T'.

Given a de Bruijn sequence we can construct the corresponding spanning tree and tree array. Now, by randomly performing a sequence of leaf node switches we can get a random spanning tree of G_n and therefore a random de Bruijn sequence.

Example 5. Consider a spanning tree, T_1, of the 3-rd order de Bruijn graph shown in Fig. 6(a). Let T_1' be any other spanning tree of G_3 rooted at the same vertex as shown in Fig. 6(g). Here $\kappa = 5$ and except 101 and 100 all other nodes in T_1 have the same outgoing edges as in T_1'. Now, 101 is a leaf node and switching its edge gives us a new spanning tree shown in Fig. 6(b). This spanning tree has 6 nodes whose outgoing edges are same as in T_1'. Now, the vertex 100 is not a leaf node. By applying depth first traversal algorithm on the sub-graph rooted at node 100 we get the list $\{111 - 110, 010, 100\}$. We now switch the nodes in the order $111 - 110, 010$ and 100. This gives us the spanning tree shown in Fig. 6(e). We now again switch the nodes 010 and $111 - 110$. This gives us the spanning tree T_1'. These switching operations are depicted in Fig. 6(a–g). The corresponding tree arrays and 4-th order de Bruijn sequences of the spanning trees T_1 and T_1' are tabulated in Table 1.

Remark 1. A sequence of Leaf node transformations can be represented by a string of vertices whose outgoing edges are switched. It is interesting to note that the set of such strings form a group under string concatenation where the zero element is the empty string and the inverse of any string is a string where the same nodes occur in the reverse order.

These results can be easily extended for k-ary de Bruijn graphs. In a k-ary de Bruijn graph of order n, $\mathcal{G}_n(k)$, every vertex $v \in \mathcal{V}$ will have k incoming edges and k outgoing edges. Let $\mathcal{T}$ be a spanning tree of $\mathcal{G}_n$. Given a vertex $v \in \mathcal{V}\backslash\{r\}$, the row L_v in the tree array has $indeg(v) = k$ elements. The last element of L_v will be the unique outgoing edge of v that occurs in $\mathcal{T}$. The remaining $k - 1$ elements in $\mathcal{G}_n(k)$ can be arranged in any order. Similarly, in case of the root

Table 1. Tree arrays and de Bruijn sequences of T_1 and T_1'

	Spanning tree (T_1)	Spanning tree (T_1')
Tree array	$l_{000} = \{(0000), (0001)\}$	$l_{000} = \{(0000), (0001)\}$
	$l_{001} = \{(0010), (0011)\}$	$l_{001} = \{(0010), (0011)\}$
	$l_{010} = \{(0101), (0100)\}$	$l_{010} = \{(0101), (0100)\}$
	$l_{011} = \{(0110), \quad \Omega \ \}$	$l_{011} = \{(0110), \quad \Omega \ \}$
	$l_{100} = \{(1001), (1000)\}$	$l_{100} = \{(1000), (1001)\}$
	$l_{101} = \{(1010), (1011)\}$	$l_{101} = \{(1010), (1011)\}$
	$l_{110} = \{(1101), (1100)\}$	$l_{110} = \{(1101), (1100)\}$
	$l_{111} = \{(1111), (1110)\}$	$l_{111} = \{(1111), (1110)\}$
de Bruijn seq.	0111101011001000	0111101011000010

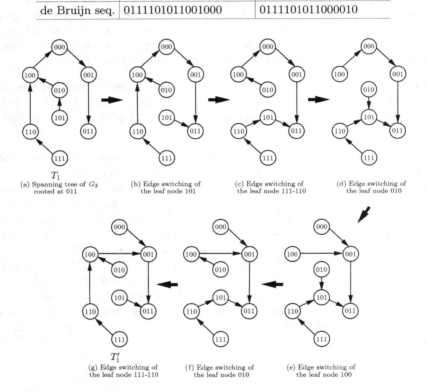

Fig. 6. Switching of leaf node edges in a spanning tree of G_3.

vertex r, we consider Ω as the last entry of L_r and the other $k-1$ entries of L_r can be chosen from any of its outgoing edges in $\mathcal{G}_n(k)$. From this tree array we can construct an Eulerian cycle of a k-ary de Bruijn graph of order n using the method given in Sect. 2. Observe that the proof of Theorem 1 would also be valid in this case.

4   Conclusion and Future Work

In this paper, we have shown that one can go from a given de Bruijn sequence to another by a sequence of leaf node edge switchings. By randomly choosing the sequence of leaf node edge switches we can generate a random spanning tree of G_n and therefore a random de Bruijn sequence.

Since every n-th order de Bruijn sequence can also be generated from a single n-th order de Bruijn sequence by using a sequence of cross joining operations, it would be interesting to find a correspondence between a sequence of leaf node switchings given in this paper and a sequence of cross joining operations. Further, it can be investigated if this method can be efficiently translated into a nonlinear feedback function for FSRs.

Acknowledgment. The authors are grateful to Prof. Harish K. Pillai, Department of Electrical Engineering, Indian Institute of Technology Bombay, without whom this work would never have been possible.

References

1. Annexstein, F.S.: Generating de Bruijn sequences: an efficient implementation. IEEE Trans. Comput. **46**(2), 198–200 (1997)
2. Bidkhori, H., Kishore, S.: A bijective proof of a theorem of Knuth. Comb. Probab. Comput. **20**(1), 11–25 (2011)
3. De Bruijn, N.G.: A Combinatorial Problem. Koninklijke Nederlandsche Akademie Van Wetenschappen, vol. 49, no. 6, pp. 758–764, June 1946
4. Etzion, T., Lempel, A.: Algorithms for the generation of full-length shift-register sequences. IEEE Trans. Inf. Theory **30**(3), 480–484 (1984)
5. Fredricksen, H.: A survey of full length nonlinear shift register cycle algorithms. SIAM Rev. **24**(2), 195–221 (1982)
6. Fredricksen, H.M.: Disjoint cycles from the de Bruijn graph. Technical report, DTIC Document (1968)
7. Golomb, S.W., et al.: Shift Register Sequences. Aegean Park Press, Laguna Hills (1982)
8. Jansen, C.J.A.: Investigations on nonlinear streamcipher systems: construction and evaluation methods. Ph.D. thesis, Technische Universiteit Delft (1989)
9. Lempel, A.: On a homomorphism of the de Bruijn graph and its applications to the design of feedback shift registers. IEEE Trans. Comput. **100**(12), 1204–1209 (1970)
10. Mykkeltveit, J., Szmidt, J.: On cross joining de Bruijn sequences. In: Topics in Finite Fields, vol. 632, pp. 335–346 (2015)
11. Schneier, B.: Applied Cryptography: Protocols, Algorithms, and Source Code in C. Wiley, New York (2007)
12. Spinsante, S., Andrenacci, S., Gambi, E.: De Bruijn sequences for spread spectrum applications: analysis and results. In: 18th International Conference on Software, Telecommunications and Computer Networks, SoftCOM 2010, pp. 365–369, September 2010

Recent Deep Learning Methods
for Melanoma Detection: A Review

Nazneen N. Sultana[✉] and N. B. Puhan

School of Electrical Sciences, IIT Bhubaneswar, Bhubaneswar 752050, India
{nns11,nbpuhan}@iitbbs.ac.in

Abstract. Melanoma is a type of skin cancer, which is not that common
like basal cell and squamous carcinoma, but it has dangerous implica-
tions since it has the tendency to migrate to other parts of body. So,
if it is detected at an early stage then we can easily treat; otherwise
it becomes fatal. Many computer-aided diagnostic methods using dermoscopy images have been proposed to assist the clinicians and dermatologists. Along with conventional methods which extract the low level
handcrafted features, nowadays researchers have focused towards deep
learning techniques which extract the deep and more generic features.
Since 2012, deep learning has been applied to classification, segmentation, localization and many other fields and made an impact. This paper
reviews about the deep learning techniques to detect melanoma cases
from the rest skin lesion in clinical and dermoscopy images.

Keywords: Melanoma · Dermoscope · Deep learning
Computer-assisted · Diagnostics

1 Introduction

There are many types of skin cancer, but the most common among them are
Basal cell carcinoma, Squamous cell carcinoma and Melanoma. Melanoma is due
to rapid growth melanin producing cells, melanocytes. The depth of penetra-
tion determines the different stages of melanoma cancer [1]. Melanoma (mainly
caused due to exposure to ultraviolet radiation) is the most dangerous and is
crucial to detect at its early stages, since it advances and spread to other parts of
the body at later stages. Early detection can significantly reduce the mortality
rate. According to the American Cancer Society (ACS), there will be an estimated 1,688,780 new diagnosed cancer cases and 600,920 cancer deaths in the
United States in 2017 out of which 87,110 new cases of melanoma will be diagnosed [2]. Though melanoma accounts for around 1% of all the skin cancer, it has
the highest mortality rate. The rate of occurrence of melanoma from 2004–2014
has increased by 2–3% per year. In 2017, an estimated 9,730 death will occur
due to melanoma. The five year survival rate is of 95% when detected early, and
this reduces to around 13% if detected at the advance stage of melanoma and
the cost of treatment is also quite high [2].

© Springer Nature Singapore Pte Ltd. 2018
D. Ghosh et al. (Eds.): ICMC 2018, CCIS 834, pp. 118–132, 2018.
https://doi.org/10.1007/978-981-13-0023-3_12

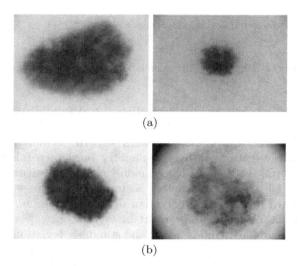

(a)

(b)

Fig. 1. Illustration of (a) benign Nevus, (b) melanoma

In recent research, various methods have been proposed by clinicians to detect melanoma from rest of the skin cancer. These methods include ABCDE method [3], Menzies method [4], pattern analysis [5], texture analysis [6] and the 7-point checklist [7]. All these methods compare the color changes, shape, symmetry and texture variations as compared to the normal skin. With the advent of dermoscopy (also referred as epiluminescence microscopy), it has been possible to assist clinicians efficiently since dermatoscope captures the dermal features and eliminates the surface glare. Dermatoscope is a non-invasive technique where a gel is applied on the surface of the skin lesion and an enhanced image is acquired using the digital imaging technique, dermatoscope. The device magnify structures which are otherwise invisible to naked eyes, thus helps in detection of melanoma from other types of skin cancer. Total body Photography is also done by dermatologists for early detection of changing lesions and avoidance of biopsy of stable lesions. Some of the examples of ISIC dataset [8], which is one of the largest collection of contact non-polarized dermoscopy images are shown in Fig. 1. The Nevus is a benign skin tumor derived from melanocytes.

The key steps in a computer-vision based diagnosis of melanoma classification are: image acquisition, preprocessing, segmentation, extraction of features and classification as shown in Fig. 2. The disease classification can be binary (malignant or benign) or n-ary (into many classes). Thus the computer vision techniques became effective to assist dermatologists and clinicians to reduce many undesired biopsies. The preprocessing techniques are applied for removal of hairs, ruler markings, air bubbles, and spurious noise. The example of such methods include DullRazor hair removal algorithm [9], median filtering [10], directional filters [11] and illumination enhancement [11]. Segmentation is the process of separating the lesion from the surrounding skin in order to perform

Fig. 2. Block diagram of melanoma detection system

lesion analysis and efficient feature extraction [12]. Visual detection by dermatologists is done by mostly studying the ABCD rule or the 7-point checklist that accurately characterizes a melanoma lesion. ABCD rule is more popular because it includes the salient features such as asymmetry, border irregularity, color variation and diameter of the lesion. The commonly extracted features are color and texture features [13] which gives better accuracy. The color statistics include mean and standard deviation, color variance, color histogram, color asymmetry and texture features include gradient histogram, gray level co-occurrence matrix (GLCM dissimilarity index, mean and standard deviation) [6,14]. Finally for lesion classification, the SVM [15–17], Naive Bayes classifier [6], KNN [18], logistic regression [19], bag-of-feature classification [20] and decision tree learning. A review on these traditional methods can be found in [6,21,22]. Due to high intra-class and low inter-class variations in melanoma, handcrafted feature-based diagnostic performance is found to be still unsatisfactory [23].

The deep neural networks take the raw image as an input while bypassing the complex procedure of pre-processing, segmentation, design of handcrafted features. Convolutional Neural Networks (CNN) has been known in the research field since few decades, but has been come to the surface quite recently after the ImageNet classification challenge which reduces the error rate significantly from 26% to 15% using the AlexNet architecture [24] which comprised of different convolution, pooling and activation layers. Deep learning has overcome the difficulty of computing handcrafted features by modulating the feature engineering step into a learning process. Instead of extracting several features in laborious steps, supervised deep learning requires only large sets of labelled data and then discovers the informative feature representation automatically.

This review paper is organized as follows. Section 2 provides the background on deep models and describes the various CNN architectures. Section 3 discusses about the various methods of deep learning for melanoma detection and their performance metrics. Section 4 concludes the paper.

2   Deep Models

2.1   Neural Networks

Neural networks are the building blocks of deep learning techniques [25]. A neural network is similar to multilayer perceptron with an input layer, hidden layers and output layer. It comprises of the activation function, weights and biases {W, b}. The output at each node can be represented as

$$a = f(W^T x + b) \tag{1}$$

Where $f(.)$ represents the non-linearity function. The typical non-linear functions that are traditionally used are sigmoid, hyperbolic and RELU (Rectified Linear Unit) [26]. RELU has mainly two advantages over the traditional sigmoid, which gives sparsity and a reduced likelihood of the vanishing gradient. Deep learning is a hierarchical feature learning process which learns features at multiple levels, thus allowing to learn a complex function without depending on the handcrafted features. The weights in the feed forward process is adjusted using the backpropagation algorithm [27] which uses the gradient descent optimization to minimize the loss at the output. At the output layer, we have the softmax function [27] as activation function to ensure that the outputs are probabilities and they add up to 1. The softmax function is a generalization of the logistic function which takes a vector of arbitrary real valued scores and squashes it to a vector of values between zero and one that sum to one [26]. It is used as a cost function for probabilistic multi-class classification and SVM is used to separate one class from rest of the classes.

2.2 Convolutional Neural Networks

Multi-layer neural networks accept input in the vector form while convolutional networks can deal with both structured and unstructured data [25]. CNN has proven to be very successful in image classification and object detection. Since 2012, the top 5 error rate of the ImageNet challenge has been reduced to 3.6% while humans have 5% error rate. CNN could recognize visual patterns directly from pixel level image representation with minimal preprocessing. As such, we do not have to extract handcrafted features, which is both difficult and domain specific. A CNN consists of a convolutional layer, a subsampling layer (either max pooling or average pooling) and optionally followed by a fully connected layer. The output at the convolution layer is given by the following equation [25].

$$A_j^l = f\left(\sum_{i=1}^{M^{l-1}} A_i^{l-1} * w_{ij}^l + b_j^l \right) \tag{2}$$

where M^{l-1} is the number of feature maps in the $(l-1)$ layer, A_j^i is the activation output at the lst layer, w_{ij}^l is the kernel weights from feature map at layer to feature map j at $(l-1)$ layer, and b_j^i is the additional bias parameter. The pooling layer (max pooling or average pooling) downsamples the feature map from previous convolutional layer. Either before or after the subsampling layer, an additional bias term is added after which a non-linearity function is added. The gradient descent with back propagation algorithm is applied to learn the trainable parameters. In the convolutional layer, CNN share the same filter weights for each receptive field in a particular layer which reduces the memory storage of different weights while giving better performance.

In the year 1998, the first convolutional neural network was proposed by LeCun et al. [28] which brought forth the deep learning architecture. In 2012, AlexNet [24] was released which incorporated the rectified linear units (RELU) in place of hyperbolic tanh function as an activation function. In 2013, ZFNet was proposed from Zeiler and Fergus in [29]. This was an improvement over the AlexNet by tweaking the hyper-parameters like using small filter size, since it helps in retaining a lot of original pixel information [26]. They explained in depth about the deconvolution concept which helped in examining a particular feature map and their relation to the input space. In 2014, the ILSVRC winner was the popular GoogLeNet [30]. The main contribution was the development of the inception module as shown in Fig. 3(a). Here, not all of the convolution and pooling layers were stacked sequentially, but they were arranged in parallel. They applied the 1×1 convolution operations before the 3×3 and 5×5 layers. This 1×1 convolution operation addressed both dimensionality reduction as well as nonlinearity addition. This architecture allows the model to recover both local features via smaller convolutions and highly abstract features with larger convolution. In VGGNet [31] as shown in Fig. 3(c), all the convolutional filter size is of 3×3. They argued that two 3×3 filters give the receptive field equivalent to one 5×5 and three 3×3 filter giving a receptive field equivalent to 7×7 filter size. Thus, they reduced the number of learnable parameters by using the smaller filter size.

In 2015, ResNet was developed which addressed the problem of gradient vanishing as we go deeper into the network [32]. It features special skip connections as shown in Fig. 3(b) and heavy use of batch normalization. Huang et al. proposed the Densely Connected Convolutional Network which concatenates outputs from the previous layers instead of using the summation [33].

2.3   Deep Generative Model

Autoencoders [34] are similar to Multi-Layer Perceptron (MLP) except that its output layer is same as its input layer, while in MLP the output layer is equal to the number of classes. MLP requires labelled data for training while autoencoders belong to the category of unsupervised learning. It is used for dimensionality reduction, reconstruction of the original image from noise corrupted version and also as a feature extractor for classification. Stacked autoencoders are formed by placing autoencoder layers on top of each other. In medical applications, autoencoder layers are often trained individually (training a 2nd hidden layer by the outputs of the 1st layer and so on) after which the full network is fine-tuned in a supervised manner. Sabbaghi et al. [35] employ stacked sparse autoencoder for skin lesion classification.

Restricted Boltzmann Machine [36] is a stochastic neural network consists of one input layer, one hidden layer and a bias unit. A deep Boltzmann machine is constructed by stacking multiple RBMs. The connection between the nodes are bidirectional; given an input, we can obtain the latent feature representation (learning lower dimensional features from high dimensional input space) and vice-versa. It is useful for dimensionality reduction, classification, regression and feature learning.

In 2006, Hinton discovered that better results could be achieved using deeper architectures when each layer (RBM) is pre-trained with an unsupervised

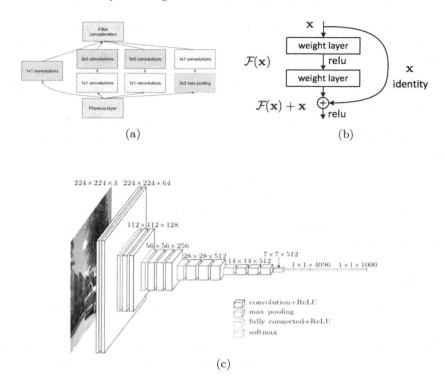

(a)   (b)

(c)

Fig. 3. (a) Inception module, (b) Residual learning, (c) VGG16 architecture

learning algorithm such as Contrastive Divergence. Deep Belief Nets [37] is similar to stacked autoencoders except that we stack RBMs here. Finally, the network is fine-tuned and trained using the backpropagation algorithm in a supervised manner.

3   Deep Learning Based Melanoma Detection

Deep learning bypasses all the complex methods of pre-processing, segmentation and low level feature extraction. It learns features important to a specific task hierarchically with minimal pre-processing. Some of the salient points towards the need of deep learning in medical imaging is as follows:

- Deep learning has achieved quite good performance in image classification, object detection, restoration, image captioning, etc. So the same can now be applied to the field of medical imaging.
- The network is trained end-to-end directly using raw pixels as input, with a single network.
- Advent of fast GPUs made it possible to have multiple hidden layers in a neural network.

- The limited number of annotated data available in medical imaging is quite a challenge, but recently many publicly available annotated dataset is available for researchers and academia's to take active participation towards the different medical challenges.
- The above challenge can be solved using unsupervised learning like training with autoencoders, or to deal with limited annotated dataset many techniques have been proposed. It includes data augmentation, input image as patches, using pretrained models. Quite recently Bayesian deep learning [38] has been proposed which is an unsupervised or semi-supervised technique to deal with limited labelled database.
- In skin lesion due to low contrasts and obscure boundaries, automated recognition tasks become harder and also due to high intra-class and low inter-class variation, classification is not accurate. Here we can apply deep learning which is a hierarchical learning process to accurately detect features which are important for classifying accurately.

3.1   Datasets

There are many datasets available for skin lesion classification. Some are publicly available and some are licensed.

- Dermofit Image library [18]: The Dermofit Image Library is a collection of 1,300 focal high quality skin lesion images collected under standardized conditions with internal colour standards. The lesions are from ten different classes: Actinic Kerato-sis, Basal Cell Carcinoma, Melanocytic Nevus (mole), Seborrhoeic Keratosis, Squa-mous Cell Carcinoma, Intraepithelial Carcinoma, Pyogenic Granuloma, Haemangi-oma, and Dermatofibroma.
- Interactive Atlas of Dermoscopy [39]: The dataset consists of 112 malignant lesion images (containing melanoma and basal cell carcinoma (BCC)), and 298 benign lesion images (containing congenital, compound, dermal, Clark, spitz, and blue nevus; dermatofibroma; and seborrheic keratosis). Each image having two modalities, dermoscopic and clinical image.
- ISIC Archive: It is a publicly available dermoscopic image dataset with 13,000 dermoscopic images.
- ISBI Challenge 2016 Dataset [8]: The Challenge Dataset contains 900 images for training (273 being melanomas) and 379 for testing (115 being melanomas).
- Dermnet [40]: This database contains over 23,000 skin lesion images separated into 23 classes of skin diseases.
- PH2 Dataset [41]: This is a dermoscopic image database acquired at the Dermatology Service of Hospital Pedro Hispano, Portugal. It consists of 200 dermoscopic images (40 are melanoma cases and 160 are non-melanoma cases).
- ISBI 2017 Challenge Dataset [42]: The official challenge dataset, with 2,000 dermoscopic images (374 melanomas, 254 seborrheic keratoses, and 1,372 benign nevi).

- IRMA Skin Lesion Dataset: With 747 dermoscopic images (186 melanomas). This dataset is unlisted, but available under special request, and the signing of a license agreement.
- MED-NODE dataset [43]: The dataset consists of 100 melanoma and 70 naevus images from the department of Dermatology of University of Medical Centre, Groningen. This dataset was used for testing in NED-MODE system for skin cancer detection using non-dermoscopic images.
- MoleMap dataset: The MoleMap NZ Ltd dataset is collected over a period of 2003 to 2015 containing both dermoscopic and clinical images. The number of images used in the paper [44] is 32,195 images from 15 disease categories of 14,754 lesions from 8,882 patients.

3.2   Performance Metrics

The model evaluation is performed using the following:

- Sensitivity - It is the ability of the test to correctly identify the diseased state (true positive rate).

$$SE = \frac{TP}{TP + FN} \tag{3}$$

- Specificity - It is the ability of the test to correctly diagnose the benign cases (true negative rate).

$$SP = \frac{TN}{TN + FP} \tag{4}$$

- Accuracy - The number of correct predictions divided by a total number of predictions. Ratio of true detected cases to all cases.

$$ACC = \frac{TP + TN}{TP + FP + FN + TN} \tag{5}$$

- Precision - Fraction of relevant instances among the retrieved instances. It is also equivalent to positive predictive value.

$$PREC = \frac{TP}{TP + FP} \tag{6}$$

- Positive Predictive Value (PPV) - It is the probability whether the subject with a positive test, truly have the disease.

$$PPV = \frac{TP}{TP + FP} \tag{7}$$

- Negative Predictive Value (NPV) - It is the probability whether the subjects with a negative screening test, truly don't have the disease.

$$NPV = \frac{TN}{TN + FN} \tag{8}$$

- ROC AUC - Area under receiver operating characteristic. It is equal to the probability that the classifier will rank a randomly chosen positive example higher than a randomly chosen negative example. It is the graph between true positive rate vs false positive rate.
- Dice coefficient - It is also known as overlap index which gives the overlap measure between the automatic and the ground truth segmentation.

$$DC = \frac{2.TP}{2.TP + FN + FP} \tag{9}$$

- Jaccard Index - It also measures the same aspect as that of Dice coefficient.

$$JA = \frac{TP}{TP + FN + FP} \tag{10}$$

3.3 Performance of Recent Deep Learning Methods for Melanoma Detection

The work of Lopez et al. [45] compares different methods of using deep learning CNN architecture. They used the VGGNet in three different ways: (1) training from scratch, (2) transfer learning from a pretrained model on a larger dataset (ImageNet), (3) transfer learning and fine tuning the CNN architecture. Zhen et al. [23] combines deep convolutional neural network with fisher vector encoding and SVM classifier. Here, instead of whole images as input to the CNN, they give samples or sub-images as the input which eliminates the small dataset problem. On the dataset of ISBI 2016, which consists of 1279 skin images, the proposed method achieved an accuracy of 83.09%. Menegola et al. [46] does classification into binary (malignant and benign) and ternary (malignant, benign and basal cell carcinoma) classes. From the results, it is shown that basal cell carcinoma is easier to diagnose than melanoma. The other medical dataset that is used for pre-training is the Kaggle Challenge for Diabetic Retinopathy Detection dataset. Esteva et al. [47] train a CNN (Inception v3 model) using a dataset of 129,450 clinical images consisting of 2,032 different diseases. The performance was measured against 21 board-certified dermatologists and it was on par with the experts, thus showing that artificial intelligence capable of classifying skin cancer with a level of competence comparable to dermatologists.

Kawahara et al. [48] proves that CNN can be efficiently used for 10-class classification and it outperforms (accuracy 81.8%) the present state-of-art techniques. This paper utilizes fully convolutional network to extract multi-scale features similar to Overfeat [49] method that is used for localization and detection problems. The full-CNN converts the fully connected layer into a convolutional layer which accepts multiscale input image. Pooling across deep features for augmented images is proposed to conserve memory and time. The method by Demyanov et al. [50] is about detection of dermosopic patterns like pigment network and regular globules. The dataset is split such that segments from the same image always belong to the same set. They implement different algorithms like CNN with data augmentation, without augmentation, k-means clustering, sparse coding and Fisher kernel. The best mean accuracy for the typical network and regular globules datasets is 88% and 83% approximately. Esfahani et al. [51] experiments on clinical images which might have illumination and noises present in the images. The images are fed to a 5-layer CNN network after removal of noise and illumination artifacts.

Yu et al. [52] integrate the FCRN (Fully Convolutional Residual Network for segmentation) and very deep residual networks (for classification) to form two-stage framework. Deep residual learning, ResNet solves the problem of degradation of accuracy (degradation problem states that simply increasing the depth by stacking layers does not contribute towards further improvement in accuracy but instead accuracy degrades rapidly). ResNet avoids this problem in being deep and still achieving quite high accuracy. The result of the experiment shows that with segmentation performs better since segmentation allows the feature extraction procedure to be conducted only on the lesion regions and thus generate more specific and representative features. For classification stage, they implemented using SVM as well as softmax classifier and found that a simple average fusion can further improve the classification accuracy. Other comparison on classification was with different models (VGG-16, GoogleNet and DRN-50); DRN-50 performed with the best result. This paper ranked 2nd among 28 teams in ISBI challenge 2016 for segmentation and first place for classification with average precision 0.637.

Kawahara et al. [53] proposed multi-tract-CNN for skin lesion classification that learns interaction across multiple image resolutions with same field of view simultaneously. The fully connected layer is converted to convolutional layer, since convolutional layers allow for variable sized inputs. Because of limited GPU memory, they used two tracts, upper tract lower resolution and lower tract higher resolution image, but it can be applied to any number. The two tract approach with auxiliary loss gives better performance (accuracy = 0.773) than without auxiliary loss (accuracy = 0.755) and further improvement in accuracy is seen if data augmentation (accuracy = 0.795) is done. Codella et al. [54] performs classification of melanoma vs non-melanoma cases and also melanoma vs atypical lesions only. The method achieves an accuracy of 93.1% for the first task and 73.9% for the second. It is observed that distinguishing melanoma from atypical cases is difficult. This method follows two parallel paths: (1) CNN with transfer learning from natural dataset, (2) Unsupervised feature learning using sparse coding. The classifiers are then trained on each using non-linear SVM using a HIK (histogram intersection kernel) and sigmoid feature normalization, and then the model is combined in late fusion (score averaging).

Ge et al. [44] deep residual network (ResNet-50) extracts the global information and bilinear pooling technique extracts the local information. The paper implemented com-pact pooling which uses Tensor Sketch to reduce the dimension in bilinear pooling. For classification, linear SVM is used after l2 normalization and concatenation of the two local and global features. 15-class disease classification using MoleMap dataset achieves 71% accuracy and using ISBI 2016 challenge dataset, an 85% accuracy. Noel et al. [55] applies fully convolutional

Table 1. Performance measures of deep learning methods

Methods	Dataset	ACC	SE	SP	PREC	ACC
Fine-tuning using VGGNet	ISBI 2016	0.813	0.7866	-	0.7974	-
Fully convolutional residual network+deep residual network	ISBI 2016	0.855	0.5407	0.941	0.637	0.804
Multi-resolution-tract CNN	Dermofit Image Library	0.795	-	-	-	-
Multi-scale feature extraction using full-CNN	Dermofit image library	0.818	-	-	-	-
CNN+Fisher Encoding	ISBI 2016	0.8309	-	-	0.535	0.7957
5 layer CNN	MED-NODE	0.81	0.81	0.80	-	-
ResNet+bilinear pooling	MoleMap	0.71	-	-	-	-
	ISBI 2016	0.85	-	-	0.625	-
Deep learning+Sparse Coding+SVM	ISIC	0.743	-	-	-	-
Deep learning ensemble	ISIC	0.807	0.693	0.836	0.649	0.843

network similar to U-Net architecture [56] for segmentation and non-linear SVM classifier to classify individual features. They combine all features like color histogram, edge histogram, multiscale color LBP, sparse coding, CNN (FC6, 4096 dimension feature vector), DRN-101 (1000 dimension extracted only from the whole image), Fully Convolutional U-Net (used as a shape descriptor extracted only from the whole image). Area under receiver operating curve (AUC) is found as 0.843 and average precision is equal to 0.649. The datasets and the results of these papers are summarized in Table 1.

4   Conclusion

In this paper, we reviewed different deep learning based methods to detect melanoma from the rest of the skin lesions. It is seen that using deep learning there is no need of complex preprocessing techniques except for normalizing the pixel value, resizing and cropping. It bypasses all the preprocessing, segmentation and handcrafted feature ex-traction. It is observed that applying deep neural networks gives better result than conventional methods. For medical imaging, the most important challenge is that of acquiring labelled images. The ISBI challenge has provided with large number of datasets (ISIC Archive), which provides a common platform for researchers and academia to evaluate their work. The future works might include: using a much larger dataset to reduce the risk of overfitting and performing additional regularization tweaks and fine-tuning of hyper-parameters. In many papers, it is shown that taking both global (deep) as well as local features into account during feature extraction provides better results. However, it is still an open issue to increase the accuracy rate. The goal should be to target highest sensitivity while optimizing to increase specificity, thereby increasing the overall accuracy for practical applications.

References

1. Satheesha, T., Satyanarayana, D., Prasad, M.G., Dhruve, K.D.: Melanoma is skin deep: a 3D reconstruction technique for computerized dermoscopic skin lesion classification. IEEE J. Trans. Eng. Health Med. **5**, 1–17 (2017)
2. Siegel, R.L., Miller, K.D., Jemal, A.: Cancer statistics. CA: A Cancer J. Clin. **67**(1), 7–30 (2017)
3. Stolz, W., Riemann, A., Cognetta, A., Pillet, L., Abmayr, W., Holzel, D., Bilek, P., Nachbar, F., Landthaler, M.: ABCD rule of dermatoscopy: a new practical method for early recognition of malignant-melanoma. Eur. J. Dermatol. **4**(7), 521–527 (1994)
4. Menzies, S.W., Ingvar, C., Crotty, K.A., McCarthy, W.H.: Frequency and morphologic characteristics of invasive melanomas lacking specific surface microscopic features. Arch. Dermatol. **132**(10), 1178–1182 (1996)
5. Sáez, A., Acha, B., Serrano, C.: Pattern analysis in dermoscopic images. In: Scharcanski, J., Celebi, M. (eds.) Computer Vision Techniques for the Diagnosis of Skin Cancer, pp. 23–48. Springer, Heidelberg (2014). https://doi.org/10.1007/978-3-642-39608-3_2
6. Maglogiannis, I., Doukas, C.N.: Overview of advanced computer vision systems for skin lesions characterization. IEEE Trans. Inf. Technol. Biomed. **13**(5), 721–733 (2009)
7. Argenziano, G., Fabbrocini, G., Carli, P., De Giorgi, V., Sammarco, E., Delfino, M.: Epiluminescence microscopy for the diagnosis of doubtful melanocytic skin lesions: comparison of the abcd rule of dermatoscopy and a new 7-point checklist based on pattern analysis. Arch. Dermatol. **134**(12), 1563–1570 (1998)
8. Gutman, D., Codella, N.C., Celebi, E., Helba, B., Marchetti, M., Mishra, N., Halpern, A.: Skin lesion analysis toward melanoma detection: a challenge at the international symposium on biomedical imaging (ISBI) 2016, hosted by the international skin imaging collaboration (ISIC). arXiv preprint arXiv:1605.01397 (2016)
9. Lee, T., Ng, V., Gallagher, R., Coldman, A., McLean, D.: Dullrazor®: a software approach to hair removal from images. Comput. Biol. Med. **27**(6), 533–543 (1997)
10. Celebi, M.E., Kingravi, H.A., Uddin, B., Iyatomi, H., Aslandogan, Y.A., Stoecker, W.V., Moss, R.H.: A methodological approach to the classification of dermoscopy images. Comput. Med. Imaging Graph. **31**(6), 362–373 (2007)
11. Barata, C., Marques, J.S., Rozeira, J.: A system for the detection of pigment network in dermoscopy images using directional filters. IEEE Trans. Biomed. Eng. **59**(10), 2744–2754 (2012)
12. Silveira, M., Nascimento, J.C., Marques, J.S., Marçal, A.R., Mendonça, T., Yamauchi, S., Maeda, J., Rozeira, J.: Comparison of segmentation methods for melanoma diagnosis in dermoscopy images. IEEE J. Sel. Top. Sig. Process. **3**(1), 35–45 (2009)

13. Marques, J.S., Barata, C., Mendonça, T.: On the role of texture and color in the classification of dermoscopy images. In: 2012 Annual International Conference of the IEEE Engineering in Medicine and Biology Society (EMBC), pp. 4402–4405. IEEE (2012)
14. Barata, C., Ruela, M., Francisco, M., Mendonça, T., Marques, J.S.: Two systems for the detection of melanomas in dermoscopy images using texture and color features. IEEE Syst. J. **8**(3), 965–979 (2014)
15. Situ, N., Wadhawan, T., Yuan, X., Zouridakis, G.: Modeling spatial relation in skin lesion images by the graph walk kernel. In: 2010 Annual International Conference of the IEEE Engineering in Medicine and Biology Society (EMBC), pp. 6130–6133. IEEE (2010)
16. Barata, C., Figueiredo, M.A.T., Celebi, M.E., Marques, J.S.: Local features applied to dermoscopy images: bag-of-features versus sparse coding. In: Alexandre, L.A., Salvador Sánchez, J., Rodrigues, J.M.F. (eds.) IbPRIA 2017. LNCS, vol. 10255, pp. 528–536. Springer, Cham (2017). https://doi.org/10.1007/978-3-319-58838-4_58
17. Jafari, M.H., Samavi, S., Karimi, N., Soroushmehr, S.M.R., Ward, K., Najarian, K.: Automatic detection of melanoma using broad extraction of features from digital images. In: 2016 IEEE 38th Annual International Conference of the Engineering in Medicine and Biology Society (EMBC), pp. 1357–1360. IEEE (2016)
18. Ballerini, L., Fisher, R.B., Aldridge, B., Rees, J.: A color and texture based hierarchical k-nn approach to the classification of non-melanoma skin lesions. In: Celebi, M., Schaefer, G. (eds.) Color Medical Image Analysis, pp. 63–86. Springer, Dordrecht (2013). https://doi.org/10.1007/978-94-007-5389-1_4
19. Blum, A., Luedtke, H., Ellwanger, U., Schwabe, R., Rassner, G., Garbe, C.: Digital image analysis for diagnosis of cutaneous melanoma. development of a highly effective computer algorithm based on analysis of 837 melanocytic lesions. Br. J. Dermatol. **151**(5), 1029–1038 (2004)
20. Situ, N., Yuan, X., Chen, J., Zouridakis, G.: Malignant melanoma detection by bag-of-features classification. In: 30th Annual International Conference of the IEEE Engineering in Medicine and Biology Society, EMBS 2008, pp. 3110–3113. IEEE (2008)
21. Mishra, N.K., Celebi, M.E.: An overview of melanoma detection in dermoscopy images using image processing and machine learning. arXiv preprint arXiv:1601.07843 (2016)
22. Korotkov, K., Garcia, R.: Computerized analysis of pigmented skin lesions: a review. Artif. Intell. Med. **56**(2), 69–90 (2012)
23. Yu, Z., Ni, D., Chen, S., Qin, J., Li, S., Wang, T., Lei, B.: Hybrid dermoscopy image classification framework based on deep convolutional neural network and fisher vector. In: 2017 IEEE 14th International Symposium on Biomedical Imaging (ISBI 2017), pp. 301–304. IEEE (2017)
24. Krizhevsky, A., Sutskever, I., Hinton, G.E.: Imagenet classification with deep convolutional neural networks. In: Advances in Neural Information Processing Systems, pp. 1097–1105 (2012)
25. Shen, D., Wu, G., Suk, H.I.: Deep learning in medical image analysis. Ann. Rev. Biomed. Eng. **19**, 221–248 (2017)
26. Karpathy, A.: Cs231n: Convolutional neural networks for visual recognition. Neural Networks 1 (2016)
27. Goodfellow, I., Bengio, Y., Courville, A.: Deep Learning. MIT Press, Cambridge (2016)
28. LeCun, Y., et al.: Lenet-5, convolutional neural networks (2015). http://yann.lecun.com/exdb/lenet

29. Zeiler, M.D., Fergus, R.: Visualizing and understanding convolutional networks. In: Fleet, D., Pajdla, T., Schiele, B., Tuytelaars, T. (eds.) ECCV 2014. LNCS, vol. 8689, pp. 818–833. Springer, Cham (2014). https://doi.org/10.1007/978-3-319-10590-1_53

30. Szegedy, C., Liu, W., Jia, Y., Sermanet, P., Reed, S., Anguelov, D., Erhan, D., Vanhoucke, V., Rabinovich, A.: Going deeper with convolutions. In: Proceedings of the IEEE Conference on Computer Vision and Pattern Recognition, pp. 1–9 (2015)

31. Simonyan, K., Zisserman, A.: Very deep convolutional networks for large-scale image recognition. arXiv preprint arXiv:1409.1556 (2014)

32. He, K., Zhang, X., Ren, S., Sun, J.: Deep residual learning for image recognition. In: Proceedings of the IEEE Conference on Computer Vision and Pattern Recognition, pp. 770–778 (2016)

33. Huang, G., Liu, Z., Weinberger, K.Q., van der Maaten, L.: Densely connected convolutional networks. arXiv preprint arXiv:1608.06993 (2016)

34. Hinton, G.E., Salakhutdinov, R.R.: Reducing the dimensionality of data with neural networks. Science 313(5786), 504–507 (2006)

35. Sabbaghi, S., Aldeen, M., Garnavi, R.: A deep bag-of-features model for the classification of melanomas in dermoscopy images. In: 2016 IEEE 38th Annual International Conference of the Engineering in Medicine and Biology Society (EMBC), pp. 1369–1372. IEEE (2016)

36. Salakhutdinov, R., Mnih, A., Hinton, G.: Restricted boltzmann machines for collaborative filtering. In: Proceedings of the 24th International Conference on Machine Learning, pp. 791–798. ACM (2007)

37. Hinton, G.E., Osindero, S., Teh, Y.W.: A fast learning algorithm for deep belief nets. Neural Comput. 18(7), 1527–1554 (2006)

38. Gal, Y., Islam, R., Ghahramani, Z.: Deep bayesian active learning with image data. arXiv preprint arXiv:1703.02910 (2017)

39. Argenziano, G., Soyer, H., De Giorgi, V., Piccolo, D., Carli, P., Delfino, M., et al.: Dermoscopy: a tutorial. EDRA, Medical Publishing & New Media, p. 16 (2002)

40. Habif, M.T.: Dermnet skin diseases Atlas (1998)

41. Mendonça, T., Ferreira, P.M., Marques, J.S., Marcal, A.R., Rozeira, J.: Ph 2-a dermoscopic image database for research and benchmarking. In: 2013 35th Annual International Conference of the IEEE Engineering in Medicine and Biology Society (EMBC), PP. 5437–5440. IEEE (2013)

42. Codella, N.C., Gutman, D., Celebi, M.E., Helba, B., Marchetti, M.A., Dusza, S.W., Kalloo, A., Liopyris, K., Mishra, N., Kittler, H., et al.: Skin lesion analysis toward melanoma detection: A challenge at the 2017 international symposium on biomedical imaging (ISBI), hosted by the international skin imaging collaboration (ISIC). arXiv preprint arXiv:1710.05006 (2017)

43. Giotis, I., Molders, N., Land, S., Biehl, M., Jonkman, M.F., Petkov, N.: Med-node: a computer-assisted melanoma diagnosis system using non-dermoscopic images. Expert Syst. Appl. 42(19), 6578–6585 (2015)

44. Ge, Z., Demyanov, S., Bozorgtabar, B., Abedini, M., Chakravorty, R., Bowling, A., Garnavi, R.: Exploiting local and generic features for accurate skin lesions classification using clinical and dermoscopy imaging. In: 2017 IEEE 14th International Symposium on Biomedical Imaging (ISBI 2017), PP. 986–990. IEEE (2017)

45. Lopez, A.R., Giro-i Nieto, X., Burdick, J., Marques, O.: Skin lesion classification from dermoscopic images using deep learning techniques. In: 2017 13th IASTED International Conference on Biomedical Engineering (BioMed), pp. 49–54. IEEE (2017)

46. Menegola, A., Fornaciali, M., Pires, R., Bittencourt, F.V., Avila, S., Valle, E.: Knowledge transfer for melanoma screening with deep learning. arXiv preprint arXiv:1703.07479 (2017)
47. Esteva, A., Kuprel, B., Novoa, R.A., Ko, J., Swetter, S.M., Blau, H.M., Thrun, S.: Dermatologist-level classification of skin cancer with deep neural networks. Nature **542**(7639), 115–118 (2017)
48. Kawahara, J., BenTaieb, A., Hamarneh, G.: Deep features to classify skin lesions. In: 2016 IEEE 13th International Symposium on Biomedical Imaging (ISBI), pp. 1397–1400. IEEE (2016)
49. Sermanet, P., Eigen, D., Zhang, X., Mathieu, M., Fergus, R., LeCun, Y.: Overfeat: Integrated recognition, localization and detection using convolutional networks. arXiv preprint arXiv:1312.6229 (2013)
50. Demyanov, S., Chakravorty, R., Abedini, M., Halpern, A., Garnavi, R.: Classification of dermoscopy patterns using deep convolutional neural networks. In: 2016 IEEE 13th International Symposium on Biomedical Imaging (ISBI), pp. 364–368. IEEE (2016)
51. Nasr-Esfahani, E., Samavi, S., Karimi, N., Soroushmehr, S.M.R., Jafari, M.H., Ward, K., Najarian, K.: Melanoma detection by analysis of clinical images using convolutional neural network. In: 2016 IEEE 38th Annual International Conference of the Engineering in Medicine and Biology Society (EMBC), pp. 1373–1376. IEEE (2016)
52. Yu, L., Chen, H., Dou, Q., Qin, J., Heng, P.A.: Automated melanoma recognition in dermoscopy images via very deep residual networks. IEEE Trans. Med. Imaging **36**(4), 994–1004 (2017)
53. Kawahara, J., Hamarneh, G.: Multi-resolution-tract CNN with hybrid pretrained and skin-lesion trained layers. In: Wang, L., Adeli, E., Wang, Q., Shi, Y., Suk, H.-I. (eds.) MLMI 2016. LNCS, vol. 10019, pp. 164–171. Springer, Cham (2016). https://doi.org/10.1007/978-3-319-47157-0_20
54. Codella, N., Cai, J., Abedini, M., Garnavi, R., Halpern, A., Smith, J.R.: Deep learning, sparse coding, and SVM for melanoma recognition in dermoscopy images. In: Zhou, L., Wang, L., Wang, Q., Shi, Y. (eds.) MLMI 2015. LNCS, vol. 9352, pp. 118–126. Springer, Cham (2015). https://doi.org/10.1007/978-3-319-24888-2_15
55. Codella, N.C., Nguyen, Q.B., Pankanti, S., Gutman, D., Helba, B., Halpern, A., Smith, J.R.: Deep learning ensembles for melanoma recognition in dermoscopy images. IBM J. Res. Dev. **61**(4), 1–5 (2017)
56. Ronneberger, O., Fischer, P., Brox, T.: U-Net: Convolutional Networks for Biomedical Image Segmentation. In: Navab, N., Hornegger, J., Wells, W.M., Frangi, A.F. (eds.) MICCAI 2015. LNCS, vol. 9351, pp. 234–241. Springer, Cham (2015). https://doi.org/10.1007/978-3-319-24574-4_28

Applied Mathematics

An Approach to Multi-criteria Decision Making Problems Using Dice Similarity Measure for Picture Fuzzy Sets

Deepa Joshi$^{(\boxtimes)}$ and Sanjay Kumar

Department of Mathematics, Statistics and Computer Science,
G. B. Pant University of Agriculture and Technology,
Pantnagar 263145, Uttarakhand, India
deepajoshi.6nov@gmail.com, skruhela@hotmail.com

Abstract. This paper presents an approach for multi criteria decision making problems based on the Dice similarity measure and weighted Dice similarity measure for picture fuzzy sets (PFSs). To illustrate the application of the proposed method, a practical problem has been considered and the results are compared and verified with an existing method. The results are well matched and the calculations are compact and much easier to analyze.

Keywords: Picture fuzzy sets · Dice similarity measure
Weighted Dice similarity measure · Multi-criteria decision making

1 Introduction

The theory of intuitionistic fuzzy sets (IFSs) proposed by *Atanassov* [1] has been successfully applied in various fields like decision making, logic programming, pattern recognition, medical diagnosis and more [2,3]. Although, IFS theory has been successfully applied in different areas, but there are situations in real life which cannot be represented by IFS [4]. Voting could be a good example of such situation as the human voters may be divided into four groups of those who; vote for, vote against, abstain and refusal of voting. Nevertheless, the IFS theory care to those who vote for or vote against, and consider those who abstain and refusal are equivalent. This concept is particularly effective in approaching the practical problems in relation to the synthesis of ideas; make decision such as voting, financial forecasting and risks in business [5].

Similarity measures are common tools used widely in measuring the deviation and closeness degree of different arguments. *Dengfeng et al.* [3] introduced the degree of similarity between IFS to propose several new similarity measures and applied those new measures into pattern recognition. A new measure of similarity for IFSs, considering the distance to its complement to analyze the extension of agreement in group of experts was proposed by *Szmidt et al.* [6]. *Ye* [7] extended the concept of the cosine similarity measure for fuzzy sets and therefore

© Springer Nature Singapore Pte Ltd. 2018
D. Ghosh et al. (Eds.): ICMC 2018, CCIS 834, pp. 135–140, 2018.
https://doi.org/10.1007/978-981-13-0023-3_13

proposed the cosine similarity measure for IFSs. He also proposed the concept of the reduct intuitionistic fuzzy set of interval valued intuitionistic fuzzy set with respect to adjustable weight vectors and the dice similarity measure based on the reduct IFS to explore the effects of optimism, neutralism and pessimism in decision making [8]. Some cosine similarity measures and weighted cosine similarity measures between picture fuzzy sets were discussed by *Wei* [9] and are applied to strategic decision making problem for selecting optimal production strategy. Cosine similarity measure is undefined when one vector is zero, since it is defined as the inner product of their lengths. In this case, the dice similarity measure for PFSs can be used which is an extension of Dice similarity measure for IFSs. Therefore, the purpose of this study is to propose the Dice similarity measure for picture fuzzy sets (PFSs) and utilize it in decision making problems.

This paper is organized as follows: In Sects. 2, 3 and 4, we review definitions of picture fuzzy set (PFS), cosine similarity measure and Dice similarity measure. In Sect. 5, we propose a Dice similarity measure for picture fuzzy sets (PFSs). The proposed Dice similarity measure based MCDM method in Sect. 6 is also implemented on a practical problem of known criteria weights in Sect. 7. Final results are compared with the result of other existing methods in Sect. 8 and finally the conclusions are presented at the end of the paper.

2   Picture Fuzzy Sets

Cuong *et al.* [2] proposed picture fuzzy sets, which is defined as follows: A picture fuzzy set A on a universe X is defined as an object of the following form;

$$A = \{ <x, \mu_A(x), \eta_A(x), \nu_A(x) > : x \in X \} \qquad (1)$$

where the functions $\mu_A(x)$, $\eta_A(x)$, $\nu_A(x)$ are respectively called the degree of positive membership, the degree of neutral membership, the degree of negative membership of x in A, and following conditions are satisfied;

$$0 \leq \mu_A(x), \eta_A(x), \nu_A(x) \leq 1$$

$$\mu_A(x) + \eta_A(x) + \nu_A(x) \leq 1 \forall x \epsilon X$$

Then, $\forall x \in X : 1 - (\mu_A(x) + \eta_A(x) + \nu_A(x))$ is called the degree of refusal membership of x in A.

3   Cosine Similarity Measure for PFSs

Wei [9] introduced similarity measures for PFSs based on the concept of cosine function. Suppose there are two PFSs given as follows:

$$A = \{ <x, \mu_A(x), \eta_A(x), \nu_A(x) > : x \in X \}$$

$$B = \{ <x, \mu_B(x), \eta_B(x), \nu_B(x) > : x \in X \}$$

in the universe of discourse $X = \{x_1 x_2 x_3 \ldots \ldots x_n\}$. Then cosine similarity measure between PFSs is as follows:

$$PFC\,(A, B) = \frac{1}{n} \sum_{j=1}^{n} \frac{\mu_A(x_j)\,\mu_B(x_j) + \eta_A(x_j)\,\eta_B(x_j) + \nu_A(x_j)\,\nu_B(x_j)}{\sqrt{\mu_A^2(x_j) + \eta_A^2(x_j) + \nu_A^2(x_j)}\sqrt{\mu_B^2(x_j) + \eta_B^2(x_j) + \nu_B^2(x_j)}} \qquad (2)$$

4   Dice Similarity Measure

Let $X = \{x_1, x_2, x_3, \ldots \ldots x_n\}$ and $Y = \{y_1, y_2, y_3, \ldots \ldots y_n\}$ be two vectors of length n where all the coordinates are positive. Then the Dice similarity measure [4] is defined as follows:

$$D = \frac{2X.Y}{\|X\|_2^2 + \|Y\|_2^2} = \frac{2\sum_{i=1}^{n} x_i y_i}{\sum_{i=1}^{n} x_i^2 + \sum_{i=1}^{n} y_i^2}$$

where $X.Y = \sum_{i=1}^{n} x_i y_i$ is the inner product of the vectors X and Y and $\|X\|_2 = \sqrt{\sum_{i=1}^{n} x^2}$ and $\|Y\|_2 = \sqrt{\sum_{i=1}^{n} y^2}$, are the Euclidean norms of X and Y. The Dice similarity measure takes value in the interval $[0, 1]$. However, it is undefined if $x_i = y_i = 0$ $(i = 1, 2, \ldots n)$.

5   Dice Similarity Measure for PFS

In this section, the dice similarity measure for PFS is proposed as a generalization of the Dice similarity measure [4] in vector space. Let A and B be two PFSs in the universe of discourse $X = \{x_1, x_2, \ldots \ldots x_n\}$. Based on the extension of the Dice similarity measure [4], the Dice similarity measure between picture fuzzy sets A and B is proposed in vector space as follows;

$$D_{PFS}\,(A, B) = \frac{1}{n} \sum_{i=1}^{n} \frac{2(\mu_{Ai}\mu_{Bi} + \eta_{Ai}\eta_{Bi} + \nu_{Ai}\nu_{Bi})}{\mu_{Ai}^2 + \eta_{Ai}^2 + \nu_{Ai}^2 + \mu_{Bi}^2 + \eta_{Bi}^2 + v_{Bi}^2} \qquad (3)$$

The Dice similarity measure between two PFSs A and B satisfies the following properties:

1. $0 \leq D_{PFS}\,(A, B) \leq 1$
2. $D_{PFS}\,(A, B) = D_{PFS}\,(B, A)$
3. $D_{PFS}\,(A, B) = 1$ if and only if $A = B$, i.e., $\mu_{Ai} = \mu_{Bi}, \nu_{Ai} = \nu_{Bi}$ and $\eta_{Ai} = \eta_{Bi}$

5.1   Proof

1. Let us consider the ith item of the summation in Eq. (3)

$$D_i\,(A_i, B_i) = \frac{2(\mu_{Ai}\mu_{Bi} + \eta_{Ai}\eta_{Bi} + \nu_{Ai}\nu_{Bi})}{\mu_{Ai}^2 + \eta_{Ai}^2 + \nu_{Ai}^2 + \mu_{Bi}^2 + \eta_{Bi}^2 + v_{Bi}^2} \qquad (4)$$

It is obvious that $D_i\,(A_i, B_i) \geq 0$ and $\mu_{Ai}^2 + \eta_{Ai}^2 + \nu_{Ai}^2 + \mu_{Bi}^2 + \eta_{Bi}^2 + v_{Bi}^2 \geq 2(\mu_{Ai}\mu_{Bi} + \eta_{Ai}\eta_{Bi} + \nu_{Ai}\nu_{Bi})$ according to the inequality $(a^2 + b^2 \geq 2ab)$. Thus $0 \leq D_i\,(A_i, B_i) \leq 1$.

2. It is obvious that the property is true.

3. When $A = B$, there are $\mu_{Ai} = \mu_{Bi}, \upsilon_{Ai} = \upsilon_{Bi}$ and $\eta_{Ai} = \eta_{Bi}$, for $i = 1, 2, \ldots n$. So there is $D_{PFS}\ (A, B) = 1$. When $D_{PFS}\ (A, B) = 1$, there are $\mu_{Ai} = \mu_{Bi}, \upsilon_{Ai} = \upsilon_{Bi}$ and $\eta_{Ai} = \eta_{Bi}$, for $i = 1, 2, \ldots n$. So there is $A = B$.

If we consider the weight of x_i, the weighted Dice similarity measure between PFSs A and B is proposed as follows:

$$W_{PFS}\,(A, B) = \sum_{i=1}^{n} w_i \frac{2(\mu_{Ai}\mu_{Bi} + \eta_{Ai}\eta_{Bi} + \upsilon_{Ai}\upsilon_{Bi})}{\mu_{Ai}^2 + \eta_{Ai}^2 + \upsilon_{Ai}^2 + \mu_{Bi}^2 + \eta_{Bi}^2 + \upsilon_{Bi}^2} \tag{5}$$

where $w_i \in [0, 1]$, $i = 1, 2, \ldots\ldots\ldots n$, and $\sum_{i=1}^{n} w_i = 1$.

6   MCDM Based on Proposed Similarity Measure

In this section, a decision making method by using above defined Dice similarity measure for PFSs has been presented followed by an illustrative example for demonstrating the approach.

Let a set of m alternatives denoted by $A = \{A_1, A_2 \ldots\ldots\ldots\ldots A_m\}$ which has been evaluated by the decision maker under the set of the different criteria $C = \{C_1, C_2 \ldots\ldots\ldots C_n\}$ whose weight vectors are $w = \{w_1, w_2 \ldots\ldots\ldots w_m\}$ such that $w_j > 0$ and $\sum_{j=1}^{n} w_j = 1$. Assume that the decision maker gave his preference in the form of PFNs $\alpha_{ij} = \langle \mu_{ij}, \upsilon_{ij}, \eta_{ij} \rangle$. Then, in the following, we develop an approach based on the proposed similarity measure for MCDM problem, which involve the following steps.

Step 1. Construct a picture fuzzy decision matrix $D = (\alpha_{ij})_{m \times n}$ by the preference given by decision maker towards the alternative A_i.

Step 2. If there are different types of criteria, namely cost (C) and benefit (B), then we normalize it by using the following equation;

$$r_{ij} = \begin{cases} \alpha_{ij}, j \in B \\ \alpha_{ij}, j \in C \end{cases} \tag{6}$$

Step 3. Define an ideal picture fuzzy set for each criterion in the ideal alternative A^* as $C_j^* = (1, 0, 0)$ for "excellence". Then applying Eq. (5), we can obtain the weighted dice similarity measure between the ideal alternative A^* and an alternative A_i $(i = 1, 2, \ldots m)$. The bigger the value of $w_i(A^*, A_i)$, the better the alternative A_i, as the alternative A_i is closer to the ideal alternative A^*.

7   Practical Applications

A practical MCDM problem has been taken from Sect. 5.1 of *Garg* [5]. Suppose a multinational company in India is planning its financial strategy for the next year, according to the group strategy objective. For this, the four alternatives are obtained after their preliminary screening and are defined as A_1: to invest in the "Southern Asian markets"; A_2: to invest in the "Eastern Asian markets";

A_3: to invest in the "Northern Asian markets"; A_4: to invest in the "Local markets". This evaluation proceeds from four aspects, namely as C_1: the growth analysis; C_2: the risk analysis; C_3: the social-political impact analysis; C_4: the environmental impact analysis whose weight vector is $w = (0.2,\ 0.3,\ 0.1,\ 0.4)$. The following steps have been performed to compute the best one;

Step 1: Picture fuzzy decision matrix given by following:

	C_1	C_2	C_3	C_4
A_1	$\langle 0.2, 0.1, 0.6 \rangle$	$\langle 0.5, 0.3, 0.1 \rangle$	$\langle 0.5, 0.1, 0.3 \rangle$	$\langle 0.4, 0.3, 0.2 \rangle$
A_2	$\langle 0.1, 0.4, 0.4 \rangle$	$\langle 0.6, 0.3, 0.1 \rangle$	$\langle 0.5, 0.2, 0.2 \rangle$	$\langle 0.2, 0.1, 0.7 \rangle$
A_3	$\langle 0.3, 0.2, 0.2 \rangle$	$\langle 0.6, 0.2, 0.1 \rangle$	$\langle 0.4, 0.1, 0.3 \rangle$	$\langle 0.3, 0.3, 0.4 \rangle$
A_4	$\langle 0.3, 0.1, 0.6 \rangle$	$\langle 0.1, 0.2, 0.6 \rangle$	$\langle 0.1, 0.3, 0.5 \rangle$	$\langle 0.2, 0.3, 0.2 \rangle$

Step 2: Since the criteria C2 and C3 are the cost criteria while C1 and C4 are benefit criteria, so we get normalized decision matrix using Eq. (6) as follows:

	C_1	C_2	C_3	C_4
A_1	$\langle 0.6, 0.1, 0.2 \rangle$	$\langle 0.5, 0.3, 0.1 \rangle$	$\langle 0.5, 0.1, 0.3 \rangle$	$\langle 0.2, 0.3, 0.4 \rangle$
A_2	$\langle 0.4, 0.4, 0.1 \rangle$	$\langle 0.6, 0.3, 0.1 \rangle$	$\langle 0.5, 0.2, 0.2 \rangle$	$\langle 0.7, 0.1, 0.2 \rangle$
A_3	$\langle 0.2, 0.2, 0.3 \rangle$	$\langle 0.6, 0.2, 0.1 \rangle$	$\langle 0.4, 0.1, 0.3 \rangle$	$\langle 0.4, 0.3, 0.3 \rangle$
A_4	$\langle 0.6, 0.1, 0.3 \rangle$	$\langle 0.1, 0.2, 0.6 \rangle$	$\langle 0.1, 0.3, 0.5 \rangle$	$\langle 0.2, 0.3, 0.2 \rangle$

Step 3: Computing the values of weighted Dice similarity measure by applying Eq. (5).

$$w\left(A^*, A_1\right) = 0.59053$$

$$w\left(A^*, A_2\right) = 0.8057$$

$$w\left(A^*, A_3\right) = 0.6259$$

$$w\left(A^*, A_4\right) = 0.3585$$

Ranking of all alternatives is obtained in accordance with the descending values of weighted Dice similarity measure, as $A_2 > A_3 > A_1 > A_4$.

Hence the best financial strategy is A_2 *i.e.* to invest in Asian market. We compare our result with method given in Sect. 5.1 of *Garg* [5], and our result is same as obtained using the PFWA operator defined by *Garg* [5] i.e. $A_2 > A_3 > A_1 > A_4$.

8   Comparison with Other Methods

1. Comparing our results with the method using cosine similarity measure given by Wie [9], we get following values of weighted cosine similarity measure

$$w^c\left(A^*, A_1\right) = 0.7835$$

$$w^c\left(A^*, A_2\right) = 0.8510$$

$$w^c\left(A^*, A_3\right) = 0.7557$$

$$w^c\left(A^*, A_4\right) = 0.5806$$

Ranking all the alternatives in accordance with the descending values of weighted cosine similarity measure, as $A_2 > A_1 > A_3 > A_4$. Hence the best alternative is A_2, which is same as our result.

2. We compare our result with method given in Sect. 5.1 of *Garg* [5], and our result is same as obtained using the PFWA operator defined by *Garg* [5] i.e. $A_2 > A_3 > A_1 > A_4$.

9   Conclusion

In this paper, we have presented a new method for handling multi criteria decision making problems, where the characteristics of the alternative are represented by picture fuzzy sets. Furthermore, a decision making method was established by the use of proposed Dice similarity measure and weighted Dice similarity measure for picture fuzzy sets. Illustrative example demonstrated the feasibility of the proposed method in practical applications. The proposed method differs from previous approaches for multi criteria decision making not only due to fact that the proposed method uses PFS theory rather than other fuzzy theories, but also due to the Dice similarity measure and weighted Dice similarity measure based on PFS, the calculations are compact and much easier to analyze.

References

1. Atanassov, K.: Intuitionistic fuzzy sets. Fuzzy Sets Syst. **20**, 87–96 (1986)
2. Coung, B.C., Kreinovich, V.: Picture fuzzy sets-a new concept for computational intelligence problems. In: Proceedings of third world congress on information and communication technologies, pp. 1–3. IEEE, Hanoi (2013)
3. Dengfeng, L., Chuntian, C.: New similarity measures of intuitionistic fuzzy sets and application to pattern recognitions. Pattern Recogn. Lett. **23**, 221–225 (2002)
4. Dice, L.R.: Measures of the amount of ecologic association between species. Ecology **26**, 297–302 (1945)
5. Garg, H.: Some picture fuzzy aggregation operators and their applications to multicriteria decision making. Arab. J. Sci. Eng. **42**, 1–16 (2017)
6. Szmidt, E., Kacprzyk, J.: A new concept of a similarity measure for intuitionistic fuzzy sets and its use in group decision making. In: Torra, V., Narukawa, Y., Miyamoto, S. (eds.) MDAI 2005. LNCS (LNAI), vol. 3558, pp. 272–282. Springer, Heidelberg (2005). https://doi.org/10.1007/11526018_27
7. Ye, J.: Cosine similarity measures fot intuitionistic fuzzy sets and their applications. Math. Comput. Model. **53**, 91–97 (2011)
8. Ye, J.: Multicriteria decision-making method using the Dice similarity measure based on the reduct intuitionistic fuzzy sets of interval-valued intuitionistic fuzzy sets. App. Math. Model. **36**, 4466–4472 (2012)
9. Wei, G.: Some similarity measures for picture fuzzy sets and their applications to strategic decision making. Informatica **28**, 547–564 (2017)

Local and Global Stability of Fractional Order HIV/AIDS Dynamics Model

Praveen Kumar Gupta[✉]

Department of Mathematics, National Institute of Technology Silchar,
Silchar 788010, Assam, India
pkguptaitbhu@gmail.com

Abstract. In this article, we discussed the dynamical behaviour of a fractional order HIV/AIDS virus dynamics model which takes account the cure of infected cells and loss of viral particles due to the fusion into uninfected cells. The local and global stability of the model is studied for disease-free equilibrium point with the help of next generation matrix method. Moreover, the numerical solutions for some particular cases are provided to verify the analytical results.

Keywords: HIV/AIDS · Local and global stability
Numerical solution

1 Introduction

The fractional derivative has been widely applied in many research areas which have been perceived an enormous growth in the last four decades. For examples, the models approaching the backgrounds of economics, physics, circuits, heat transfer, diffusion, electro-chemistry, and even biology are always apprehensive with fractional derivative [1–5]. In fact, fractional derivative based approaches establish more advanced and updated models of engineering systems than the ordinary derivative based approaches do in many applications. The theories of fractional derivatives generalize the idea of ordinary derivatives to some extent. The literature shows that there is no field that has remained untouched by fractional derivatives. However, development still needs to be achieved before the ordinary derivatives could be truly interpreted as a subset of the fractional derivatives [6–8].

In 2012, Safiel et al. [9] examines the effect of screening and treatment on the transmission of HIV/AIDS infection in a population and shows the screening of HIV infectives and treatment of screened HIV infectives has the effect of reducing the transmission of the disease. Kaur et al. [10] studied the transmission of infectives and counselling on the spread of HIV infection. In 2009, Ding and Ye [11] introduce a fractional-order HIV infection of CD4+ T-cells model, which determined the non-negative solutions, and carry out a detailed analysis on the stability of equilibrium. Gkdogan et al. [12] have applied the multi-step

© Springer Nature Singapore Pte Ltd. 2018
D. Ghosh et al. (Eds.): ICMC 2018, CCIS 834, pp. 141–148, 2018.
https://doi.org/10.1007/978-981-13-0023-3_14

differential transform method to present an analytical solution of nonlinear fractional order HIV model for infection of CD4$^+$ T cells. Recently, Arafa et al. [13] describes the fractional-order model for HIV infection of CD4$^+$ T cells with therapy effect, and they employed Generalized Euler Method to find the numerical solution of such problem.

More precisely, we reflect on a HIV/AIDS virus dynamic model describing the interaction between the host susceptible CD4$^+$ T cells (H), infected CD4$^+$ T cells (I) and virus (V), and it is formulated by the following non-linear system of fractional differential equations in Caputo sense

$$D_0^\alpha H = \mu - (\delta + \gamma)HV - d_1 H + \sigma I \tag{1a}$$

$$D_0^\alpha I = \delta HV - (d_2 + \sigma)I \tag{1b}$$

$$D_0^\alpha V = \beta I - \gamma HV - d_3 V \tag{1c}$$

and the initial conditions are

$$H(0) = H_0, I(0) = I_0, V(0) = V_0, \tag{2}$$

where the formation rate of susceptible host cells is μ, die at a rate $d_1 H$ and turn into infected δHV by virus, recovered or cured at a rate σI and destroy at a rate γHV due to fusion. Infected cells might be killed because of virion in their nucleus. The loss rate of infected cells is given by $(d_2 + \sigma)I$, where $d_2 I$ is the death rate of infected cells and σI is the cure rate into the susceptible cells. Finally, virions are produced by infected cells at a rate βI, decays at a rate $d_3 V$, and destroy at a rate γHV due to fusion.

In this study, we analysed a HIV/AIDS dynamical model with effect of fusion. One more vital feature of the model is the fact that we incorporate also a cure rate of the infected cells to the susceptible cells.

2   Analysis of the Model

2.1   Positivity and Boundedness

Denote $\mathbb{R}_+^3 = \{x \in \mathbb{R}^3 : x \geq 0\}$ and let $x(t) = [H(t), I(t), V(t)]^T$. To prove the main theorem, we need the following generalized mean value theorem and corollary [9, 12].

Lemma 1 ([11]). *Suppose that $f(x) \in C[a, b]$ and Caputo derivative $D_a^\alpha f(x) \in C[a, b]$ for $0 < \alpha \leq 1$, then we have*

$$f(x) = f(a) + \frac{1}{\Gamma(\alpha)}(D_a^\alpha f)(\tau)(x - a)^\alpha \tag{3}$$

with $a \leq \tau \leq x$, $\forall x \in (a, b]$.

Corollary 2. *Let $f(x) \in C[a, b]$ and Caputo derivative $D_a^\alpha f(x) \in C[a, b]$ for $0 < \alpha \leq 1$.*

- If $D_a^\alpha f(x) \geq 0$, $\forall x \in (a,b)$, then $f(x)$ is non-decreasing function for each $x \in [a,b]$.
- If $D_a^\alpha f(x) \leq 0$, $\forall x \in (a,b)$, then $f(x)$ is non-increasing function for each $x \in [a,b]$.

Theorem 3. *There is a unique solution $x(t) = [H(t), I(t), V(t)]^T$ to the system (1) and initial condition (2) on $t \geq 0$ and the solution will remain in $\mathbb{R}_+^3$. Furthermore, $H(t)$ and $I(t)$ are all bounded.*

Proof. According to Lin [14], we can determine the solution on $(0, +\infty)$, by solving the model (1) and initial conditions (2), which is not only existent but also unique. Subsequently, we have to explain the non-negative octant $\mathbb{R}_+^3$ is a positively invariant region. From Eq. (1), we find

$$D_0^\alpha H = \mu > 0, \ D_0^\alpha I = \delta HV \geq 0, \ D_0^\alpha V = \beta I \geq 0. \tag{4}$$

By Corollary 2, the solution of model (1) will be remain in $\mathbb{R}_+^3$. Furthermore, from equation (1) we make out that

$$D_0^\alpha T_{total} = \mu - \gamma HV - d_1 H - d_2 I,$$

where, $T_{total} = H + I$.

Death by infected CD4+ T cells occurs faster than death by natural means; that is, $d_2 > d_1$. Therefore,

$$D_0^\alpha T_{total} + d_1 T_{total} < \mu. \tag{5}$$

Thus, by Corollary 2, in the case of HIV infection, the total T-cell population, T_{total}, i.e., the sub populations $H(t)$ and $I(t)$, are bounded.

2.2 Equilibrium Points, Reproduction Number and Local Stability

Equilibrium Points. To evaluate the equilibrium points of model (1), let

$$D_0^\alpha H = 0, \ D_0^\alpha I = 0, \ D_0^\alpha V = 0. \tag{6}$$

Then $E_0 = (\frac{\mu}{d_1}, 0, 0)$ and $E^* = (H^*, I^*, V^*)$ are the infection-free and endemic equilibrium points, respectively, where

$$H^* = \frac{d_3(d_2 + \sigma)}{\beta\delta - \gamma(d_2 + \sigma)},$$

$$I^* = \frac{\delta(d_1 d_3(d_2 + \sigma) + \mu(\gamma(d_2 + \sigma) - \beta\delta)}{(\gamma(d_2 + \sigma) - \beta\delta)(d_2(\gamma + \delta) + \gamma\sigma)},$$

$$V^* = \frac{\mu(\beta\delta - \gamma(d_2 + \sigma)) - d_1 d_3(d_2 + \sigma)}{d_2 d_3(\gamma + \delta) + d_3\gamma\sigma}.$$

Reproduction Number. Now, we compute the reproduction number ($\Re_0$) for the model (1). $\Re_0$ is defined as the number of secondary infections due to a single infection in a completely susceptible population, and it is

$$\Re_0 = \frac{\beta\delta\mu}{(d_1 d_3 + \gamma\mu)(d_2 + \sigma)}. \tag{7}$$

Local Stability of Equilibria. The Jacobian matrix of model (1) at a general point is given by

$$J = \begin{pmatrix} -d_1 - (\gamma + \delta)V & \sigma & -(\gamma + \delta)H \\ \delta V & -d_2 - \sigma & \delta H \\ -\gamma V & \beta & -d_3 - \gamma H \end{pmatrix} \tag{8}$$

Based on Jacobian matrix approach by evaluating (8) at infection-free equilibrium point E_0, we can obtain the following results:

Lemma 4. *The infection-free equilibrium point E_0 is locally asymptotically stable if all eigenvalues λ_i of the Jacobian matrix $J(E_0)$ for model (1), satisfy $|arg(\lambda_i)| > \alpha\frac{\pi}{2}$.*

Proof. The Jacobian matrix $J(E_0)$ for model (1) evaluated at the infection-free equilibrium steady state E_0, is given by

$$J(E_0) = \begin{pmatrix} -d_1 & \sigma & -(\gamma + \delta)\frac{\mu}{d_1} \\ 0 & -d_2 - \sigma & \delta\frac{\mu}{d_1} \\ 0 & \beta & -d_3 - \gamma\frac{\mu}{d_1} \end{pmatrix} \tag{9}$$

The characteristic equation of the Jacobian matrix $J(E_0)$ is

$$(\lambda + d_1)(\lambda^2 + a_1\lambda + a_2) = 0,$$

where, $a_1 = (\sigma + d_2 + d_3 + \gamma\frac{\mu}{d_1}) > 0$, and,

$$a_2 = \frac{(d_1 d_3 + \mu\gamma)(d_2 + \sigma) - \beta\delta\mu}{d_1} = \frac{1 - \Re_0}{d_1(d_1 d_3 + \mu\gamma)(d_2 + \sigma)}. \tag{10}$$

Many researchers studied the Routh-Hurwitz stability conditions for fractional order systems [12–15], and describe the necessary and sufficient condition $|arg(\lambda_i)| > \alpha\frac{\pi}{2}$, for various models. Routh–Hurwitz criteria states that all roots of the characteristic equation $(\lambda + d_1)(\lambda^2 + a_1\lambda + a_2) = 0$ have negative real parts if and only if $a_1 > 0$ and $a_2 > 0$. Therefore, Eq. (10) implies that if $\Re_0 < 1$ then all roots will be negative and for this condition the necessary and sufficient condition will satisfy. Hence, a sufficient condition for the local asymptotic stability of the equilibrium points is that the eigenvalues λ_i of the Jacobian matrix of $J(E_0)$ satisfy the condition $|arg(\lambda_i)| > \alpha\frac{\pi}{2}$. This confirms that fractional-order differential equations are, at least, as stable as their integer order counterpart.

The global existence of the solution of the fractional differential equation always becomes a most important concern, which is carry out in the next section.

2.3   Global Stability of Equilibria

Lemma 5 ([14]).   *Assume that the function $G : \mathbb{R}_+ \times \mathbb{R}^3 \to \mathbb{R}^3$ satisfies the following conditions in the global space:*

(I)   The function $G(t, x(t))$ is Lebesgue measurable with respect to t on $\mathbb{R}$.
(II)   The function $G(t, x(t))$ is continuous with respect to $x(t)$ on $\mathbb{R}^3$.
(III) $\dfrac{\partial G(t, x(t))}{\partial x}$ is continuous with respect $x(t)$ on $\mathbb{R}^3$.
(IV) $\|G(t, x(t))\| \leq \omega + \lambda \|x\|$, for almost every $t \in \mathbb{R}$ and all $x \in \mathbb{R}^3$.

Here, ω, λ are two positive constants and $x(t) = [H(t), I(t), V(t)]^T$. Then, the initial value problem

$$\begin{cases} D_t^\alpha x(t) = G(t, x(t)), \\ x(t_0) = x_0, \end{cases} \tag{11}$$

has a unique solution.

Theorem 6. *There is a unique solution for system (1) and solution remains in $\mathbb{R}_+^3$.*

Proof. From Lemma 5, we obtain the unique solution on $(0, \infty)$ by solving the system (1). Firstly, Lin [14] discussed the proof of theorem and shows that the solution is not only exist but also unique. In Theorem 3, we already proof that the solution of model (1) will be remain in $\mathbb{R}_+^3$. The global stability of the model also verified with the help of Fig. 1, which shows that after some time the susceptible population is going to constant while the number of infected population and virions are tends to zero, i.e., we achieve the infection-free stage.

3   Numerical Results and Discussion

In this article, we will solve the system (1) by using Mathematica 9. Consider that $\mu = 10 \; mm^{-3} day^{-1}$, $\beta = 160 \; day^{-1}$, $\delta = 0.000024 \; mm^3 day^{-1}$, $\gamma = 0.00001 \; mm^3 day^{-1}$, $\sigma = 0.2 \; day^{-1}$, $d_3 = 3.4 \; day^{-1}$ [13]. We choose $d_1 = 0.05 \; day^{-1}$ and $d_2 = 0.6 \; day^{-1}$ (since death rate of host infected cells by virions will be slightly higher than those of susceptible cells) with initial conditions $H(0) = 1000$, $I(0) = 10$ and $V(0) = 10$. The recovery rate σ will vary for situation of the patient, such as availability of the drugs, etc.

Figure 1(a) shows that the host susceptible CD4$^+$ T cell population decreases with increase of time and tends to positive equilibrium point $\frac{\mu}{d_1}$. Figure 1(b) verify that infected CD4$^+$ T cell population is increases in the first ten days, after that it decreases drastically with increase of time and it tends to zero. Similarly, Fig. 1(c) exhibit that first ten days the virus population is increases rapidly compare to infected population later on decreased radically and it tends to zero.

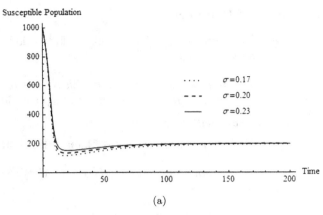

(a)

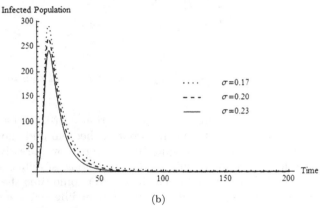

(b)

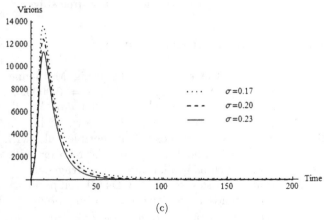

(c)

Fig. 1. The densities of the host susceptible population H(t), infected population I(t) and virions V(t) when $\alpha = 1$, $\mu = 10$, $\beta = 160$, $\delta = 0.000024$, $\gamma = 0.00001$, $d_1 = 0.05$, $d_2 = 0.06$, $d_3 = 3.4$. The solid line ($\sigma = 0.23$), the dashed line ($\sigma = 0.20$), and the dotted line ($\sigma = 0.17$).

4 Conclusion

In this paper, we establish a system of Caputo sense fractional-order HIV/AIDS dynamics model with the help of Srivastava and Hattaf et al. [16, 17]. The author explained the non-negative solutions and boundedness as an essential part of any population dynamics model. The authors have defined the equilibrium points and reproduction number for the proposed model. By using stability analysis on an anticipated fractional order system, we obtained a sufficient condition on the parameters for the stability of the infection-free steady state. The recent appearance of fractional differential equations as models in some fields of applied mathematics makes it necessary to investigate analysis of solution for such equations and we hope that this work is a step in this direction. The numerical solutions have performed for different values of σ.

References

1. Debnath, L.: Recent applications of fractional calculus to science and engineering. Int. J. Math. Math. Sci. **2003**(54), 3413–3442 (2003)
2. Machado, J.A.T., Silva, M.F., Barbosa, R.M., Jesus, I.S., Reis, C.M., Marcos, M.G., Gal-hano, A.F.: Some applications of fractional calculus in engineering. Math. Probl. Eng. **2010**, 1–34 (2010)
3. Das, S., Gupta, P.K.: A mathematical model on fractional Lotka-Volterra equations. J. Theor. Biol. **277**(1), 1–6 (2011)
4. Gupta, P.K.: Approximate analytical solutions of fractional BenneyLin equation by reduced differential transform method and the homotopy perturbation method. Comput. Math. Appl. **61**(9), 2829–2842 (2011)
5. Ali, M.F., Sharma, M., Jain, R.: An application of fractional calculus in electrical engineering. Adv. Eng. Technol. Appl. **5**(2), 41–45 (2016)
6. Atangana, A., Baleanu, D.: New fractional derivatives with nonlocal and non-singular kernel: theory and applications to heat transfer model. Therm. Sci. **20**(2), 763–769 (2016)
7. Atangana, A., Koca, I.: Chaos in a simple nonlinear system with AtanganaBaleanu derivatives with fractional order. Chaos Solitons Fractals **89**, 447–454 (2016)
8. Atangana, A.: Fractal-fractional differentiation and integration: connecting fractal calculus and fractional calculus to predict complex system. Chaos Solitons Fractals **102**, 396–406 (2017)
9. Safiel, R., Massawe, E.S., Makinde, D.O.: Modelling the effect of screening and treatment on transmission of HIV/AIDS infection in a population. Am. J. Math. Stat. **2**(4), 75–88 (2012)
10. Kaur, N., Ghosh, M., Bhatia, S.S.: Mathematical analysis of the transmission dynamics of HIV/AIDS: role of female sex workers. Appl. Math. Inf. Sci. **8**(5), 2491–2501 (2014)
11. Ding, Y., Ye, H.: A fractional-order differential equation model of HIV infection of CD4+ T-cells. Math. Comput. Model. **50**, 386–392 (2009)
12. Gkdogan, A., Yildirim, A., Merdan, M.: Solving a fractional order model of HIV infection of CD4+ T cells. Math. Comput. Model. **54**, 2132–2138 (2011)
13. Arafa, A.A.M., Rida, S.Z., Khalil, M.: A fractional-order model of HIV infection with drug therapy effect. J. Egypt. Math. Soc. **22**, 538–543 (2014)

14. Lin, W.: Global existence theory and chaos control of fractional differential equations. J. Math. Anal. Appl. **332**, 709–726 (2007)
15. Ahmed, E., El-Sayed, A.M.A., El-Saka, H.A.A.: Equilibrium points, stability and numerical solutions of fractional-order predator-prey and rabies models. J. Math. Anal. Appl. **325**, 542–553 (2007)
16. Srivastava, P.K., Chandra, P.: Modelling the dynamics of HIV and CD4+ T cells during primary infection. Nonlinear Anal. Real World Appl. **11**(2), 612–618 (2010)
17. Hattaf, K., Yousfi, N.: Global stability of a virus dynamics model with cure rate and absorption. J. Egypt. Math. Soc. **22**, 386–389 (2014)

A Study of an EOQ Model Under Cloudy Fuzzy Demand Rate

Snigdha Karmakar[1(✉)], Sujit Kumar De[2], and A. Goswami[1]

[1] Department of Mathematics, IIT Kharagpur, Kharagpur, West Bengal, India
mathsnigdha.316@gmail.com
[2] Department of Mathematics, Midnapore College,
West Midnapore 721101, West Bengal, India

Abstract. This paper deals with a new fuzzy number namely, cloudy fuzzy number and its new defuzzification method for a classical economic order quantity (EOQ) inventory management problem. In fuzzy system, the measures of ambiguity depend upon the area of applicability and the observations of experimenters. The lack of insight over the set consideration causes the invention of new fuzzy set "cloudy fuzzy set". The traditional assumptions over fuzziness were fixed over time, but in this study we see fuzziness can be removed as time progresses. Here the crisp model is solved first then taking the demand rate as general fuzzy as well as cloudy fuzzy number we have solved the problem under usual Yager's index method and extension of Yager's index method respectively. With the help of numerical example we have compared the objective values for all cases and the implication of the cloudy fuzzy number has been discussed exclusively. Graphical illustrations, sensitivity analysis are given for better justification of the model. Finally, a conclusion is made.

Keywords: Inventory · Cloudy fuzzy number · Cloud index
Extension of Yager's index method · Optimization

1 Introduction

The classical EOQ model was developed with some limitations in the early stage of 20th century. Till date it has been expanded in many dimensions towards the creation of deterministic models. Harris (1915) developed the concept of modeling in the management problem. In the threshold period of developing inventory management Hanssmann (1962), Hadley and Whitin (1963) worked a lot over inventory system. By this time Ghare and Schrader (1963) studied a model for exponentially decaying inventory. Haneveld and Teunter (1998) discussed the effects of discount and demand rate variability on the EOQ model. Hariga (1996) studied an EOQ model for deteriorating items with time-varying demand. Fuzzy sets was first developed by Zadeh (1965), subsequently, it was applied by Bellman and Zadeh (1970) in decision making problem. Since then several researchers were being engaged to characterize the actual nature of the fuzzy

© Springer Nature Singapore Pte Ltd. 2018
D. Ghosh et al. (Eds.): ICMC 2018, CCIS 834, pp. 149–163, 2018.
https://doi.org/10.1007/978-981-13-0023-3_15

sets [Dubois and Prade (1978); Kaufmann and Gupta (1992); Báez-Sáncheza et al. (2012); Ban and Coroianu (2014); De and Sana (2015)]. Goetschel and Voxman (1985) studied over eigen fuzzy number sets.

In fuzzy environment, the inventory system itself have been developing in its natural way under great thinkers. In its process, Lee and Yao (1999) developed a fuzzy EOQ model without back order. De et al. (2003, 2008) studied the inventory model considering fuzzy demand rate, fuzzy deterioration rate and fuzzy cost co-efficient respectively. De (2013) developed an EOQ model with natural idle time and wrongly measured demand rate. De and Sana (2013) studied fuzzy order quantity inventory model with fuzzy shortage quantity and fuzzy promotional index. Kao and Hsu (2002) developed a lot size-reorder point inventory model with fuzzy demands. Kumar et al. (2012) developed a fuzzy model for ramp type demand and partial backlogging.

However in defuzzification analysis, specially, on ranking fuzzy numbers the contribution of Yager (1981) kept a unquestionable destination. Few years later a huge number of researchers were began to study over the ranking methods and finally invented numerous formulae over the subject. Researchers like Chu and Tsao (2002), Ramli and Mohamad (2009), Allahviranloo and Saneifard (2012), Ezzati et al. (2012), Deng (2014) etc. discussed the methods for ranking fuzzy numbers based on center of gravity. Cheng (1998) discussed a new approach for ranking fuzzy numbers by distance method. Buckley and Chanas (1989) discussed a fast method of ranking alternatives using fuzzy numbers. Ezatti and Saneifard (2010) developed a method of continuous weighted quasi-arithmetic means. Based on deviation degree, the extensive works over L-R fuzzy numbers by Wang et al. (2009), Kumar et al. (2011), Hajjari and Abbasbandy (2011), Xu et al. (2012) etc. kept a new milestone in the subject. Yu et al. (2013) developed fuzzy ranking generalized fuzzy numbers in fuzzy decision making based on the left and right transfer coefficients and areas. Zhang et al. (2014) studied a new method for ranking fuzzy numbers and its application to group decision making problems. Rezvani (2015) developed ranking generalized exponential trapezoidal fuzzy numbers based on variance. De and Beg (2016) used the learning experience over fuzzy set and introduced the triangular dense fuzzy set (TDFS). Karmakar et al. (2017) applied the TDFS on a pollution sensitive EOQ model. De (2017) extended the TDFS and developed triangular dense fuzzy lock set. Very recent De and Mahata (2017) have studied the cloudy fuzzy numbers on an backordered inventory model.

In our present study, we have utilized a new fuzzy number namely cloudy fuzzy number and its corresponding new defuzzification technique via the extension of Yager (1981)'s ranking index method. By this study we have shown that, a cloud indicator exists and it cannot be removed in any inventory process. A simple EOQ model has been considered and the model itself has been solved by crisp, general fuzzy and cloudy fuzzy environments also. A numerical study followed a comparative graphical illustrations and a sensitivity analysis as well. Finally decision is made over the applicable region by realistic feasibility of the model itself.

2 Preliminary

2.1 Normalized General Triangular Fuzzy Number (NGTFN) [De and Mahata (2017)]

Let $\tilde{D}$ be a NGTFN having the form $\tilde{D} = \langle D_1, D_2, D_3 \rangle$. Then its membership function (Fig. 1) is defined by

$$\mu(x) = \begin{cases} 0 & if\, D < D_1 \text{ and } D > D_3 \\ \frac{D-D_1}{D_2-D_1} & if\, D_1 < D < D_2 \\ \frac{D_3-D}{D_3-D_2} & if\, D_2 < D < D_3 \end{cases} \quad (1)$$

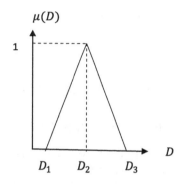

Fig. 1. Membership function of NGTFN

Now, the left and right α-cuts of $\mu(D)$ are given by

$$L(\alpha) = D_1 + \alpha(D_2 - D_1) \text{ and } R(\alpha) = D_3 - \alpha(D_3 - D_2) \quad (2)$$

Note that the measures of fuzziness can be obtained from the following formula.

2.2 Yager's (1981) Ranking Index

If $L(\alpha)$ and $R(\alpha)$ are the left and right α-cuts of a fuzzy number $\tilde{D}$ then the defuzzification rule under Yager's ranking index is given by

$$I(\tilde{D}) = \frac{1}{2} \int_0^1 \{L(\alpha) + R(\alpha)\} d\alpha = \frac{1}{4}(D_1 + 2D_2 + D_3) \quad (3)$$

Note that the measures of fuzziness (degree of fuzziness d_f) can be obtained from the formula $d_f = \frac{1}{2m}(U_b - L_b)$ where L_b and U_b are the lower bounds and upper bounds of the fuzzy numbers respectively and m be their respective mode.

2.3 Cloudy Normalized Triangular Fuzzy Number (CNTFN) [De and Mahata (2017)]

A fuzzy number of the form $\tilde{A} = \langle a_1, a_2, a_3 \rangle$ is said to be cloudy triangular fuzzy number if after infinite time the set itself converges to a crisp singleton. This means that, as t tends to infinity, both $a_1, a_3 \rightarrow a_2$.

Let us consider the fuzzy number

$$\tilde{A} = \langle a_2(1 - \frac{\rho}{1+t}), a_2, a_2(1 + \frac{\sigma}{1+t}) \rangle, \text{ for } 0 < \rho, \sigma < 1 \tag{4}$$

Note that, $\lim\limits_{t \to \infty} a_2(1 - \frac{\rho}{1+t}) \rightarrow a_2$ and $\lim\limits_{t \to \infty} a_2(1 + \frac{\sigma}{1+t}) \rightarrow a_2$, so $\tilde{A} \rightarrow a_2$. Then the membership function for $0 \leq t$ as follows:

$$\mu(x,t) = \begin{cases} 0 & if\, x < a_2(1 - \frac{\rho}{1+t}) \text{ and } x > a_2(1 + \frac{\sigma}{1+t}) \\ \frac{x - a_2(1 - \frac{\rho}{1+t})}{\frac{\rho a_2}{1+t}} & if\ a_2(1 - \frac{\rho}{1+t}) < x < a_2 \\ \frac{a_2(1 + \frac{\sigma}{1+t}) - x}{\frac{\sigma a_2}{1+t}} & if\ a_2 < x < a_2(1 + \frac{\sigma}{1+t}) \end{cases} \tag{5}$$

The graphical representation of CNTFN is given in Fig. 2.

2.4 Defuzzification Method of CNTFN [De and Mahata (2017)]

Let $\tilde{A}$ be a CNTFN stated in (4). We take left and right α-cuts of $\mu(x,t)$ from (5) noted as $L(\alpha,t)$ and $R(\alpha,t)$ respectively. Then the defuzzification formula under time extension of Yager's ranking index is given by

$$I(\tilde{A}) = \frac{1}{2T} \int_{\alpha=0}^{\alpha=1} \int_{t=0}^{t=T} \{L(\alpha,t) + R(\alpha,t)\} d\alpha dt \tag{6}$$

Note that, α and t are independent variables. Now by using (6) for CNTFN (4), we get

$$I(\tilde{A}) = \frac{a_2}{2T} [2T + \frac{(\sigma - \rho)}{2} Log(1+T)] \tag{7}$$

Again (7) can be rewritten as

$$I(\tilde{A}) = a_2[1 + \frac{(\sigma - \rho)}{4} \frac{Log(1+T)}{T}] \tag{8}$$

Obviously $\lim\limits_{T \to \infty} \frac{Log(1+T)}{T} \rightarrow 0$ and therefore $I(\tilde{A}) \rightarrow a_2$.

Now, we may call the factor $\frac{Log(1+T)}{T}$ as cloudy index (CI) \qquad (9)

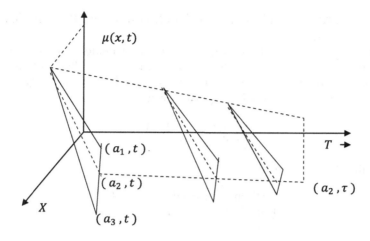

Fig. 2. Membership function of CNTFN

Here also we notice that, since for practical purpose the time horizon cannot be infinite so after defuzzification the indexed values do not come back to its crisp original even the restrictions have been removed in our assumptions.

3 Assumptions and Notations

The following notations and assumptions are used to develop the model.

3.1 Assumptions

1. Replenishments are instantaneous.
2. The time horizon is infinite (months).
3. Shortages are not allowed.

3.2 Notations

q The order quantity per cycle.
d Demand rate per year.
k Setup cost per cycle (\$).
h Inventory holding cost per unit quantity per cycle (\$).
c Purchasing price of unit item (\$).
T Cycle time (days).
z Average total cost of the inventory (\$).

4   Crisp Mathematical Model (Classical EOQ Model)

Let the inventory starts with order quantity q and it depletes with the constant demand rate d. Then upto the cycle time T it reaches zero. The costs associated with it are unit purchasing price, unit holding cost and set up cost only. Therefore the average total cost is given by (shown in Fig. 3.)

$$\begin{cases} z = cd + \frac{k}{T} + \frac{hdT}{2} \\ \text{Subject to } q = dT \end{cases} \tag{10}$$

Thus our problem is given by

$$\begin{cases} \text{Minimize } z = cd + \frac{k}{T} + \frac{hdT}{2} \\ \text{Subject to } q = dT \end{cases} \tag{11}$$

5   Formulation of the Fuzzy Mathematical Model

Let the demand rate follows the fuzzy flexibility during the inventory run time. Then fuzzifying (11) we have

$$\begin{cases} \text{Minimize } \tilde{z} = c\tilde{d} + \frac{k}{T} + \frac{h\tilde{d}T}{2} \\ \text{Subject to } \tilde{q} = \tilde{d}T \end{cases} \tag{12}$$

Now let us consider the fuzzy number $\tilde{d}$ as follows:

$$\tilde{d} = \begin{cases} \langle d_1, d_2, d_3 \rangle \text{ for } NGTFN \\ \langle d(1 - \frac{\rho}{1+t}), d, d(1 + \frac{\sigma}{1+t}) \rangle \text{ for } CNTFN \\ \text{for } 0 < \rho, \sigma < 1 \text{ and } T > 0 \end{cases} \tag{13}$$

Therefore, using (1) the membership function for the fuzzy objective and order quantity under NGTFN are given by

$$\mu_1(z) = \begin{cases} 0 & if\ z \leq z_1 \text{ and } z \geq z_3 \\ \frac{z - z_1}{z_2 - z_1} & if\ z_1 \leq z \leq z_2 \\ \frac{z_3 - z}{z_3 - z_2} & if\ z_2 \leq z \leq z_3 \end{cases} \tag{14}$$

where $z_1 = (c + \frac{hT}{2})d_1 + \frac{k}{T}$, $z_2 = (c + \frac{hT}{2})d_2 + \frac{k}{T}$ and $z_3 = (c + \frac{hT}{2})d_3 + \frac{k}{T}$

$$\mu_2(q) = \begin{cases} 0 & if\ q \leq q_1 \text{ and } q \geq q_3 \\ \frac{q - q_1}{q_2 - q_1} & if\ q_1 \leq q \leq q_2 \\ \frac{q_3 - q}{q_3 - q_2} & if\ q_2 \leq q \leq q_3 \end{cases} \tag{15}$$

for $q_1 = d_1 T$, $q_2 = d_2 T$ and $q_3 = d_3 T$

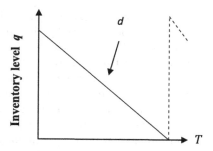

Fig. 3. The Classical EOQ model

Using (2)–(3) the index values of fuzzy objective and order quantities are respectively given by

$$\begin{cases} I(\tilde{z}) = \frac{1}{4}(z_1 + 2z_2 + z_3) = \frac{1}{4}(d_1 + 2d_2 + d_3)(c + \frac{hT}{2}) + \frac{k}{T} \\ I(\tilde{q}) = \frac{1}{4}(q_1 + 2q_2 + q_3) = \frac{1}{4}(d_1 + 2d_2 + d_3)T \end{cases} \tag{16}$$

Again, using (5) the membership function for the fuzzy objective and order quantity under CNTFN are given by

$$w_1(z, T) = \begin{cases} 0 & \text{if } z \leq z_1 \text{ and } z \geq z_3 \\ \frac{z - z_1}{z_2 - z_1} & \text{if } z_1 \leq z \leq z_2 \\ \frac{z_3 - z}{z_3 - z_2} & \text{if } z_2 \leq z \leq z_3 \end{cases} \tag{17}$$

where $z_1 = (c + \frac{hT}{2})d(1 - \frac{\rho}{1+T}) + \frac{k}{T}, z_2 = (c + \frac{hT}{2})d + \frac{k}{T}$ and $z_3 = (c + \frac{hT}{2})d(1 + \frac{\sigma}{1+T}) + \frac{k}{T}$

$$w_2(q, T) = \begin{cases} 0 & \text{if } q \leq q_1 \text{ and } q \geq q_3 \\ \frac{q - q_1}{q_2 - q_1} & \text{if } q_1 \leq q \leq q_2 \\ \frac{q_3 - q}{q_3 - q_2} & \text{if } q_2 \leq q \leq q_3 \end{cases} \tag{18}$$

for $q_1 = d(1 - \frac{\rho}{1+T})T, q_2 = dT$ and $q_3 = d(1 + \frac{\sigma}{1+T})T$
Using (6) the index values of fuzzy objective and order quantities are respectively given by

$$\begin{cases} I(\tilde{z}) = cd + \frac{hdT}{4} + \frac{k}{T}Log|\frac{T}{\varepsilon}| + \frac{cd(\sigma - \rho)}{4}\frac{Log(1+T)}{T} + \frac{hd(\sigma - \rho)}{8}\left\{1 - \frac{Log(1+T)}{T}\right\} \\ I(\tilde{q}) = \frac{dT}{2} + \frac{d(\sigma - \rho)}{4}\left\{1 - \frac{Log(1+T)}{T}\right\} \end{cases}$$

$$\tag{19}$$

Here we take ε as sufficiently small so that the given objective converges at a finite value.

6   Implication of Cloudy Fuzzy Environment in Inventory Process

The concept of fuzziness usually arises from the uncertainties over a particular subject. Meanwhile suppose x be a crisp number then under uncertainties it will be "around x". Our focus of attention in this uncertainty depends upon the situation over time. In any inventory process, initially the uncertainties viewed are high as there is no information available about the production process, demand and customer's behavior. But, as the time progresses the decision-maker gains more information, analyzes the situation and the uncertainties begin to decrease. With time, experience and knowledge the prediction becomes precise and after a long period of time it becomes more and more accurate. In such way the ambiguities underlying in the inventory system can be removed after a long time and it is experiencing from the very ancient stage of any management system. When the inventory cycle time is low the ambiguity is high and reversely if the cycle time is high then uncertainty is low; it is the most common phenomenon of day to day life. Let us consider about the ambiguity over the demand rate, a most vital parameter of an inventory process. Here at the beginning the ambiguity over demand rate is high because, the people will usually take much time (no matter what offers have been declared or how attractive the getup of the system be) to accept and adopt the process. If the cycle time ends at the "fully adopted" time period then the cost becomes high. The basic insight in the public opinion is that 'the system is less reliable' as because the DM is hesitating to run the process more time. This feeling must affect directly to the customers' satisfactions as well as on demand rate. However as the cycle time becomes more the customers are began to satisfy. A saturation on adoptability and reliability reaches. So the ambiguities have been removed from the process and a grand paradigm shift on progress (financial development, cost minimization, achievement of large customer) of that system has been viewed. Since the cloud of uncertainties removes over time from the process so such uncertainties we have named as 'cloudy fuzzy'. Hence the concept is noble.

7   Numerical Example-1

Let us consider $c = 10, h = 2.5, k = 155, d = 20, \langle d_1, d_2, d_3 \rangle = \langle 16, 20, 22 \rangle$, $\sigma = .15, \rho = .12, \varepsilon = .05$ in the Eqs. (11), (16) and (19) then we get the following results.

Table 1. Optimal solution of the classical EOQ model

Model	T (days)	q^*	z^* (\$)	$d_f = \frac{U_b - L_b}{2m}$	$CI = \frac{Log(1+T)}{T}$
Crisp	2.49	49.80	324.50	...	...
Fuzzy	2.52	49.17	317.93	0.141	...
Cloudy fuzzy	7.00	70.18	397.50	...	0.129

Table 2. Objective values under several cycle times

Cycle time	Crisp model		Fuzzy model		Cloudy fuzzy model	
T (days)	q^*	z^*	q^*	z^*	q^*	z^*
2	40	327.50	39.00	321.25	20.07	511.80
3	60	326.67	58.50	319.79	30.08	449.84
4	80	338.75	78.00	331.25	40.09	420.52
5	100	356.00	97.50	347.88	50.10	405.92
6	120	375.83	117.00	367.08	60.10	399.29
7	140	397.14	136.50	387.77	70.10	397.50
8	160	419.38	156.00	409.38	80.10	398.88
9	180	442.22	175.00	431.60	90.10	402.46
10	200	465.50	195.00	454.25	100.10	407.63

8   Discussion on Tables 1 and 2

From Table 1 we see that, the total average cost ($317.93 only) is minimum with the order quantity 49.17 units and the replenishment time (cycle time) 2.52 days only whenever a general fuzzy solution is considered. But the cloudy fuzzy solution is giving the inventory cost something more values ($397.50) with respect to other solutions. Again Table 2 shows, at cycle time 3 days the crisp and fuzzy objective giving better solution but among them fuzzy solution is better. Moreover we see when the cycle time reaches 7 days the crisp as well as cloudy fuzzy objective behaves almost same but the order quantity is minimum due to cloudy fuzzy (70.11 units only) where the fuzzy objective gets something less value throughout.

Also the degree of fuzziness (d_f) under general fuzzy numbers is 0.140 (see Appendix) for all the time and the cloudy fuzzy index at the optimum cycle time is 0.129 $(<d_f)$.

However we may take cloudy fuzzy solution as a final decision whenever (situation basis) it is impossible to replenish the items each after 3 days exclusively. By this study we may draw the following merits:

1. **Merits of Fuzzy Solutions:** It can be accepted if the items are easy to replenish and a huge number of items are available in the market as per instant order, it can be applicable in the market situating at the geographically plane region provinces.
2. **Merits of Cloudy Fuzzy Solutions:** It can be chosen when the items are not easy to replenish due to some transportation problem as well as unavailability in a huge lot in the market as per instant order, it can be applicable in the market situating at the geographically hilly region provinces.

9   Sensitivity Analysis

Let us take the sensitivity of the parameters $c, h, d, k, \varepsilon, \rho$ and σ from $(-50\%$ to $+50\%)$ in Table 3.

Table 3. Sensitivity analysis of the cloudy fuzzy objective

Parameters	% change	T^* (days)	q^*	z_*	$\frac{z_* - z^*}{z^*} 100\%$
c	+50%	7.00	70.17	497.72	53.37
	+30%	7.00	70.14	457.63	41.02
	−30%	7.00	70.07	337.37	3.97
	−50%	7.00	70.05	297.28	−8.38
h	+50%	5.54	55.53	436.28	34.44
	+30%	6.02	60.30	421.83	29.98
	−30%	8.58	85.93	368.52	13.55
	−50%	10.39	103.98	344.97	6.31
d	+50%	5.55	83.35	536.54	63.34
	+30%	6.02	78.43	481.97	48.53
	−30%	8.58	60.12	308.40	−4.96
	−50%	10.38	51.94	244.80	−24.56
k	+50%	...	...	...	...
	+30%	...	...	...	...
	−30%	5.71	57.15	362.03	11.56
	−50%	4.70	47.06	334.34	3.03
ε	+50%	6.57	65.62	388.24	19.64
	+30%	6.73	67.37	391.57	20.66
	−30%	7.35	73.62	405.20	24.87
	−50%	7.66	76.74	412.15	27.01
ρ	+50%	...	...	...	...
	+30%	7.00	69.86	396.81	22.28
	−30%	7.00	70.36	398.19	22.71
	−50%	7.00	70.53	398.65	22.86
σ	+50%	7.00	70.63	398.94	22.93
	+30%	7.00	70.42	398.37	22.76
	−30%	7.00	60.79	396.63	22.22
	−50%	7.00	69.59	396.06	22.04

10   Discussion on Sensitivity Analysis Table 3

From Table 3 we see that, the purchasing cost, the holding cost and the demand rate are highly sensitive parameters. When c and h are increased by $+50\%$ then the objective values are increased by +53.37% and +63.34% respectively. In other changes they are sensitive on average. Whenever the demand rate decreases to −50% the objective value reaches at −24.56% only giving the minimum cost among all the parametric changes in the whole table. When the ordering cost changes +50% and +30% and the parameter ρ changes +50% then the objective function gives no feasible solution. At −30% and −50% changes of the ordering cost k, a negligible sensitivity occurs for the objective itself. In case of model convergence parameter ε and fuzzy denoting parameters (ρ, σ), for all changes the objective function behaves average sensitivity as well.

11 Graphical Illustrations of the Model

Here we draw the figures of the proposed model under three different cases.

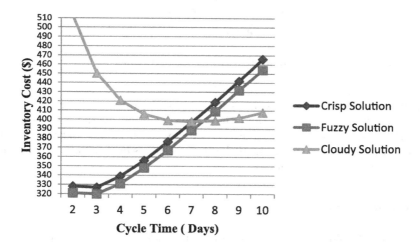

Fig. 4. Total average cost vs. cycle time

12 Discussion on Figs. 4 and 5

From the Fig. 4 we see that, when the fuzzy and crisp objective functions follow an exponential path but the cloudy fuzzy follows a hyperbolic path and they meet at some points. In cloudy fuzzy objective, at the very beginning its value was high (due to high ambiguity) and it began to decrease (due to disappearance of the clouds) with cycle time and finally gets the minimum value at 7 days cycle time only. However, the fuzzy objective path meets below the cloudy objective path, it indicates that fuzzy solution giving better solution (due to fuzzy constancy) for the same cycle time. The crisp objective path passes above side by side with the fuzzy path without intersecting each other. This shows that they are

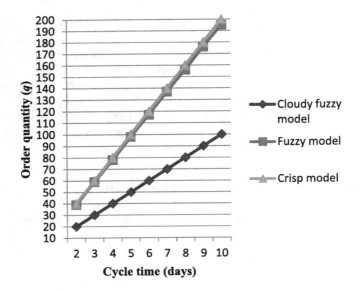

Fig. 5. Order quantity vs. cycle time

independent. Also the Fig. 5 explores that, the order quantities for crisp as well as fuzzy model are very high (and they are very often overlapping over cycle time) with respect to that of cloudy fuzzy model. In cloudy fuzzy model the order quantity is restricted within 10 daysts for 10 days cycle time but it becomes doubled for the other cases also.

13   Conclusion

Here we have developed a classical EOQ model under cloudy fuzzy enviroment. In our work, the fuzzification by cloudy fuzzy set theory has been discussed elaborately. The proposed method is simple and easy for computation. We have seen some cases the crisp as well as general fuzzy solution is better but they do not obey the reality. Also, we have observed the cost function of cloudy fuzzy cuts those crisp cost and the fuzzy cost near the same points. So the true solution will be there (no matter whether any one of these curves getting minimum values beyond these intersecting points or not). The basic drawback of the general fuzzy model is that it keeps the same uncertainties in all the time which is an unrealistic assumption. Thus DM's view point the following decision can be achieved:

1. Choose a solution so that replenishment time obeys the practice.
2. Choose a solution where the uncertainty/ambiguity is less.
3. Choose the final solution where the curves for fuzzy and cloudy fuzzy meet at a point.

Finally, "A birth follows a death"-so nothing is immortal, the concept of infinite time is vague and hence time is finite. Therefore, solution must come within cloudy fuzzy environment. Further, we are warning that, in no cases total cloud can be removed in practice. This is the central focus of attraction of this article. In future, we can apply the cloudy fuzzy in more complex inventory model, stochastic model as well as in supply chain models.

Appendix

We have the demand rate as fuzzy number $\langle d_1, d_2, d_3 \rangle = \langle 16, 20, 22 \rangle$. The lower and upper bounds are $L_b = 16$ and $U_b = 22$ respectively. The mean is 19.333, the median is 20, so the mode (m) is $3 \times median - 2 \times mean = 60 - 38.666 = 21.334$. Therefore, $d_f = \frac{1}{2m}(U_b - L_b) = \frac{6}{42.668} = 0.141$ and $CI = \frac{Log(1+T)}{T} = \frac{Log(1+7)}{7} = 0.129$.

References

Allahviranloo, T., Saneifard, R.: Defuzzification method for ranking fuzzy numbers based on center of gravity. Iran. J. Fuzzy Syst. 9(6), 57–67 (2012)

Báez-Sáncheza, A.D., Morettib, A.C., Rojas-Medarc, M.A.: On polygonal fuzzy sets and numbers. Fuzzy Sets Syst. **209**, 54–65 (2012)

Ban, A.I., Coroianu, L.: Existence, uniqueness and continuity of trapezoidal approximations of fuzzy numbers under a general condition. Fuzzy Sets Syst. **257**, 3–22 (2014)

Bellman, R.E., Zadeh, L.A.: Decision making in a fuzzy environment. Manag. Sci. **17**, B141–B164 (1970)

Buckley, J.J., Chanas, S.: A fast method of ranking alternatives using fuzzy numbers. Fuzzy Set Syst. **30**, 337–339 (1989)

Cheng, C.H.: A new approach for ranking fuzzy numbers by distance method. Fuzzy Sets Syst. **95**, 307–317 (1998)

Chu, T., Tsao, C.: Ranking fuzzy numbers with an area between the centroid point and original point. Comput. Math. Appl. **43**, 111–117 (2002)

De, S.K.: Triangular dense fuzzy lock sets. Soft Comput. 1–12 (2017). https://doi.org/10.1007/s00500-017-2726-0

De, S.K., Beg, I.: Triangular dense fuzzy sets and new defuzzification methods. Int. J. Intell. Fuzzy Syst. **31**, 469–477 (2016)

De, S.K., Kundu, P.K., Goswami, A.: Economic ordering policy of deteriorated items with shortage and fuzzy cost co-efficients for vendor and buyer. Int. J. Fuzzy Syst. Rough Syst. **1**(2), 69–76 (2008)

De, S.K., Kundu, P.K., Goswami, A.: An economic production quantity inventory model involving fuzzy demand rate and fuzzy deterioration rate. J. Appl. Math. Comput. **12**(1), 251–260 (2003)

De, S.K.: EOQ model with natural idle time and wrongly measured demand rate. Int. J. Inventory Control Manag. **3**(1–2), 329–354 (2013)

De, S.K., Mahata, G.: Decision of a fuzzy inventory with fuzzy backorder model under cloudy fuzzy demand rate. Int. J. Appl. Comput. Math. **3**(3), 2593–2609 (2017). https://doi.org/10.1007/s40819-016-0258-4

De, S.K., Sana, S.S.: Fuzzy order quantity inventory model with fuzzy shortage quantity and fuzzy promotional index. Econ. Model. **31**, 351–358 (2013)

De, S.K., Sana, S.S.: An EOQ model with backlogging. Int. J. Manag. Sci. Eng. Manag. **11**(3), 143–154 (2015). https://doi.org/10.1080/17509653.2014.995736

Deng, H.: Comparing and ranking fuzzy numbers using ideal solutions. Appl. Math. Model. **38**, 1638–1646 (2014)

Dubois, D., Prade, H.: Operations on fuzzy numbers. Int. J. Syst. Sci. **30**, 613–626 (1978)

Ezatti, R., Saneifard, R.: A new approach for ranking of fuzzy numbers with continuous weighted quasi-arithmetic means. Math. Sci. **4**, 143–158 (2010)

Ezzati, R., Allahviranloo, T., Khezerloo, S., Khezerloo, M.: An approach for ranking of fuzzy numbers. Expert Syst. Appl. **39**, 690–695 (2012)

Ghare, P.M., Schrader, G.P.: A model for exponentially decaying inventory. J. Ind. Eng. **14**, 238–243 (1963)

Goetschel Jr., R., Voxman, W.: Eigen Fuzzy number sets. Fuzzy Sets Syst. **16**, 75–85 (1985)

Hadley, G., Whitin, T.M.: Analysis of Inventory Systems. Prentice Hall, Englewood Cliffs (1963)

Hajjari, T., Abbasbandy, S.: A note on "the revised method of ranking LR fuzzy number based on deviation degree". Expert Syst. Appl. **39**, 13491–13492 (2011)

Haneveld, W.K., Teunter, R.H.: Effects of discounting and demand rate variability on the EOQ. Int. J. Prod. Econ. **54**, 173–192 (1998)

Hanssmann, F.: Operations Research in Production and Inventory Control. Wiley, New York (1962)

Hariga, M.A.: Optimal EOQ models for deteriorating items with time-varying demand. J. Oper. Res. Soc. **47**(10), 1228–1246 (1996)

Harris, F.: Operations and Cost. Factory Management Series. Chicago (1915)

Kao, C., Hsu, W.K.: Lot size reorder point inventory model with fuzzy demands. Comput. Math. Appl. **43**, 1291–1302 (2002)

Karmakar, S., De, S.K., Goswami, A.: A pollution sensitive dense fuzzy economic production quantity model with cycle time dependent production rate. Int. J. Clean. Prod. **154**, 139–150 (2017)

Kaufmann, A., Gupta, M.M.: Introduction to Fuzzy Arithmetic Theory and Applications. Van Nostrand Reinhold, New York (1992)

Kumar, A., Singh, P., Kaur, P., Kaur, A.: A new approach for ranking of L-R type generalized fuzzy numbers. Expert Syst. Appl. **38**, 10906–10910 (2011)

Kumar, R.S., De, S.K., Goswami, A.: Fuzzy EOQ models with ramp type demand rate, partial backlogging and time dependent deterioration rate. Int. J. Math. Oper. Res. **4**, 473–502 (2012)

Lee, H.M., Yao, J.S.: Economic order quantity in fuzzy sense for inventory without back order model. Fuzzy Sets Syst. **105**, 13–31 (1999)

Ramli, N., Mohamad, D.: A comparative analysis of centroid methods in ranking fuzzy numbers. Eur. J. Sci. Res. **28**(3), 492–501 (2009)

Rezvani, S.: Ranking generalized exponential trapezoidal fuzzy numbers based on variance. Appl. Math. Comput. **262**, 191–198 (2015)

Wang, Z.X., Liu, Y.J., Fan, Z.P., Feng, B.: Ranking L-R fuzzy number based on deviation degree. Inf. Sci. **179**, 2070–2077 (2009)

Xu, P., Su, X., Wu, J., Sun, X., Zhang, Y., Deng, Y.: A note on ranking generalized fuzzy numbers. Expert Syst. Appl. **39**, 6454–6457 (2012)

Yager, R.R.: A procedure for ordering fuzzy subsets of the unit interval. Inf. Sci. **24**, 143–161 (1981)

Yu, V.F., Chi, H.T.X., Dat, L.Q., Phuc, P.N.K., Shen, C.W.: Ranking generalized fuzzy numbers in fuzzy decision making based on the left and right transfer coefficients and areas. Appl. Math. Model. **37**, 8106–8117 (2013)

Zadeh, L.A.: Fuzzy sets. Inf. Control **8**, 338–356 (1965)

Zhang, F., Ignatius, J., Lim, C.P., Zhao, Y.: A new method for ranking fuzzy numbers and its application to group decision making. Appl. Math. Model. **38**, 1563–1582 (2014)

A Delayed Non-autonomous Predator-Prey Model with Crowley-Martin Functional Response

Jai Prakash Tripathi[1] and Vandana Tiwari[2(✉)]

[1] Department of Mathematics, Central University of Rajasthan,
Bandarsindri, Ajmer 305817, Rajasthan, India
jtripathi85@gmail.com
[2] Department of Mathematical Sciences, Indian Institute of Technology (BHU),
Varanasi 221005, UP, India
vandanpp@gmail.com

Abstract. In this work, we propose a delayed non-autonomous prey-predator system with Crowley-Martin functional response (CMFR). Mutual interference by predators at high prey density differentiate between Beddington-DeAngelis functional response and CMFR. We discuss the permanence, extinction, stability, existence and uniqueness of a globally attractive almost periodic solution (APS). In addition to effect of Crowley-Martin parameter, we also show that the intrinsic growth rate leaves positive effect on the permanence of the considered model system. Some numerical examples are also presented to support obtained analytical results.

Keywords: Almost periodic solution · Delay
Crowley-Martin functional response · Global stability · Permanence
Periodic solution

1 Introduction

Understanding the relationship between predator and prey is a central goal in ecology [1,2,5,6,11,14,16]. Almost periodicity and periodicity play significant role in several branches of science and engineering. There are several reasons due to which environment varies: like, mating habits, food supplies, harvesting and therefore several vital rates of populations such as death rates, birth rates and several others, alter in time [4,9]. When temporal inhomogeneity is incorporated into a model system, the system becomes a non-autonomous system (a system with time dependent coefficients). Main objective of the present study is the analysis of a non-autonomous model with delay [17,22] and CMFR [3,7,8].

A non-autonomous (systems with time dependent periodic or almost periodic parameters) model system indicates the presence of temporal irregularity in the environment. In the recent years, almost periodicity [10,27] in ecological modelling has been extensively studied by several authors [17,20,21,23–26]. Here we

© Springer Nature Singapore Pte Ltd. 2018
D. Ghosh et al. (Eds.): ICMC 2018, CCIS 834, pp. 164–173, 2018.
https://doi.org/10.1007/978-981-13-0023-3_16

investigate the following prey-predator model system with mutual interference [15, 18]:

$$\frac{dx(t)}{dt} = x(t)\big(a_1(t) - b_1(t)x(t - \tau(t))\big) - \frac{r_1(t)x(t)}{a(t) + b(t)x(t) + c(t)y(t) + e(t)x(t)y(t)}y^m(t),$$

(1)

$$\frac{dy(t)}{dt} = y(t)\big(-a_2(t) - b_2(t)y(t)\big) + \frac{r_2(t)x(t)}{a(t) + b(t)x(t) + c(t)y(t) + e(t)x(t)y(t)}y^m(t),$$

with the initial conditions

$$x(\theta) = \phi(\theta), \quad \theta \in [-\tau, 0], \quad \phi(\theta) \in C([-\tau, 0], R_+),$$
$$y(0) > 0, \, y(\theta) \geq 0, \quad \theta \in [-\tau, 0],$$

Here $x(t)$ is the size of prey population and $y(t)$ is the size of predator population; $0 < m < 1$; $\tau(t)$ is nonnegative and continuously differentiable APF on R, and $\tau = \max_{t \in R}\{\tau(t)\}$, $\min_{t \in R}\{1 - \dot{\tau}(t)\} > 0$. The coefficients $a_i(t), b_i(t), r_i(t), a(t), b(t), c(t), e(t)$ $(i = 1, 2)$ are continuous positive APFs. For the ecological meaning of these coefficients, one can refer [19].

2   Boundedness and Permanence

Note that the existence of solutions of the model system (1) can be shown for all $t \geq 0$. In this section, we establish the positive invariance, boundedness, permanence and global asymptotic stability. Let $R_+^2 = \{(x, y) \in R^2 | x \geq 0, y \geq 0\}$. The positive invariance of the positive half cone can be easily proved.

Now we state a theorem that will help us to show the boundedness and permanence of the model system (1).

Theorem 1. *If* $a_1^l > \dfrac{r_1^u M_2^{m-1}}{e^l}$, *then the set defined by*

$$\kappa := \big\{(x, y) \in R^2 | m_1 \leq x \leq M_1, m_2 \leq y \leq M_2\big\}$$

is positively invariant with respect to the system (1), where

$$M_1 := \frac{a_1^u}{b_1^l e^{-a_1^u \tau}}, \qquad m_1 := \left(\frac{a_1^l}{b_1^u} - \frac{r_1^u M_2^{m-1}}{e^l b_1^u}\right) \exp\left\{\left(a_1^l - b_1^u M_1 - \frac{r_1^u M_2^{m-1}}{e^l}\right)\tau\right\},$$

$$M_2 := \left(\frac{3r_2^u}{2a_2^l}\right)^{\frac{1}{1-m}}, \qquad m_2 := \left[\frac{r_2^l m_1}{2(a^u + b^u M_1 + c^u M_2 + e^u M_1 M_2)(a_2^u + b_2 M_2)}\right]^{\frac{1}{1-m}}.$$

Theorem 2. *The system (1) becomes permanent and the set κ defined in Theorem 1 is an ultimately bounded region (1) if* $a_1^l > \dfrac{r_1^u M_2^{m-1}}{e^l}$ *hold.*

Proof. For the proof of the above theorems (Theorems 1 and 2), one can refer, [17, 19].

3 Global Attractivity and Almost Periodicity

For relevant definitions and results related to almost periodic functions, we refer to [24, 27].

Theorem 3. *Let*

$$P_1(t) = b_1(t) - \frac{r_1(t)M_2^m(b(t) + e(t)\eta M_2 c(t))}{\zeta^2(t, m_1, m_2)} - \int_t^{\xi^{-1}(t)} b_1(u)du\Big(a_1(t) + b_1(t)M_1 +$$

$$\frac{r_1(s)M_2^m}{\zeta(t, m_1, m_2)}\Big) - \frac{M_1 b_1(\xi^{-1}(t))}{1 - \dot{\tau}(\xi^{-1}(t))} \int_{\xi^{-1}(t)}^{\xi^{-1}(\xi^{-1}(t))} b_1(u)du - \frac{c(t)r_2(t)M_2^m}{\eta\zeta(t, m_1, m_2)},$$

$$P_2(t) = -\frac{\eta^m r_1(t)}{\zeta(t, m_1, m_2)} - \frac{\eta(c(t) + e(t)M_1)}{\zeta^2(t, m_1, m_2)} - \Big(\frac{M_1^2\eta^m r_1(t)\Big(M_1 b(t) + e(t)\frac{M_2}{\eta^{m-1}}\Big) + r_1(t)M_1^3 M_2^m)}{(b(t)m_1(s) + \eta e(t)m_1 m_2)^2}$$

$$\int_t^{\xi^{-1}(t)} b_1(u)du + \eta b_2(t) - r_2(t)M_2^{m-1}M_1\Big\{\frac{c(t) + e(t)M_1}{\zeta(t, m_1, m_2)}\Big\}. \text{ If all the conditions}$$

of Theorems 1 and 2 and $\liminf\limits_{t\to\infty} P_1(t) > 0$, $\liminf\limits_{t\to\infty} P_2(t) > 0$ *hold, then for any two positive solutions* $X(t) = (x_1(t), y_1(t))$ *and* $Y(t) = (x_2(t), y_2(t))$ *of system (1), we obtain*

$$\lim_{t\to\infty} |X(t) - Y(t)| = 0$$

Proof. If $(x(t), y(t))$ be a positive solution of the model system (1) then we have $m_1 < x(t) < M_1$, $m_2 < y(t) < M_2$. We define $\eta = \min\{y(t)\}$, then $\eta > m_2 > 0$. Let us consider any two positive solution $X(t) = (x_1(t), y_1(t))$ and $Y(t) = (x_2(t), y_2(t))$ of (1) and let

$$x_i(t) = x(t), y_i(t) = \frac{y(t)}{\eta} \ \forall \ i = 1, 2. \tag{2}$$

Furthermore, define $\zeta(t, x(t), y(t)) = a(t) + b(t)x(t) + c(t)y(t) + e(t)x(t)y(t)$. Define

$$V_{11}(t) = |\ln x_1(t) - \ln x_2(t)|.$$

Calculating the upper right derivative of $V_{11}(t)$ along the solution of (1) it follows that

$$D^+ V_{11}(t) = sgn(x_1(t) - x_2(t))\Big(\frac{\dot{x}_1(t)}{x_1(t)} - \frac{\dot{x}_2(t)}{x_2(t)}\Big)$$

$$= sgn(x_1(t) - x_2(t))\Big[-b_1(t)x_1(t - \tau(t)) - x_2(t - \tau))$$

$$-\eta^m r_1(t)\Big(\frac{y_1^m(t)}{\zeta(t, x_1(t), y_1(t))} - \frac{y_2^m(t)}{\zeta(t, x_2(t), y_2(t))}\Big)\Big]$$

$$\leq -b_1(t)|x_1(t) - x_2(t)| + b_1(t)\Big|\int_{t-\tau(t)}^t (\dot{x}_1(s) - \dot{x}_2(s))ds\Big| + \frac{\eta^m r_1(t)|y_1^m(t) - y_2^m(t)|}{\zeta(t, x_1(t), y_1(t))}$$

$$+ \frac{\eta^m r_1(t)\Big[(b(t) + e(t)\eta c(t)y_2(t))|x_1(t) - x_2(t)| + \eta(c(t) + ex_1(t))|y_1(t) - y_2(t)|\Big]}{\zeta(t, x_1(t), y_1(t))\zeta(t, x_2(t), y_2(t))}.$$

Putting the suitable values of $\dot{x}_1(t)$ and $\dot{x}_2(t)$, we have

$$D^+V_{11}(t) \leq -b_1(t)|x_1(t) - x_2(t)| + b_1(t)\left|\int_{t-\tau(t)}^t \left\{x_1(s)\left(a_1(s) - b_1(s)x_1(s - \tau(s))\right.\right.\right.$$

$$-\frac{\eta^m r_1(s)y_1^m(s)}{\zeta(s, x_1(s), \eta y_1(s))}\right) - x_2(s)\left(a_1(s) - b_1(s)x_2(s - \tau(s))\right.$$

$$\left.\left.-\frac{\eta^m r_1(s)y_2^m(s)}{\zeta(s, x_2(s), \eta y_2(s))}\right)\right\}ds\right| + \frac{\eta^m r_1(t)|y_1^m(t) - y_2^m(t)|}{\zeta(t, x_1(t), y_1(t))}$$

$$+\frac{\eta^m r_1(t)\left[(b(t) + e(t)\eta c(t)y_2(t))|x_1(t) - x_2(t)| + \eta(c(t) + ex_1(t))|y_1(t) - y_2(t)|\right]}{\zeta(t, x_1(t), y_1(t))\zeta(t, x_2(t), y_2(t))}$$

$$\leq -b_1(t)|x_1(t) - x_2(t)| + b_1(t)\int_{t-\tau(t)}^t \left\{x_1(s)\left(a_1(s) + b_1(s)x_1(s - \tau(s))\right.\right.$$

$$+\frac{\eta^m r_1(s)y_1^m(s)}{\zeta(s, x_1(s), \eta y_1(s))}\right)|x_1(s) - x_2(s)| + x_2(s)b_1(s)|x_1(s - \tau(s)) - x_2(s - \tau(s))|$$

$$+\frac{x_1(s)x_2(s)\eta^m r_1(s)(x_2(s)b(s) + \eta e(s)y_2(s)|y_2^m(s) - y_1^m(s)|)}{(b(s)x_1(s) + \eta e(s)x_1(s)y_1(s))(b(s)x_2(s) + \eta e(s)x_2(s)y_2(s))}$$

$$\left.+\frac{x_2(s)y_2^m|y_2(s) - y_1(s)|)}{(b(s)x_1(s) + \eta e(s)x_1(s)y_1(s))(b(s)x_2(s) + \eta e(s)x_2(s)y_2(s))}\right\}ds$$

$$+\frac{\eta^m r_1(t)|y_1^m(t) - y_2^m(t)|}{\zeta(t, x_1(t), y_1(t))}$$

$$+\frac{\eta^m r_1(t)\left[(b(t) + e(t)\eta c(t)y_2(t))|x_1(t) - x_2(t)| + \eta(c(t) + ex_1(t))|y_1(t) - y_2(t)|\right]}{\zeta(t, x_1(t), y_1(t))\zeta(t, x_2(t), y_2(t))}.$$

$$(3)$$

Now from (2) and (3), we have for $t \geq T + \tau$

$$D^+V_{11}(t) \leq -b_1(t)|x_1(t) - x_2(t)| + b_1(t)\int_{t-\tau(t)}^t \zeta(s)ds + \frac{\eta^m r_1(t)|y_1^m(t) - y_2^m(t)|}{\zeta(t, m_1, m_2)}$$

$$+\frac{r_1(t)M_2^m\left[(b(t) + e(t)\eta M_2 c(t))|x_2(t) - x_1(t)| + \eta(c(t) + e(t)M_1)|y_1(t) - y_2(t)|\right]}{\zeta^2(t, m_1, m_2)},$$

$$(4)$$

where $\zeta(s) = \left(a_1(s) + b_1(s)M_1 + \frac{r_1(s)M_2^m}{\zeta(t, m_1, m_2)}\right)|x_1(s) - x_2(s)| + M_1 b_1(s)|x_1(s - \tau(s)) - x_2(s - \tau(s))|$

$$+\frac{M_1^2\eta^m r_1(s)\left(M_1 b(s) + e(s)\frac{M_2}{\eta^{m-1}}|y_2^m(s) - y_1^m(s)|\right) + r_1(s)M_1^3 M_2^m|y_2(s) - y_1(s)|}{(b(s)m_1(s) + \eta e(s)m_1 m_2)^2}$$

Define

$$V_{12} = \int_t^{\xi^{-1}(t)}\int_{\xi(t)}^t b_1(u)\zeta(s)dsdu, \quad V_{13} = \int_{\xi(t)}^t\int_{\xi^{-1}(v)}^{\xi^{-1}(\xi^{-1}(v))} \frac{b_1(u)b_1(\xi^{-1}(v))}{1 - \dot{\tau}(\xi^{-1}(v))}|x_1(v) - x_2(v)|dudv.$$

$$(5)$$

Thus we define first component of Lyapunov function by

$$V_1(t) = V_{11}(t) + V_{12}(t).$$

$$(6)$$

From (4) and (5) and $|y_1^m(t) - y_2^m(t)| \leq |y_1(t) - y_2(t)|$, we have

$$D^+V_1(t) \leq -b_1(t)|x_1(t) - x_2(t)| + \frac{\eta^m r_1(t)|y_1(t) - y_2(t)|}{\zeta(t, m_1, m_2)}$$

$$+ \frac{r_1(t)M_2^m\left[(b(t) + e(t)\eta M_2 c(t))|x_2(t) - x_1(t)| + \eta(c(t) + e(t)M_1)|y_1(t) - y_2(t)|\right]}{\zeta^2(t, m_1, m_2)}$$

$$+ \int_t^{\xi^{-1}(t)} b_1(u)du|x_1(t) - x_2(t)|\left(a_1(t) + b_1(t)M_1 + \frac{r_1(s)M_2^m}{\zeta(t, m_1, m_2)}\right)$$

$$\left(\frac{M_1^2\eta^m r_1(t)\left(M_1 b(t) + e(t)\frac{M_2}{\eta^{m-1}}|y_1(t) - y_2(t)|\right) + r_1(t)M_1^3 M_2^m|y_1(t) - y_2(t)|}{(b(t)m_1(s) + \eta e(t)m_1 m_2)^2}\right)$$

$$\int_t^{\xi^{-1}(t)} b_1(u)du + \frac{M_1 b_1(\xi^{-1}(t))}{1 - \dot{\tau}(\xi^{-1}(t))} \int_{\xi^{-1}(t)}^{\xi^{-1}(\xi^{-1}(t))} b_1(u)du|x_1(t) - x_2(t)|.$$

Again, define

$V_2(t) = |\ln y_1(t) - \ln y_2(t)|$, which gives

$$D^+V_2(t) = sgn(y_1(t) - y_2(t))\left(\frac{\dot{y}_1(t)}{y_1(t)} - \frac{\dot{y}_2(t)}{y_2(t)}\right)$$

$$= sgn(y_1(t) - y_2(t))\left[-\eta b_2(t)(y_1(t) - y_2(t)) + \eta^{m-1}r_2(t)\left(\frac{x_1(t)y_1^{m-1}(t)}{\zeta(t, x_1(t), y_1(t))}\right.\right.$$

$$\left.\left. - \frac{x_2(t)y_2^{m-1}(t)}{\zeta(t, x_2(t), y_2(t))}\right)\right]$$

$$\leq -\eta b_2(t)|y_1(t) - y_2(t)| + r_2(t)M_2^{m-1}M_1\left(y_1(t) - y_2(t)\right)$$

$$\left\{\frac{c(t)x_1(t)(y_1(t) - y_2(t)) + y_2(t)c(t)(x_1(t) - x_2(t)) + e(t)x_1(t)x_2(t)(y_1(t) - y_2(t))}{\zeta(t, m_1, m_2)^2}\right\}$$

$$\leq -\eta b_2(t)|y_1(t) - y_2(t))| + r_2(t)M_2^{m-1}M_1\left\{\frac{c(t) + e(t)M_1}{\zeta(t, m_1, m_2)}\right\}|y_1(t) - y_2(t)|$$

$$+ \frac{c(t)r_2(t)M_2^m}{\eta\zeta(t, m_1, m_2)}\left|x_1(t) - x_2(t)\right|.$$

Define

$V(t) = V_1(t) + V_2(t)$ then, we obtain $D^+V(t) = -P_1(t)|x_1(t) - x_2(t)| - P_2(t)|y_1(t) - y_2(t)|$.

Define $\rho(t) = \min\{P_1(t), P_2(t)\}$ where $P_1(t)$ and $P_2(t)$ are defined in the Theorem 3. For $t \geq T + \tau$, we have

$$D^+V(t) = -\rho(t)[|x_1(t) - x_2(t)| + |y_1(t) - y_2(t)|]. \tag{7}$$

Thus under the condition of the Theorem 3, there must exist a constant ω and $T_0 > T + \tau$ such that

$$D^+V(t) = -\omega[|x_1(t) - x_2(t)| + |y_1(t) - y_2(t)|]. \tag{8}$$

Integrating Eq. (8) from T_0 to t, we have

$$V(t) + \omega \int_{T_0}^t \left[|x_1(s) - x_2(s)| + |y_1(s) - y_2(s)|\right]ds < V(T_0) < +\infty. \tag{9}$$

Thus, we find

$$\limsup_{t \to \infty} \int_{T_0}^t \Big[|x_1(s) - x_2(s)| + |y_1(s) - y_2(s)| \Big] ds < \frac{V(T)}{\omega} < +\infty. \qquad (10)$$

Finally we conclude that

$$\lim_{t \to \infty} |x_1(t) - x_2(t)| = 0, \ \lim_{t \to \infty} |y_1(t) - y_2(t)| = 0.$$

Thus the existence of a globally attractive solution of (1) is established. Furthermore, the uniqueness of an APS can easily be proved [19].

4   Numerical Example

In the support of analytical results established in the previous sections, we consider and numerically simulate the following examples:

Example 1. Let $a_1(t) = 2 + 0.1 \sin \sqrt{3}t, b_1(t) = 2 - 0.1 \sin t, a_2(t) = 0.04 + 0.01 \sin \sqrt{5}t, b_2(t) = 0.05 - 0.01 \sin t, r_1(t) = 0.05 - 0.01 \sin t, r_2(t) = 0.02 - 0.01 \sin t, a = 1, b = 1.5, c = 0.001, e = 0.01, m = 0.2$ then the system (1) becomes

$$\frac{dx(t)}{dt} = x(t)\big(2 + 0.1 \sin \sqrt{7}t - 2 - 0.1 \sin t x(t - 0.01)\big)$$

$$- \frac{0.05 - 0.01 \sin t x(t)}{1 + 1.5x(t) + 0.001y(t) + 0.01x(t)y(t)} y^{0.2}(t), \qquad (11)$$

$$\frac{dy(t)}{dt} = y(t)\big(-0.02 + 0.01 \sin \sqrt{3}t - 0.05 - 0.01 \sin t y(t)\big)$$

$$+ \frac{0.02 - 0.01 \sin t x(t)}{1 + 1.5x(t) + 0.001y(t) + 0.01x(t)y(t)} y^{0.2}(t),$$

with the initial conditions $\phi(\theta) = 0.6, y(\theta) = 0.7$. Here the bounds of the coefficients are: $a_1^l = 1.90, a_1^u = 2.1, b_1^l = 1.99, b_1^u = 2.1, a_2^l = 0.03, a_2^u = 0.05, b_2^l = 0.04, b_2^u = 0.06, r_1^l = 0.04, r_1^u = 0.06, r_2^l = 0.01, r_2^u = 0.03, c^l = c^u = 0.001, e^l = e^u = 0.1$. Furthermore, using these bounds, we obtain

$$M_2 = \left(\frac{3r_2^u}{2a_2^l}\right)^{\frac{1}{1-m}} = \left(\frac{3}{2}\right)^{\frac{5}{4}} = 1.66, \quad a_1^l = 1.90 > 0.40 = \frac{0.6 \times (1.66)^{-0.8}}{0.1} = \frac{r_1^u M_2^{m-1}}{e^l}.$$

Figure 1 shows the existence of unique globally attractive APS. The orbits for predator-prey are shown in the Fig. 2.

Fixing all the parametric values same as in the Example 1 except $a_1 = 0.99 + 0.1 \sin \sqrt{7}t$ and $e(t) = 0.01$, we see that parametric condition of Theorem 2 does

not hold as $a_1^l = 1 < 3.96 = \frac{0.06 \times (1.66)^{-0.8}}{0.01} = \frac{r_1^u M_2^{m-1}}{e^l}$. However the Fig. 3 ensures the permanence of the model system (1). This ensures that condition obtained in the Theorem 2 is only sufficient.

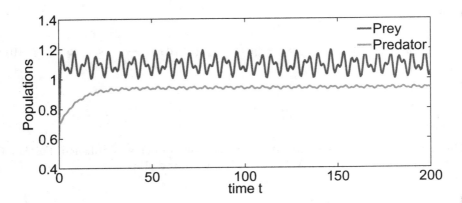

Fig. 1. Solution curves for the model system (11). The predator-y and prey-x are persistent.

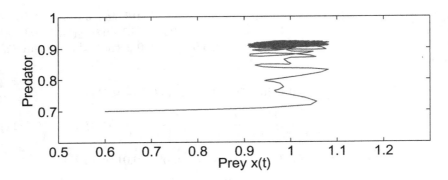

Fig. 2. Orbit of the predator-prey system (11)

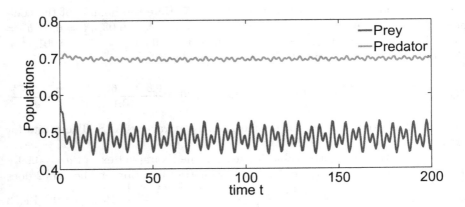

Fig. 3. The predator-y and prey-x are persistent.

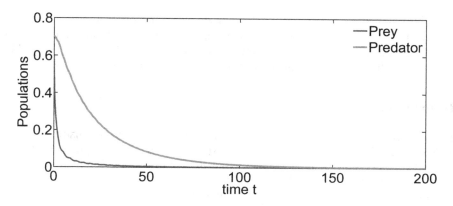

Fig. 4. The predator-y and prey-x are not persistent.

Moreover, if we consider $a_1 = 0.01$ and all other parametric values same as in Example 1, these parametric values again do not satisfy the parametric condition obtained in Theorem 2 as $a_1^l = 0.01 < 0.40 = \dfrac{r_1^u M_2^{m-1}}{e^l}$. The Fig. 4 shows that model system (1) is not permanent.

5 Concluding Remarks

Variability in environmental factors plays an important role in shaping the intrinsic population dynamics. Predator-prey model is one of the basic links among populations which determine population dynamics and trophic structure. In this work, environmental variability is captured in the model parameters with time dependent APFs, which makes the model non-autonomous. Here we proposed and studied a non-autonomous system with almost periodic coefficients and delay in competition among preys. We summarize the conclusion as under:

(i) The results (refer the Theorems 2 and 3) obtained for almost periodic model system (1) also hold for the corresponding periodic system. In other words, the oscillatory coexistence of all the population involved in the model system is ensured under the sufficient condition for persistence.

(ii) The sufficient condition obtained for permanence (see, Theorem 2) ensure that the permanence of the model system (1) depends on Crowley-Martin parameter e (mutual interference among predator at high prey density) and intrinsic growth rate a_1 of prey species. When the intrinsic growth rate a_1 takes value below a threshold level, both the species become extinct. (refer the Theorem 2 and Fig. 4).

(iii) It is interesting to note that in our illustrative examples, only prey growth rate and predator natural death rate are assumed to be APFs. However, it can be observed that consideration of other coefficients as APFs will give similar dynamics. The mutual interference leaves stabilizing effect.

The delayed system come back in a stable situation from delay induced instability for a threshold value of m. Existence and uniqueness of APS of the model system (1) is established using Bochner's definition of APF. The time evolution of prey and predator are almost periodic.

References

1. Lotka, A.: Elements of Mathematical Biology. Dover, New York (1956)
2. Tripathi, J.P., Meghwani, S.S., Thakur, M., Abbas, S.: A modified Leslie–Gower predator-prey interaction model and parameter identifiability. Commun. Nonlinear Sci. Numer. Simulat. **54**, 331–346 (2018)
3. Crowley, P.H., Martin, E.K.: Functional responses and interference within and between year classes of a dragonfly population. J. N. Am. Benthol. Soc. **8**, 211–221 (1989)
4. Fan, M., Kuang, Y.: Dynamics of a non-autonomous predator-prey system with Beddington-DeAngelis functional response. J. Math. Anal. Appl. **295**, 15–39 (2004)
5. Jana, D., Tripathi, J.P.: Impact of generalist type sexually reproductive top predator interference on the dynamics of a food chain model. Int. J. Dynam. Control. 80 (2015) https://doi.org/10.1007/s40435-016-0255-9
6. Cui, J., Takeuchi, Y.: Permanence, extinction and periodic solution of predator-prey system with Beddinton-DeAngelis functional response. J. Math. Anal. Appl. **317**, 464–474 (2006)
7. Sklaski, G.T., Gilliam, J.F.: Functional responses with predator interference: viable alternatives to the Holling type II model. Ecology **82**(11), 3083–3092 (2001)
8. Dong, Q., Ma, W., Sun, M.: The asymptotic behaviour of a chemostat model with Crowley Martin type functional response and time delays. J. Math. Chem. **51**, 1231–1248 (2013)
9. Chesson, P.: Understanding the role of environment variation in population and community dynamics. Theor. Popul. Biol. **64**, 253–254 (2003)
10. Bohr, H.: On the theory of almost periodic functions. Acta Math. **45**, 101–214 (1925)
11. Parshad, R.D., Basheer, A., Jana, D., Tripathi, J.P.: Do prey handling predators really matter: subtle effects of a Crowley-Martin functional response. Chaos, Solitons Fractals **103**, 410–421 (2017)
12. Bohr H.: Almost periodic functions, Chelsea (1947)
13. Lin, X., Chen, F.: Almost periodic solution for a Volterra model with mutual interference and Beddington-DeAngelis functional response. Appl. Math. Comput. **214**, 548–556 (2009)
14. Abrams, P.A., Ginzburg, L.R.: The nature of predation: prey dependent, ratio dependent or neither? Trends Ecol. Evol. **15**, 337–341 (2000)
15. Hassell, M.P.: Mutual interference between searching insect parasites. J. Anim. Ecol. **40**, 473–486 (1971)
16. Abbas, S., Tripathi, J.P., Neha, A.A.: Dynamical analysis of a model of social behaviour: criminal versus non-criminal. Chaos, Solitons Fractals **98**, 121–129 (2017)
17. Du, Z., Lv, Y.: Permanence and almost periodic solution of a Lotka–Volterra model with mutual interference and time delays. Appl. Math. Model. **37**, 1054–1068 (2013)

18. Hassell, M.P.: Density-dependence in single-species populations. J. Anim. Ecol. 283–295 (1975)
19. Tripathi, J.P., Abbas, S.: Almost periodicity of a modified Leslie-Gower Predator-Prey system with Crowley-Martin functional response. In: Agrawal, P., Mohapatra, R., Singh, U., Srivastava, H. (eds.) Mathematical Analysis and Its Applications, pp. 309–317. Springer, New Delhi (2015). https://doi.org/10.1007/978-81-322-2485-3_25
20. Guo, H., Chen, X.: Existence and global attractivity of positive periodic solution for a Volterra model with mutual interference and Beddington–DeAngelis functional response. Appl. Math. Comput. **217**(12), 5830–5837 (2011)
21. Abbas, S., Sen, M., Banerjee, M.: Almost periodic solution of a non-autonomous model of phytoplankton allelopathy. Nonlinear Dyn. **67**, 203–214 (2012)
22. Gopalasamy, K.: Stability and Oscillation in Delay Equation of Population Dynamics. Kluwer Academic Publishers, Dordrect (1992)
23. Wang, Q., Dai, B.X.: Almost periodic solution for n-species Lotka-Volterra competitive systems and feedback controls. Appl. Math. Comput. **200**(1), 133–146 (2008)
24. Fink, A.M.: Almost Periodic Differential Equations. Lecture Notes in Mathematics, vol. 377. Springer, Berlin (1974). https://doi.org/10.1007/BFb0070324
25. Tripathi, J.P.: Almost periodic solution and global attractivity for a density dependent predator-prey system with mutual interference and Crowley-Martin response function. Differ. Eqn. Dyn. Syst. (2016). https://doi.org/10.1007/s12591-016-0298-6
26. Tripathi, J.P., Abbas, S.: Global dynamics of autonomous and nonautonomous SI epidemic models with nonlinear incidence rate and feedback controls. Nonlinear Dyn. **86**(1), 337–351 (2016)
27. Besicovitch, A.S.: Almost Periodic Functions. Dover Publications Inc., New York (1954)

Cauchy Poisson Problem for Water with a Porous Bottom

Piyali Kundu[1($\boxtimes$)], Sudeshna Banerjea[1], and B. N. Mandal[2]

[1] Jadavpur University, Kolkata 700032, India
kundupiyali92@gmail.com, sudeshna.banerjea@yahoo.co.in
[2] Physics and Applied Mathematics Unit, Indian Statistical Institute,
203 B.T. Road, Kolkata 700108, India
bnm2006@rediffmail.com

Abstract. This paper is concerned with generation of surface waves in an ocean with porous bottom due to initial disturbances at free surface. Assuming linear theory the problem is formulated as an initial value problem for the velocity potential describing the motion in the fluid. Laplace transform in time and Fourier transform in space have been utilized in the mathematical analysis to obtain the form of the free surface in terms of an integral. This integral is then evaluated asymptotically for large time and distance by the method of stationary phase for prescribed initial disturbance at the free surface in the form of depression of the free surface or an impulse at the free surface concentrated at the origin. The form of the free surface is depicted graphically for these two types of initial conditions in a number of figures to demonstrate the effect of the porosity at the bottom.

Keywords: Cauchy Poisson problem · Porous bottom
Laplace and Fourier transform · Method of stationary phase
Free surface depression

1 Introduction

Problems of unsteady motion in water due to any initial disturbance at the free surface are known as Cauchy Poisson problems. The disturbance may be either in the form of prescribed elevation or depression of the free surface or a distribution of impulse or a combination of these. These problems are in general difficult to tackle mathematically. However in two dimensions the problems can be formulated in a comparatively simple manner assuming linear theory and solution can be obtained easily. For example, the Fourier transform techniques was used to obtain the form of the free surface in the form of an infinite integral for the two dimensional Cauchy Poisson problem in deep water in which the motion is generated due to initial surface disturbances in the form of initial depression or impulse concentrated at a point on the free surface (cf. Lamb (1945); Stoker (1957)). The infinite integral was evaluated asymptotically by

© Springer Nature Singapore Pte Ltd. 2018
D. Ghosh et al. (Eds.): ICMC 2018, CCIS 834, pp. 174–185, 2018.
https://doi.org/10.1007/978-981-13-0023-3_17

the method of stationary phase for large time and distance so that the form of the free surface could be found. Kranzer and Keller (1959) obtained explicit forms of the free surface due to axially symmetric initial disturbances in water of finite depth. They compared their theoretical results with experimental results and found that agreement was fairly good. Very recently Baek et al. (2017) investigated the Cauchy Poisson problem with measured initial profile taking into account the effect of the surface tension at the free surface and compared the theoretical results with those obtained by experiment carried out in water tanks wherein the waves were generated by dropping water droplets and high speed camera was used to record video clips for wave field. The experimental results agree quite well with the theoretical results. These validate the use of linear theory in the study of Cauchy Poisson problem.

The 2D Cauchy Poisson problems could easily be solved for uniform finite depth water wherein the bottom is assumed to be rigid (cf. Chaudhuri (1968); Wen (1982)). However, there is not much work involving Cauchy Poisson problems, if one considers a porous bottom instead of a rigid bottom. An ocean bottom is far from rigid and in reality it is actually permeable or porous. Thus it may be interesting to study the effect of bottom porosity on the wave motion due to initial disturbances at the free surface. Gangopadhyay and Basu (2013) considered a porous bottom while studying the problem wave generation due to initial depression of the free surface concentrated at the origin. However, their mathematical analysis leading to the asymptotic form of the free surface appears to be incomplete. As in Martha et al. (2007) and Maiti and Mandal (2014), here a special type of porous bottom is considered for which the porosity parameter is taken to be only real and has the dimension of $(\text{length})^{-1}$.

In the present paper we consider Cauchy Poisson problem in water with the porous bottom. Fourier and Laplace transform techniques are employed to solve the problem formulated as an initial value problem assuming linear theory. The form of the free surface is obtained in terms of an integral which is evaluated asymptotically for large time and distance from the origin, assuming of course that the initial disturbance on the free surface is concentrated at the origin. The asymptotic form of the free surface is depicted graphically against non-dimensional distance for fixed time and against non-dimensional time for fixed distance and for different values of the porosity parameter (non-dimensionalised) leas than unity.

2   Mathematical Formulation

A Cartesian co-ordinate system is used wherein the y-axis is chosen vertically downwards in the fluid region and the xz-plane is the undisturbed free surface and $y = h$ corresponds to the bottom composed of some specific kind of porous materials (cf. Fig. 1). The porous bottom is characterized by a real quantity G which has the dimension of inverse of length. The fluid is assumed to be inviscid and the motion in the fluid starts from rest so that it is irrotational and thus can

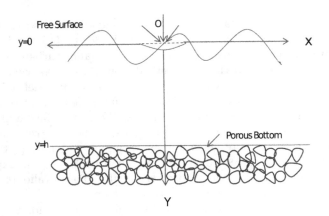

Fig. 1. Sketch of physical model.

be described by a velocity potential $\phi(x, y, t)$ for two dimensional case. Assuming linear theory it is easy to see that ϕ satisfies the initial value problem described by,

$$\nabla^2 \phi = 0, \quad 0 \le y \le h, \quad -\infty < x < \infty,$$

$$\frac{\partial^2 \phi}{\partial t^2} - g\frac{\partial \phi}{\partial y} = 0 \quad \text{on} \quad y = 0,$$

$$\frac{\partial \phi}{\partial y} - G\phi = 0 \quad \text{on} \quad y = h, \quad -\infty < x < \infty$$

where G is the porosity parameter of the fluid bottom (cf. Martha et al. (2007)). When the initial depression of the free surface is prescribed then the initial conditions are,

$$\phi(x, 0, 0) = 0,$$

$$\phi_t(x, 0, 0) = gf(x)$$

where $f(x)$ denotes the prescribed initial depression of the free surface, g is the acceleration due to gravity. We will assume that the initial depression is concentrated at the origin.

When the initial disturbance is in the form of an impulsive pressure $I(x)$ per unit area applied to the free surface, then initial conditions are,

$$\phi(x, 0, 0) = -\frac{I(x)}{\rho},$$

$$\phi_t(x, 0, 0) = 0$$

where ρ is the density of water. We will also assume that the initial impulse is also concentrated at the origin. The expression for free surface depression is given by

$$\eta(x, t) = \frac{1}{g}\phi_t(x, 0, t).$$

Introducing a characteristic length l, characteristic time $\sqrt{\frac{l}{g}}$, and characteristic mass m. We define the dimensionless quantities as,

$$\bar{x} = \frac{x}{l}, \ \bar{y} = \frac{y}{l}, \ \bar{t} = t\sqrt{\frac{l}{g}}, \ \bar{\eta} = \frac{\eta}{l}, \ \bar{h} = \frac{h}{l}, \ \bar{\phi} = \frac{\phi}{l\sqrt{gl}}, \ \bar{G} = Gl, \ \bar{I} = \frac{lI}{m\sqrt{\frac{l}{g}}}, \ \bar{\rho} = \frac{\rho l^3}{m}.$$

Removing the bars the dimensionless quantities satisfy,

$$\nabla^2\phi = 0, \quad 0 \leq y \leq h, \quad -\infty < x < \infty, \tag{2.1}$$

$$\frac{\partial^2\phi}{\partial t^2} - \frac{\partial\phi}{\partial y} = 0 \quad \text{on} \quad y = 0, \tag{2.2}$$

$$\frac{\partial\phi}{\partial y} - G\phi = 0 \quad \text{on} \quad y = h, \quad -\infty < x < \infty. \tag{2.3}$$

Initial conditions for the case of prescribed initial depression of the free surface are,

$$\phi(x, 0, 0) = 0, \tag{2.4}$$

$$\phi_t(x, 0, 0) = \delta(x) \tag{2.5}$$

where $\delta(x)$ is the Dirac Delta function.

Initial conditions in case of impulsive pressure applied to the free surface are,

$$\phi(x, 0, 0) = -\frac{\delta(x)}{\rho}, \tag{2.6}$$

$$\phi_t(x, 0, 0) = 0. \tag{2.7}$$

Once $\phi(x, y, t)$ is obtained by solving the initial value problem, the corresponding non dimensional depression of the free surface is obtained as,

$$\eta(x, t) = \phi_t(x, 0, t). \tag{2.8}$$

3   Method of Solution

We use the Fourier and Laplace transform techniques to solve the aforesaid initial value problems. Fourier transform of $\phi(x, y, t)$ with respect to the variable x is,

$$\bar{\phi}(s, y, t) = \frac{1}{\sqrt{2\pi}} \int_{-\infty}^{\infty} \phi(x, y, t)e^{-isx} \, dx.$$

Laplace transform of $\bar{\phi}(s, y, t)$ with respect to the variable t is,

$$\bar{\bar{\phi}}(s, y, p) = \int_{0}^{\infty} \bar{\phi}(s, y, t)e^{-pt} \, dt.$$

Taking Fourier transform with respect to the variable x of the Eqs. (2.1), (2.2), (2.3), (2.4), (2.5), (2.6), (2.7) and (2.8) we have,

$$\bar{\phi}_{yy} - s^2\bar{\phi} = 0, \quad 0 \le y \le h, \tag{3.1}$$

$$\bar{\phi}_{tt} - \bar{\phi}_y = 0, \quad \text{on} \quad y = 0, \tag{3.2}$$

$$\bar{\phi}_y - G\bar{\phi} = 0, \quad \text{on} \quad y = h. \tag{3.3}$$

Initial conditions for the case of initial depression of the free surface are,

$$\bar{\phi}(s,0,0) = 0, \tag{3.4}$$

$$\bar{\phi}_t(s,0,0) = 1. \tag{3.5}$$

Initial conditions in case of impulsive pressure applied to the free surface are,

$$\bar{\phi}(s,0,0) = -\frac{1}{\rho}, \tag{3.6}$$

$$\bar{\phi}_t(s,0,0) = 0. \tag{3.7}$$

Depression of free surface at time t is,

$$\bar{\eta}(s,t) = \bar{\phi}_t(x,0,t). \tag{3.8}$$

Taking Laplace transform of the Eqs. (3.1), (3.2), (3.3) and using the conditions (3.4), (3.5) we get,

$$\frac{d^2\bar{\bar{\phi}}}{dy^2} - s^2\bar{\bar{\phi}} = 0, \quad 0 \le y \le h, \tag{3.9}$$

$$\frac{d\bar{\bar{\phi}}}{dy} - p^2\bar{\bar{\phi}} = -1, \quad \text{on} \quad y = 0, \tag{3.10}$$

$$\frac{d\bar{\bar{\phi}}}{dy} - G\bar{\bar{\phi}} = 0, \quad \text{on} \quad y = h. \tag{3.11}$$

Solution of (3.9) satisfying (3.10) and (3.11) is,

$$\bar{\bar{\phi}}(s,y,p) = \frac{A(s,y,h)}{p^2 + \mu(s)}, \tag{3.12}$$

where

$$A(s,y,h) = \frac{s\cosh(s(y-h)) + G\sinh(s(y-h))}{\cosh(sh)(s - G\tanh(sh))}$$

and

$$\mu(s) = \frac{s(G - s\tanh(sh))}{(G\tanh(sh) - s)}.$$

As $s \to 0$ the function $\mu(s)$ has limiting value $\frac{G}{Gh-1}$.

When $Gh < 1$ the graph of, $s(G - s\tanh(sh))$ and $(G\tanh(sh) - s)$ are shown in Fig. 2. There are three roots of

$$s(G - s\tanh(sh)) = 0$$

say $s = \pm\lambda_1$ and $s = 0$. Again at $s = 0$ both numerator and denominator of $\mu(s)$ vanish and the limit of $\mu(s)$ as $s \to 0$ exists and its value is negative.
Graph of $\mu(s)$ for $Gh < 1$ is shown in Fig. 3. From Fig. 3 it is observed that $\mu(s) \geq 0$ for $|s| \geq \lambda_1$ and $\mu(s) < 0$ for $|s| < \lambda_1$.

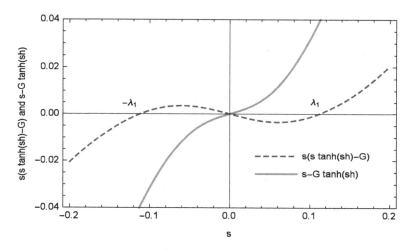

Fig. 2. Graph of $s(G - s\tanh(sh))$ and $(G\tanh(sh) - s), Gh = 0.9$

Let

$$\mu(s) = \begin{cases} \mu_1^2(s) & \text{for} & |s| \geq \lambda_1, \\ -\mu_2^2(s) & \text{for} & |s| < \lambda_1. \end{cases}$$

Then,

$$\bar{\bar{\phi}}(s, y, p) = \begin{cases} \dfrac{A(s, y, h)}{p^2 + \mu_1^2(s)} & \text{for} \quad |s| \geq \lambda_1, \\[3mm] \dfrac{A(s, y, h)}{p^2 - \mu_2^2(s)} & \text{for} \quad |s| < \lambda_1. \end{cases}$$

Taking Laplace inversion we obtain,

$$\bar{\phi}(s, y, t) = \begin{cases} \dfrac{A(s, y, h)}{\mu_1(s)} \sin(\mu_1 t) & \text{for} \quad |s| \geq \lambda_1, \\[3mm] \dfrac{A(s, y, h)}{\mu_2(s)} \sinh(\mu_2 t) & \text{for} \quad |s| < \lambda_1. \end{cases}$$

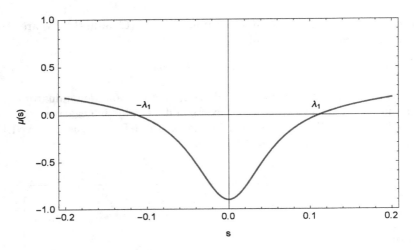

Fig. 3. Graph of $\mu(s)$ for $Gh = 0.9$

Then,

$$\bar{\eta}_1(s,t) = \begin{cases} cos(\mu_1 t) & \text{for} \quad |s| \geq \lambda_1, \\ cosh(\mu_2 t) & \text{for} \quad |s| < \lambda_1. \end{cases}$$

Taking inverse Fourier transform we find,

$$\eta_1(x,t) = \frac{1}{\sqrt{2\pi}} \left[\int_0^{\lambda_1} 2\cosh(\mu_2 t) \cos(sx) \, ds + \int_{\lambda_1}^{\infty} 2\cos(\mu_1 t) \cos(sx) \, ds \right] \quad (3.13)$$

$$= I_1 + I_2 (say)$$

where

$$I_1 = \frac{1}{\sqrt{2\pi}} \int_0^{\lambda_1} 2\cosh(\mu_2 t) \cos(sx) \, ds$$

and

$$I_2 = \frac{1}{\sqrt{2\pi}} \int_{\lambda_1}^{\infty} 2\cos(\mu_1 t) \cos(sx) \, ds.$$

If initial impulse is assumed to be concentrated at the origin, then expression for depression of free surface at time t is obtained by a similar procedure as

$$\eta_2(x,t) = \frac{1}{\rho\sqrt{2\pi}} \left[-\int_0^{\lambda_1} 2\mu_2 \sinh(\mu_2 t) \cos(sx) \, ds + \int_{\lambda_1}^{\infty} 2\mu_1 \sin(\mu_1 t) \cos(sx) \, ds \right]$$

$$(3.14)$$

$$= I_3 + I_4 (say)$$

where,

$$I_3 = -\frac{1}{\rho\sqrt{2\pi}} \int_0^{\lambda_1} 2\mu_2 \sinh(\mu_2 t) \cos(sx) \, ds$$

and

$$I_4 = \frac{1}{\rho\sqrt{2\pi}} \int_{\lambda_1}^{\infty} 2\mu_1 \sin(\mu_1 t)\cos(sx)\ ds.$$

4   Asymptotic Expansion

We are interested in the waves after a long lapse of time and at a large distance from the origin. For this we use the method of stationary phase to evaluate the integral (3.13) for large x and t such that $\frac{x}{t} > 1$. Integral I_1 does not contribute to $\eta_1(x,t)$. Since,

$$I_1 = \frac{1}{\sqrt{2\pi}} \int_0^{\lambda_1} 2\cosh(\mu_2 t)\cos(sx)\ ds$$

$$= \frac{1}{\sqrt{2\pi}} \left[\int_0^{\lambda_1} \frac{e^{\mu_2 t} + e^{-\mu_2 t}}{2} e^{isx} ds + \int_0^{\lambda_1} \frac{e^{\mu_2 t} + e^{-\mu_2 t}}{2} e^{-isx} ds \right]$$

For the first integral, we write,

$$f_1(s) = s$$

and

$$g_1(s) = e^{\mu_2 t} + e^{-\mu_2 t}$$

and x is large. Now,

$$f_1'(s) = 1$$

which has no zero in the range of integration.

Similarly for the second integral we write

$$f_2(s) = -s, g_2(s) = e^{\mu_2 t} + e^{-\mu_2 t}$$

and x is large. Thus I_1 does not contribute to $\eta_1(x,t)$ as

$$f_2'(s) = -1$$

which has no zero in the range of integration.

For I_2 we have,

$$\eta_1(x,t) = \frac{1}{\sqrt{2\pi}} \int_{\lambda_1}^{\infty} \left[\cos(sx + \mu_1 t) + \cos(sx - \mu_1 t) \right] ds$$

$$= \frac{1}{\sqrt{2\pi}} \Re \left[\int_{\lambda_1}^{\infty} e^{it(\mu_1 + s\frac{x}{t})}\ ds + \int_{\lambda_1}^{\infty} e^{it(\mu_1 - s\frac{x}{t})}\ ds \right]. \qquad (4.1)$$

In (4.1) the first integral does not have any stationary point in the range of the integration. So this integral does not contribute to $\eta_1(x,t)$. For the second integral let

$$f_3(s) = \mu_1(s) - s\frac{x}{t},$$

$$g_3(s) = 1$$

and t is large. Now

$$f_3'(s) = \frac{1}{2}\left[\frac{(sh-1)G\tanh(sh)}{s-G\tanh(sh)}\mu_1(s) + \frac{s(\frac{sh}{\cosh^2(sh)}+\tanh(sh))}{s-G\tanh(sh)}\mu_1^{-1}(s)\right] - \frac{x}{t}.$$

We see that $f_3'(s)$ is monotonically decreasing in (λ_1, ∞) and has only one zero in the range of integration for $\frac{x}{t} > 1$. Let α be the root of

$$f_3'(s) = 0$$

then

$$\eta_1(x,t) \approx \frac{1}{\sqrt{2\pi}}\left(\frac{2\pi}{t|f_3''(\alpha)|}\right)^{1/2}\cos(tf_3(\alpha) - \frac{\pi}{4}), \tag{4.2}$$

negative sign of $\frac{\pi}{4}$ being taken since the sign of $f_3''(\alpha)$ is negative.

Similarly in case of (3.14) the integral I_3 does not contribute to $\eta_2(x,t)$ and by similar calculation for large x and t such that $\frac{x}{t} > 1$ asymptotic expansion of $\eta_2(x,t)$ is

$$\eta_2(x,t) \approx \frac{1}{\rho\sqrt{2\pi}}\left(\frac{2\pi}{t|f_4''(\beta)|}\right)^{1/2}g_4(\beta)\sin(tf_4(\beta) - \frac{\pi}{4}), \tag{4.3}$$

where

$$f_4(s) = \mu_1(s) - s\frac{x}{t}, \quad g_4(s) = \mu_1(s)$$

and β is the unique real root of

$$f_4'(s) = 0$$

in the range of integration and negative sign is taken since the sign of $f_4''(\beta)$ is negative.

5   Numerical Results

To display the effect of free surface elevation in an ocean with porous bottom due to the initial disturbances at the free surface, the non dimensional asymptotic form of $\eta_1(x,t)$ are depicted graphically against x for fixed t and against t for fixed x in a number of figures. To visualize the nature of the wave motion due to prescribed initial disturbance at the free surface $\eta_1(x,t)$ is plotted in Fig. 4 against t for fixed $x = 300$ and t ranging from 200 to 250 and porosity parameter $G = 0, 0.05, 0.09$ and $h = 10$.

From Fig. 4 it is observed that the amplitude of the wave profile increases as time increases. Also $\eta_1(x,t)$ is plotted in Fig. 5 for fixed time and variable distance from the origin. In Fig. 5 $\eta_1(x,t)$ is plotted for fixed $t = 180$ and x ranging from 220 to 280 and the porosity parameter $G = 0, 0.05, 0.09$ and $h = 10$.

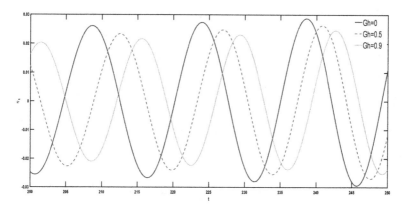

Fig. 4. Wave motion due to initial displacement for fixed distance $x = 300$.

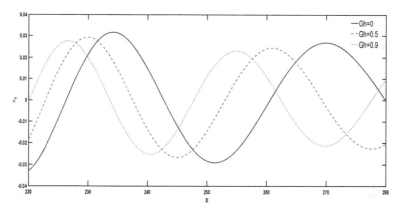

Fig. 5. Wave motion due to initial displacement for fixed time $t = 180$.

From Fig. 5 it is observed that as x increases amplitude of wave motion decreases.

Similarly $\eta_2(x, t)$ obtained from Eq. (4.3) due to an initial disturbance in the form of an impulse concentrated at the origin, is plotted in Figs. 6 and 7. In Fig. 6 $\rho\eta_2(x, t)$ is plotted against t for fixed $x = 300$ and t ranging from 200 *to* 250 and porosity parameter $G = 0, 0.05, 0.09$ and $h = 10$. From Fig. 6 it is observed that the amplitude of the wave profile increases as time increases. In Fig. 7 $\rho\eta_2(x, t)$ is plotted against x for fixed $t = 180$ and x ranging from 220 *to* 280 and porosity parameter $G = 0, 0.05, 0.09$ and $h = 10$. From Fig. 7 it is observed that the amplitude of the wave profile decreases as distance increases. When $G = 0$ the Eqs. (4.2) and (4.3) coincide with the corresponding result when the ocean bottom is rigid. In the Figs. 4, 5, 6 and 7 the curves for $G = 0$ match well with the corresponding curves for the case of rigid ocean bottom.

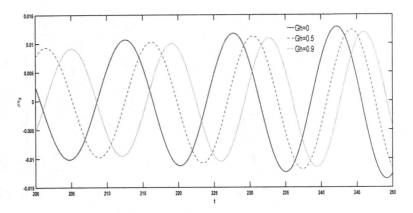

Fig. 6. Wave motion due to impulse for fixed distance $x = 300$.

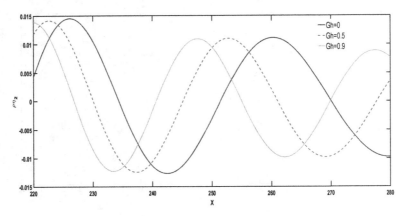

Fig. 7. Wave motion due to impulse for fixed time $t = 180$.

6   Conclusion

Cauchy Poisson problem for water in an ocean with porous bottom is considered due to two types of initial disturbances at the free surface, one is prescribed initial depression of the free surface at time $t = 0$ and the other one is the initial disturbance of the free surface due to impulsive pressure. In the present problem we consider the porosity parameter G to be real and consider the case only when $Gh < 1$. It is observed that when $G = 0$ the figures are almost same when ocean bottom is rigid.

The case $Gh > 1$ needs to be studied somewhat carefully which will be done in future.

Acknowledgments. The authors thank the reviewers for their comments to modify the paper in the present form. This work is carried out under CSIR research project No. 25(0253)/16/EMR-II.

References

Baek, H.M., Kim, Y.J., Lee, I.J., Kwon, S.H.: Revisit of Cauchy Poisson problem in unsteady water wave problem. In: 32 IWWWFB, Dalian, China, pp. 13–16 (2017)

Chaudhuri, K.: Waves in shallow water due to arbitrary surface disturbances. App. Sci. Res. **19**, 274–284 (1968)

Gangopadhyay, S., Basu, U.: Water wave generation due to initial disturbance at the free surface in an ocean with porous bed. Int. J. Sci. Eng. Res. **4**(2), 1–4 (2013)

Kranzer, H.C., Keller, J.B.: Water waves produced by explosions. J. Appl. Phys. **30**, 398–407 (1959)

Lamb, H.: Hydrodynamics. Dover, New York (1945)

Maiti, P., Mandal, B.N.: Water wave scattering by an elastic plate floating in an ocean with a porous bed. Appl. Ocean Res. **47**, 73–84 (2014)

Martha, S.C., Bora, S.N., Chakraborti, A.: Oblique water wave scattering by small undulation on a porous sea bed. Appl. Ocean Res. **29**, 86–90 (2007)

Stoker, J.J.: Water Waves. Interscience Publishers, New York (1957)

Wen, S.L.: Int. J. Math. Educ. Sci. Technol. **13**, 55–58 (1982)

Semi-frames and Fusion Semi-frames

N. K. Sahu[1(✉)] and R. N. Mohapatra[2]

[1] Dhirubhai Ambani Institute of Information and Communication Technology,
Gandhinagar, India
nabindaiict@gmail.com
[2] Department of Mathematics, University of Central Florida, Orlando, USA
ramm1627@gmail.com

Abstract. This paper is a short survey of the theory of semi-frames and fusion semi-frames in Hilbert and Banach spaces.

1 Introduction

A frame is a redundant representation of a basis. In general the elements in a frame need not be linearly independent unlike the elements in a basis. The concept of frame was introduced by Duffin and Schaeffer [7] in 1952 while studying nonharmonic Fourier series. There was a gap of nearly thirty years, where there is no work in that area. In early 90's the work was revived by Daubechies et al. [6]. They reintroduced it while studying Wavelet theory, thereafter it became popular in Gabor and Wavelet analysis. A good frame in a Hilbert space is almost as good as an orthonormal basis for expanding arbitrary elements of a signal and is often easier to construct.

Definition 1.1. Let H be a Hilbert space. Let I be an index set. A sequence of elements $\{f_k\}_{k\in I}$, in H is called a frame for H if there exist positive constants A and B such that

$$A\|f\|^2 \le \sum_{k\in I} |\langle f_k, f\rangle|^2 \le B\|f\|^2, \quad \forall f \in H.$$

A and B are called lower and upper frame bound respectively.

If $A = B$ then the frame is called a tight frame, and if $A = B = 1$ then the frame is called a Parseval frame. A frame is called a normalized frame if each frame element has unit norm. To get more familiar with the theory of frames one may refer to Christensen [5] and the references there in.

Let H be a Hilbert space. Consider the discrete sequence space l^2. A sequence $\{f_k\}_{k\in I}$ in H is called a Bessel sequence for the Hilbert space H, if there exists a constant $M > 0$ such that

$$\sum_{k\in I} |\langle f_k, f\rangle|^2 \le M\|f\|^2, \forall f \in H. \tag{1}$$

© Springer Nature Singapore Pte Ltd. 2018
D. Ghosh et al. (Eds.): ICMC 2018, CCIS 834, pp. 186–193, 2018.
https://doi.org/10.1007/978-981-13-0023-3_18

Let $T : H \to l^2$ be an operator defined by

$$Tf = \{\langle f_k, f \rangle\}_{k \in I}. \tag{2}$$

The adjoint operator $T^* : l^2 \to H$ is given by

$$T^*(C) = \sum_{k \in I} c_k f_k, \text{ where } C = \{c_k\} \in l^2. \tag{3}$$

Let S be the composition operator $T^*T : H \to H$ given by

$$S(f) = T^*T(f) = \sum_{k \in I} \langle f_k, f \rangle f_k. \tag{4}$$

When the sequence $\{f_k\}_{k \in I}$ in H is a frame that is if there exist positive constants A and B such that

$$A\|f\|^2 \le \sum_{k \in I} |\langle f_k, f \rangle|^2 \le B\|f\|^2, \quad \forall f \in H,$$

then the operator T^* is called the analysis operator, the operator T is called the synthesis operator, the operator S is called the frame operator. In this paper we shall introduce several important concepts and state results without proof. We state the first result below:

Theorem 1.1. Let $\{f_k\}_{k \in I}$ be a frame for H, with analysis operator T^*, synthesis operator T, and the frame operator S, then
(i) $\{f_k\}_{k \in I}$ is total in H. The operator S has a bounded inverse $S^{-1} : H \to H$, and the vectors in H can be reconstructed by

$$f = S^{-1}Sf = \sum_{k \in I} \langle f_k, f \rangle S^{-1}f_k, \ \forall f \in H,$$

and

$$f = SS^{-1}f = \sum_{k \in I} \langle S^{-1}f_k, f \rangle f_k, \ \forall f \in H.$$

(ii) $Rng(T)$ is a closed subspace of l^2. The projection P from l^2 onto $Rng(T)$ is given by $P = TS^{-1}T^* = TT^\dagger$, where $T^\dagger$ is the pseudo inverse of T.

Frames in Banach spaces have been studied in Casazza and Christensen [3], Christensen and Heil [4], Gröchenig [9], Kaushik [10] and Stoeva [15]. Stoeva [16] introduced X_d-frames and their dual in Banach spaces. By using the semi-inner product structure of Banach spaces, Zhang and Zhang [17] have studied X_d-frame, X_d^*-frame in Banach spaces.

2   Semi-frames in Hilbert Space

2.1   Upper Semi-frames

The upper and lower frame bounds are more strict restrictions for a frame sequence. To loosen that restriction semi-frame was introduced. The notion of semi-frame in Hilbert space was studied by Antoine and Balazs [1].

Definition 2.1. A sequence of elements $\{f_k\}_{k \in I}$ in a Hilbert space H is said to be a upper semi-frame if there exists a positive constant M such that

$$0 < \sum_{k \in I} |\langle f_k, f \rangle|^2 \leq M\|f\|^2, \quad \forall f \in H, \ f \neq 0.$$

Antoine and Balazs [1] proved the following result:

Theorem 2.1. $\{f_k\}_{k \in I}$ is an upper semi-frame if and only if it is a total Bessel sequence.

If $\{f_k\}_{k \in I}$ is a frame in a Hilbert space H, then there exists some sequence $\{g_k\}_{k \in I}$ in H such that

$$f = \sum_{k \in I} \langle f_k, f \rangle g_k = \sum_{k \in I} \langle g_k, f \rangle f_k, \quad \forall f \in H. \tag{5}$$

If the above equality (5) holds true then $\{f_k\}_{k \in I}$ and $\{g_k\}_{k \in I}$ are said to be dual of each other.

However there exist an upper semi-frame $\{f_k\}_{k \in I}$ in H which are not necessarily frame for which (5) still holds true for some sequence $\{g_k\}_{k \in I}$. We see this in the following example:

Example 2.1. Let $H = l^2$ be the Hilbert space and $\{e_k\}_{k \in I}$ is canonical basis for l^2. Let

$$\{f_k\}_{k \in I} = \{\frac{1}{k} e_k\}_{k \in I}$$
$$\{g_k\}_{k \in I} = \{k e_k\}_{k \in I}.$$

One can easily see that $\{f_k\}_{k \in I}$ is an upper semi-frame but not a frame as the lower frame bound is missing. But we have

$$f = \sum_{k \in I} \langle f_k, f \rangle g_k = \sum_{k \in I} \langle g_k, f \rangle f_k, \quad \forall f \in l^2.$$

Antoine and Balazs [1] has proved the following theorem for upper semi-frame.

Theorem 2.2. Let $\{f_k\}_{k \in I}$ be an upper semi-frame for the Hilbert space H. Let T^* be the analysis operator as defined in (3), T be the synthesis operator as defined in (2), and $S = T^*T$ be the operator as defined in (4). Then

(i) The analysis operator T^* is an injective bounded operator, and the synthesis operator T is a bounded operator with dense range. The operator S is a bounded, self adjoint and positive operator with dense range. Its inverse S^{-1} is densely defined and self adjoint.

(ii) $T(Rng(S)) \subseteq Rng(T) \subseteq \overline{Rng(T)}$.

For an upper semi-frame, the frame operator is bounded, but has an unbounded inverse.

Definition 2.2. An upper semi-frame $\{f_k\}_{k\in I}$ is said to be regular if $\{f_k\}_{k\in I} \in Dom(S^{-1}) = Rng(S)$.

Theorem 2.3. If $\{f_k\}_{k\in I}$ is an upper semi-frame for the Hilbert space H, then

$$f = S^{-1}Sf = \sum_{k\in I}\langle f_k, S^{-1}f\rangle f_k, \quad \forall f \in Rng(S).$$

Note that if the upper semi-frame is regular, then we can write the reconstruction formula using a dual sequence.

Theorem 2.4. Let $\{f_k\}_{k\in I}$ be a regular upper semi-frame for H. Then

$$f = SS^{-1}f = \sum_{k\in I}\langle S^{-1}f_k, f\rangle f_k.$$

2.2   Lower Semi-frames

Definition 2.3. A sequence $\{f_k\}_{k\in I}$ in a Hilbert space H is said to be a lower semi-frame if there exists a positive constant m such that

$$m\|f\|^2 \le \sum_{k\in I}|\langle f_k, f\rangle|^2, \quad \forall f \in H.$$

In upper semi-frame, the positivity requirement on the left hand side ensures that the sequence $\{f_k\}_{k\in I}$ is total, whereas in lower semi-frame $\{f_k\}_{k\in I}$ is total automatically. A lower semi frame has an unbounded frame operator, with bounded inverse.

Example 2.2. Let $\{e_k\}_{k\in I}$ be orthonormal basis for the Hilbert space l^2. Let $f_k = \frac{e_k}{k}$, $k \in I$. We see that for every $f \in l^2$,

$$0 < \sum_{k\in I}|\langle f_k, f\rangle|^2 = \sum_{k\in I}|\frac{1}{k}\langle e_k, f\rangle|^2 = \sum_{k\in I}\frac{1}{k^2}|\langle e_k, f\rangle|^2$$

$$\le \sum_{k\in I}|\langle e_k, f\rangle|^2 = \sum_{k\in I}|f_k|^2 = \|f\|^2.$$

Hence the sequence $\{\frac{e_k}{k}\}_{k\in I}$ is an upper semi-frame for l^2.

On the other hand the sequence $\{\frac{e_k}{k}\}_{k\in I}$ is not a lower semi-frame for l^2, as it is failed to satisfy the lower frame bound condition. We can see this by taking $f = e_p$,

$$\sum_{k\in I}|\langle f_k, f\rangle|^2 = \sum_{k\in I}|\langle \frac{e_k}{k}, e_p\rangle|^2 = \frac{1}{p^2}.$$

Example 2.3. Let $\{e_k\}_{k\in I}$ be orthonormal basis for the Hilbert space l^2. Let $f_k = \frac{e_k}{k}$, $k \in I$ and $\phi_k = ke_k$, $k \in I$. In the previous example, we have seen that the sequence $\{f_k\}_{k\in I}$ is an upper semi-frame for l^2. Again we observe that for every $f \in l^2$,

$$\sum_{k\in I}\langle f_k, f\rangle \phi_k = \sum_{k\in I}\langle \frac{e_k}{k}, f\rangle ke_k = \sum_{k\in I}\langle e_k, f\rangle e_k = \sum_{k\in I} f_k e_k = f.$$

Hence $\{\phi_k\}_{k\in I}$ is a dual semi-frame for $\{f_k\}_{k\in I}$.

Moreover, for every $f \in l^2$,

$$\sum_{k\in I}|\langle \phi_k, f\rangle|^2 = \sum_{k\in I}|\langle ke_k, f\rangle|^2 = \sum_{k\in I} k^2|\langle e_k, f\rangle|^2$$
$$\geq \sum_{k\in I}|\langle e_k, f\rangle|^2 = \|f\|^2.$$

Hence $\{\phi_k\}_{k\in I}$ is a lower semi-frame which is dual to the upper semi-frame $\{f_k\}_{k\in I}$.

3   Semi-frames in Banach Spaces

Semi-frames in Banach spaces have been introduced by Antoine and Balazs [1]. Sahu and Mohapatra [14] introduced semi-X_d-frames in Banach spaces. They have used the semi-inner product structure available in Banach spaces to define semi-frames. The notion of semi-inner product was introduced by Lumer [13], and further investigation was done by Giles [8] and Koehler [12].

Definition 3.1. Let X be a Banach space with a compatible semi-inner product $[.,.]$ and norm $\|.\|_X$. Let X_d be an associated BK-space with norm $\|.\|_{X_d}$. A sequence of elements $\{f_j\} \subseteq X$ is called upper semi-X_d-frame for X if
(i) $\{f_j\}$ is total in X;
(ii) $\{[f, f_j]\} \in X_d$ for all $f \in X$;
(iii) there exists a positive constant B such that

$$0 \leq \|\{[f, f_j]\}\|_{X_d} \leq B\|f\|_X \quad \text{for all } f \in X.$$

Definition 3.2. Let X be a Banach space with a compatible semi-inner product $[.,.]$ and norm $\|.\|_X$. Let X_d be an associated BK-space with norm $\|.\|_{X_d}$. A sequence of elements $\{f_j\} \subseteq X$ is called lower semi-X_d-frame for X if

(i) $\{f_j\}$ is total in X;
(ii) $\{[f, f_j]\} \in X_d$ for all $f \in X$;
(iii) there exists a positive constants A such that

$$A\|f\|_X \le \|\{[f, f_j]\}\|_{X_d} \text{ for all } f \in X.$$

Similarly, we define upper semi-X_d^*-frame and lower semi-X_d^*-frame for the dual space X^*.

Definition 3.3. Let X be a Banach space with a compatible semi-inner product $[.,.]$ and norm $\|.\|_X$. Let X^* be the dual space of X. Let X_d be an associated BK-space with norm $\|.\|_{X_d}$, and X_d^* be the dual space of X_d. A sequence of elements $\{f_j^*\} \subseteq X^*$ is upper semi-X_d^*-frame for X^* if
(i) $\{f_j^*\}$ is total in X^*;
(ii) $\{[f_j, f]\} \in X_d^*$ for all $f \in X$;
(iii) there exists a positive constants B such that

$$0 \le \|\{[f_j, f]\}\|_{X_d^*} \le B\|f\|_X \text{ for all } f \in X.$$

Definition 3.4. Let X be a Banach space with a compatible semi-inner product $[.,.]$ and norm $\|.\|_X$. Let X^* be the dual space of X. Let X_d be an associated BK-space with norm $\|.\|_{X_d}$, and X_d^* be the dual space of X_d. A sequence of elements $\{f_j^*\} \subseteq X^*$ is lower semi-X_d^*-frame for X^* if
(i) $\{f_j^*\}$ is total in X^*;
(ii) $\{[f_j, f]\} \in X_d^*$ for all $f \in X$;
(iii) there exists a positive constants A such that

$$A\|f\|_X \le \|\{[f_j, f]\}\|_{X_d^*} \text{ for all } f \in X.$$

4   Fusion Semi-frames

The fusion frame also called as frame of subspaces, is a natural generalization of frame theory and related to the construction of global frames from local frames in Hilbert spaces. The fusion frame in Hilbert spaces was introduced by Casazza and Kutyniok [2].

Definition 4.1 [2].   Let H be a Hilbert space and $\{W_i\}_{i \in I}$ be sequence of closed subspaces of H. Let $\{w_i\}_{i \in I}$ be a family of positive weights. Let $\pi_{W_i} : H \to W_i$ be the orthogonal projections onto the subspaces W_i. Then the sequence $\{(W_i, w_i)\}_{i \in I}$ is said to be a fusion frame for H, if there exist positive constants A and B such that

$$A\|f\|^2 \le \sum_{i \in I} w_i^2 \|\pi_{W_i}(f)\|^2 \le B\|f\|^2, \quad \forall f \in H. \tag{6}$$

Fusion semi-frames in Hilbert space was introduced by Antoine and Balazs [1].

Definition 4.2 [1].   Let H be a Hilbert space and $\{W_i\}_{i \in I}$ be sequence of closed subspaces of H. Let $\{w_i\}_{i \in I}$ be a family of positive weights. Let $\pi_{W_i} :$ $H \to W_i$ be the orthogonal projections onto the subspaces W_i. Then the sequence $\{(W_i, w_i)\}_{i \in I}$ is said to be a fusion upper semi-frame for H, if there exists positive constant M such that

$$0 \leq \sum_{i \in I} w_i^2 \|\pi_{W_i}(f)\|^2 \leq M \|f\|^2, \quad \forall f \in H. \tag{7}$$

Fusion frame in Banach space was introduced by Khosravi and Khosravi [11].

Definition 4.3 [11].   Let X be a Banach space, $\{\pi_i\}_{i \in I}$ be a sequence of continuous linear projections on X, $W_i = \pi_i(X)$ for each $i \in I$. Let $\{w_i\}_{i \in I}$ be a sequence of positive weights. The sequence $\{(w_i, W_i)\}_{i \in I}$ is said to be a fusion frame for X if there exists a sequence $\{q_i \in B(W_i, X) : i \in I\}$ and an invertible operator $S \in B(X)$ such that
(i) The series $\sum_{i \in I} w_i^2 q_i \ o \ p_i(x)$ converges to $S(x)$;
(ii) There exists a solid $BK-$space X_d and positive constants A and B such that for every $x \in X$

$$A\|x\|_X \leq \|\{\|\pi_i(x)\|\}\|_{X_d} \leq B\|x\|. \tag{8}$$

Now we are in a position to define fusion semi-frame in Banach spaces.

Definition 4.4. Let X be a Banach space, $\{\pi_i\}_{i \in I}$ be a sequence of continuous linear projections on X, $W_i = \pi_i(X)$ for each $i \in I$. Let $\{w_i\}_{i \in I}$ be a sequence of positive weights. The sequence $\{(w_i, W_i)\}_{i \in I}$ is said to be a fusion upper semi-frame for X if there exists a sequence $\{q_i \in B(W_i, X) : i \in I\}$ and a bounded operator $S \in B(X)$ such that
(i) The series $\sum_{i \in I} w_i^2 q_i \ o \ p_i(x)$ converges to $S(x)$;
(ii) There exists a solid $BK-$space X_d and positive constants A and B such that for every $x \in X$

$$0 \leq \|\{\|\pi_i(x)\|\}\|_{X_d} \leq B\|x\|. \tag{9}$$

5   Conclusion

So far there is no work on fusion semi-frames in Banach spaces. This opens up a new direction of research. Presently, the authors are working on this area. This paper is a short survey to get the readers into this area of research and the papers in our list of references will be useful to the readers.

References

1. Antoine, J.P., Balazs, P.: Frames and semi-frames. J. Phys. A Math. Theor. **44**(20), 205201 (2011)
2. Casazza, P.G., Kutyniok, G.: Frames of subspaces. Contemp. Math. **345**, 87–114 (2004)
3. Casazza, P.G., Christensen, O.: The reconstruction property in Banach spaces and perturbation theorem. Can. Math. Bull. **51**, 348–358 (2008)
4. Christensen, O., Heil, C.: Perturbations of Banach frames and atomic decompositions. Math. Nachr. **158**, 33–47 (1997)
5. Christensen, O.: An Introduction to Frames and Riesz Bases. Birkhäuser, Basel (2003)
6. Daubechies, I., Grossman, A., Meyer, Y.: Painless nonorthogonal expansions. J. Math. Phys. **27**, 1271–1283 (1986)
7. Duffin, R.J., Schaeffer, A.C.: A class of nonharmonic Fourier series. Trans. Am. Math. Soc. **72**, 341–366 (1952)
8. Giles, J.R.: Classes of semi-inner product spaces. Trans. Am. Math. Soc. **129**, 436–446 (1967)
9. Gröchenig, K.: Localization of frames, Banach frames, and the invertibility of the frame operator. J. Fourier Anal. Appl. **10**, 105–132 (2004)
10. Kaushik, S.K.: A generalization of frames in Banach spaces. J. Contemp. Math. Anal. **44**, 212–218 (2009)
11. Khosravi, A., Khosravi, B.: Fusion frames and G-frames in Banach spaces. Proc. Indian Acad. Sci. (Math. Sci.) **121**, 155–164 (2011)
12. Koehler, D.O.: A note on some operator theory in certain semi-inner product spaces. Proc. Am. Math. Soc. **30**, 363–366 (1971)
13. Lumer, G.: Semi-inner product spaces. Trans. Am. Math. Soc. **100**, 29–43 (1961)
14. Sahu, N.K., Mohapatra, R.N.: Frames in semi-inner product spaces. Mathematical Analysis and its Applications. PROMS, vol. 143, pp. 149–158. Springer, Heidelberg (2015). https://doi.org/10.1007/978-81-322-2485-3_11
15. Stoeva, D.T.: On p-frames and reconstruction series in separable Banach spaces. Integral Transforms Spec. Funct. **17**, 127–133 (2006)
16. Stoeva, D.T.: X_d frames in Banach spaces and their duals. Int. J. Pure Appl. Math. **52**, 1–14 (2009)
17. Zhang, H., Zhang, J.: Frames, Riesz bases, and sampling expansions in Banach spaces via semi-inner products. Appl. Comput. Harmon. Anal. **31**, 1–25 (2011)

A Study on Complexity Measure
of Diamond Tile Self-assembly System

M. Nithya Kalyani⬥, P. Helen Chandra$^{(\boxtimes)}$⬥, and S. M. Saroja T. Kalavathy⬥

Jayaraj Annapackiam College for Women (Autonomous),
Periyakulam, Theni District, Tamilnadu, India
rnithraj@gmail.com, chanrajac@yahoo.com, kalaoliver@gmail.com

Abstract. Molecular self-assembly gives rise to a great diversity of complex forms from crystals and DNA helices to microtubules and holoenzymes. We study a formal self-assembly model called the Diamond Tile Assembly System in which a diamond tile may be added to the growing object when the total interaction strength with its neighbours exceeds a parameter $\mathcal{T}$. Self-assembled objects can also be studied from the point of view of computational complexity. Here, we define the program-size complexity of an $N \times N$ diamond to be the minimum number of distinct tiles required to self-assemble the diamond. We study this complexity under the Diamond Tile Assembly Model and find a dramatic decrease in complexity from N^2 tiles to $O(logN)$ tiles, as $\mathcal{T}$ is increased from 1 where bonding is non co-operative to 2 allowing co-operative bonding. Further, we observe that the size of the largest diamond uniquely produced by a set of n tiles grows faster than any computable function.

Keywords: Self-assembly · Diamond Tile Assembly
Program-size complexity

1 Introduction

Self-assembly is the process by which a collection of relatively simple components, beginning in a disorganized state, spontaneously and without external guidance coalesce to form more complex structures. The process is guided by only local interactions between the components, which typically follow a basic set of rules. Despite the seemingly simplistic nature of self-assembly, its power can be harnessed to form structures of incredible complexity and intricacy. In order to model such systems, theoretical models have been developed and one of the most popular among these is the Tile Assembly Model introduced by Erik Winfree in his Ph.D. thesis [Wi2]. The complexity of self-assembled shapes is investigated in [LL1, SE1, Su1].

Branched DNA molecules [Se1] provide a direct physical motivation for the Tile Assembly Model. DNA double-crossover molecules, each bearing four *sticky ends* analogous to the four sides of a Wang tile, have been designed to self-assemble into a periodic two dimensional lattice [WL1]. The binding interactions between double-crossover molecules may be redesigned by changing the base

© Springer Nature Singapore Pte Ltd. 2018
D. Ghosh et al. (Eds.): ICMC 2018, CCIS 834, pp. 194–204, 2018.
https://doi.org/10.1007/978-981-13-0023-3_19

sequence of their sticky ends, thus allowing arbitrary sets of molecular Wang tiles. From a physically-based stochastic model of such a system, the Tile Assembly Model is obtained in the limit of strong binding domains and low monomer concentrations [Ra1,Wi1]. This model is an extension of the theory of Wang tiles [Wa1] to include a specific mechanism for growth based on the physics of molecular self-assembly.

A *program* consists of a finite set of unit diamond tiles with sides having molecular binding domain and thus each side has an associated *binding strength*, which in our model must be an integer. Starting from a chosen seed tile, growth occurs by addition of single tiles. Tiles bind a growing assembly only if their binding interactions are of sufficient strength as determined by the *temperature* parameter T. T measures the *co-operativity* of the binding interactions. It is interesting to observe that cooperative effects play a major role in gene regulation and many other biological systems.

In this paper, we introduce a new model called Diamond Tile Self-assembly System. It is a formal model for the self-assembly of molecules, such as protein or DNA, constrained to self-assembly on a diamond lattice. We measure the complexity of self-assembly by considering diamond instead of square [RE1]. Standard complexity measures in computer science are based on time, space, program size and decidability. Here, we discuss the program-size complexity of self-assembled diamonds, where complexity is measured by the number of distinct tile types involved.

2   Diamond Tile Self-assembly System

In this section, we introduce a new model called Diamond Tile Self-assembly System.

Definition 1. *A **Diamond Tile Self-assembly System** D_{TAS} is defined by the quadruple*

$$\mathbb{T} = <T, S, g, T>$$

where T is a finite set of diamond tile types containing empty, S is a seed assembly with finite domain, g is a strength function and $T \geq 0$ is the temperature. We consider only $|S| = 1$, where $S = A_s^{(0,0)}$.

*Diamond tile self-assembly is defined by a relation between configurations: $A \rightarrow_{\mathbb{T}} B$ if there exists a diamond tile $t \in T$ and a site (x, y) such that $B = A + A_t^{(x,y)}$ and B is T-stable. In particular, at $T = 1$, a diamond tile may be added if it makes any bond to a neighbour, whereas at $T = 2$, the diamond tile to be added must either make two weak bonds or a single strong bond. $\rightarrow_{\mathbb{T}}^*$ is the reflexive and transitive closure of $\rightarrow_{\mathbb{T}}$. The diamond tile self-assembly system defines a partially ordered set, the **produced** assemblies $D_{Prod(\mathbb{T})}$ where*

$$D_{Prod(\mathbb{T})} = \{A, \exists S \in T \ s.t. \ S \rightarrow_{\mathbb{T}}^* A\} \ and \ A \leq B \ if \ A \rightarrow_{\mathbb{T}}^* B.$$

*Another set, the **terminal** assemblies $D_{term(\mathbb{T})}$ is defined as the maximal elements of $D_{Prod(\mathbb{T})}$:*

$$D_{Term(\mathbb{T})} = \{A \in D_{Prod(\mathbb{T})}, \nexists B \ s.t. \ A < B\}.$$

The produced assemblies include intermediate products of the self-assembly process, whereas the terminal assemblies are just the end products and may be considered as the output. If

$$A \in D_{Prod(\mathbb{T})} \Rightarrow \exists B \in D_{Term(\mathbb{T})} \ s.t. \ A \rightarrow_{\mathbb{T}}^{*} B$$

then $\mathbb{T}$ *is said to be* **haltable***, in the sense that every path of self-assembly can eventually terminate. If* $\mathbb{T}$ *is haltable and* $D_{Term(\mathbb{T})}$ *is finite,* $\mathbb{T}$ *is said to be* **halting** *in the sense that every path of self-assembly does eventually terminate. In general, if* $D_{Prod(\mathbb{T})}$ *is a lattice, we say that* $\mathbb{T}$ *produces a* **unique pattern-**$\mathbb{T}$ *need not be halting nor even haltable.*

Example 1. Consider the Diamond Tile Self-assembly System $\mathbb{T} = <T, S, g, \mathcal{T}>$ where

$$T = \left\{ \langle S \rangle, \langle SW \rangle, \langle SE \rangle, \langle 0 \rangle, \langle 1 \rangle, \langle 0 \rangle, \langle 1 \rangle \right\}, \ S = \left\{ \langle S \rangle \right\},$$

$g = Strength \ Function, \mathcal{T} = 2.$

The tile set T, consists of four diamond rule tiles with strength-1 binding domains, two border diamond tiles with strength-1 and 2 binding domains and one seed diamond tile with strength-2 binding domains. At $\mathcal{T} = 2$, these tiles

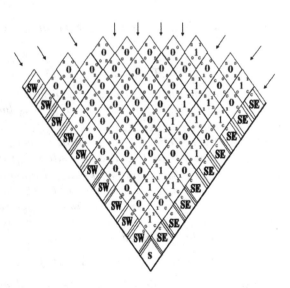

Fig. 1. Simulating a binary counter with diamond self-assembly

count in binary; the nth row above the origin represents the integer n (which is rotated 45° anticlockwise). This self-assembly *program* is analogous to an infinite loop-there are no terminal assemblies. Diamond rule tiles may be added only if both their southwest and southeast neighbors are already in place and there is a unique diamond rule tile for each possible pair of binding domains the neighbors could present; furthermore, the property that only northwest and northeast sides exposed in the assembly is preserved from step to step. The computation is possible when the system temperature $= 2$ and at least two strength -1 bonds must cooperate for a tile to be added to an assembly. The assembly is not terminal and arrows indicate positions at which may grow. The picture pattern generated by D_{TAS} is shown in Fig. 1.

3   Complexity of Diamond Self-assembly

In this section, we introduce Complexity of Diamond Self-Assembly. We will be measuring program-size complexity using asymptotic notion.

All functions will be from $\mathbb{N} \to \mathbb{N}$. A function $f(n)$ is **non-decreasing** iff $\forall n$, $f(n) \leq f(n+1)$. A function $f(n)$ is **un bounded** iff $\forall c$, $\exists n$ s.t. $f(n) \geq c$. We say $f(n) = O(g(n))$ iff $\exists c, n_0$ s.t. $\forall n > n_0$, $f(n) \leq cg(n)$. We say $f(n) = \Omega(g(n))$ iff $\exists c, n_0$ s.t. $\forall n > n_0$, $f(n) \geq cg(n)$. We assert proposition $P(n)$ **infinitely often** iff $\forall n_0 > 0$, $\exists n > n_0$ s.t. $P(n)$. Define $O_{i.o.}$ (big-O infinitely often) such that $f(n) = O_{i.o.}(g(n))$ iff $\exists c$ s.t. $f(n) \leq cg(n)$ infinitely often. We assert proposition $P(n)$ **for almost all** n iff $lim_{n_0 \to \infty} \frac{|1 \leq n \leq n_0 s.t. P(n)|}{n_0} = 1$. Define $\Omega_{a.a}$ (big-Ω almost always) such that $f(n) = \Omega_{a.a}(g(n))$ iff $\exists c$ s.t. $f(n) \geq cg(n)$ for almost all n.

We can now formally describe the program-size complexity of an $N \times N$ diamond. An assembly A is an $N \times N$ **diamond** if there exists a site (x_0, y_0) such that $(x, y) \in A$ iff $x \geq x_0$ and $x < x_0 + N$ and $y \geq y_0$ and $y < y_0 + N$. In other words the choice of tiles may be arbitrary, so long as they are there. Diamond A is a **full diamond** if for all (x, y) and $(x', y') \in A$ such that (x, y) and (x', y') are neighbours (x, y) and (x', y') bind with non-zero strength. In other words, every adjacent pair of tiles must have non-zero interaction strength. We are interested in which diamonds can be self-assembled by tile systems:

$$D^{\mathcal{T}} = \{(N, n) \in \mathbb{N} \times \mathbb{N} \text{ s.t. there exists a tile system}$$
$$\mathbb{T} = <T, S, g, \mathcal{T}>, |T| = n + 1 \text{ and } \mathbb{T}$$
$$\text{uniquely produces an } N \times N \text{ full diamond}\}.$$

We define the program size complexity $\mathcal{K}_{DA}^{\mathcal{T}}(N)$ of a diamond to be the minimum number of distinct non-empty tiles required to uniquely produce the diamond-physically the number of distinct types of molecules that must be prepared.

$$\mathcal{K}_{DA}^{\mathcal{T}}(N) = min\{n \, s.t. \, (N, n) \in D^{\mathcal{T}}\}$$

Our investigations rely on several constructions. We need an easy way to verify that these constructions do indeed *uniquely* produce the target structure. For each construction, the argument is an elaboration of the argument given for the binary counter tiles, only now an assembly may have more than one diagonal growth front. Specifically, the property that is preserved from step to step is that the assembly is *stop-sign*-shaped: the orientations of the exposed sides along the (clockwise) perimeter are of the form

$$NE^*\{NE, SE\}^* SE^*\{SE, SW\}^* SW^*\{SW, NW\}^* NW^*\{NW, NE\}^*.$$

These arguments rely on showing that there is exactly one strength-2 bond joining each row and each column.

We begin by studying $\mathcal{K}^{\mathcal{T}}_{DA}(N)$ for $\mathcal{T} = 1$ and obtain the following theorem.

Theorem 1. $\mathcal{K}^1_{DA}(N) = N^2$.

Proof. *To show $\mathcal{K}^1_{DA}(N) \leq N^2$, we construct N^2 diamond tiles, one for each position in the diamond, with a unique strength-1 binding domain for each adjacent pair of diamond tiles as in Fig. 2. In Fig. 2 (a): $N^2 = 16$ tiles with unique side labels uniquely produce a terminal 4×4 full diamond at $\mathcal{T} = 1$. (b): $2N - 1 = 7$ tiles uniquely produce a 4×4 diamond (but this is not a full diamond since thick sides have strength 0). Except for the sides labels with a circle, each interacting pair of tiles share a unique side label. This construction is conjectured to be minimal for diamonds assembled at $\mathcal{T} = 1$.*

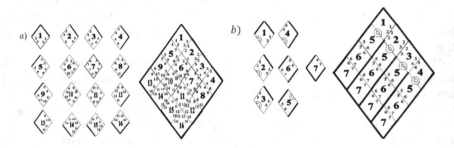

Fig. 2. Formation of diamonds at $\mathcal{T} = 1$.

To show $\mathcal{K}^1_{DA}(N) \geq N^2$, suppose a diamond tile set T with $|T| < N^2$ produces an $N \times N$ full diamond A (Fig. 3). In Fig. 3, a full $N \times N$ diamond with fewer than N^2 diamond tiles must have some tile i present at two sites. Consider the assembly W (the white diamond tiles) which includes an assembly V (bounded diamond tile i), the seed tile S and a tile that connects the seed tile to V. W can be extended indefinitely with the addition of translated segments of V (e.g. V^2_{+1} shown in gray). Then some tile i is present at two sites in A, say (x_1, y_1) and (x_2, y_2).

Let V be the V-shaped (or possibly linear) assembly consisting only of the tiles at $(x_1, y_1), \ldots, (x_2, y_2)$; let V^1 be the assembly such that $V^1 + (x_2, y_2) = V$; let V^2 be the assembly such that $V^2 + (x_1, y_1) = V$; let $V^k_n = [V^k(x + n * (x_2 -$

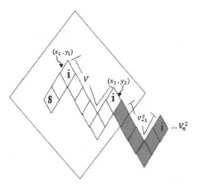

Fig. 3. No $\mathcal{T} = 1$ tile system with fewer than N^2 diamond tiles can uniquely produce an $N \times N$ diamond.

$x_1), y + n * (y_2 - y_1))$ be a translated version of V^k for $k = 1, 2$ and let W consist of V, S and the fewest diamond tiles in A required to connect S to V. Because W is contained in A and A is a full diamond, all adjacent pairs of diamond tiles interact on a strength-(at least)-1 side and therefore $S \to_{\mathbb{T}}^* W$. At least one of $\{V_{-1}^1, V_{+1}^1, V_{-1}^2, V_{+1}^2\}$, say V_s^r, can be added to W, resulting in a larger assembly also produced by $\mathbb{T}$. This can be continued indefinitely: if $s = +1$ then for all n, $W + \Sigma_{i=+1}^n V_i^r$ is in $D_{Prod(\mathbb{T})}$; if $s = -1$ then for all n, $W + \Sigma_{i=-n}^{-1} V_i^r$ is in $D_{Prod(\mathbb{T})}$. This contradicts the assumption that $\mathbb{T}$ is halting and terminates in $N \times N$ full diamonds.   ●

At $\mathcal{T} = 2$ the situation is markedly different.

Theorem 2. $\mathcal{K}_{DA}^2(N) = O(N)$.

Proof. *Figure 4 shows two constructions for an $N \times N$ full diamond using $2N$ (Fig. 4a) and $N + 4$ (Fig. 4b) diamond tiles respectively. Diamond tile self-assembly from the seed diamond tile A expands initially by single strength-2 interactions creating the northeast and northwest borders with the alphabetic diamond tiles. As the border grows, two cooperative strength-1 interactions allow the blank tile to fill in and complete the diamond.*

In Fig. 4b, diamond tile self-assembly from the seed diamond tile A expands initially by single strength-2 interactions creating the northeast border with the numbered diamond tiles. The U and V diamond tiles proceeds in the diagonal sides from west to east by their strength-2 interactions. Thus allowing the rest of the column to be filled with blanks. The $N \times N$ full diamond can be easily verified to be a terminal assembly.   ●

This is only the beginning. The construction in Fig. 4b can be combined with a fixed-width version of the binary counter of Fig. 1 to obtain a set of tiles that produce the full diamond by counting in binary instead of by counting in unary.

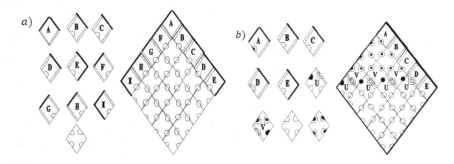

Fig. 4. Formation of full diamonds at $T = 2$.

Theorem 3. $\mathcal{K}_{DA}^2(N) = O(logN)$.

Proof. *Figure 5 constructs an $N \times N$ full diamond using $n + 22$ diamond tiles, where $n = \lceil logN \rceil$. $n + 2$ diamond tiles, including the diamond seed tile, produce an $(n - 1) \times (n - 1)$ diamond as in the previous construct (Fig. 4b). Here $N = 52, n = 6$ and 28 tiles are used. Additionally, the $n - 1$ diamond tiles in the seed row have northwest sides encoding the bits of the integer $c = 1 + 2^{n-1} - (N - n)/2$, the initial value of the counter. We must use a fixed-width version of the counter diamond tiles of Fig. 1; this requires a special set of diamond tiles for the southwest and southeast columns of bits. The counter counts from c to 2^{n-1} using two rows for each integer. In order to detect when the counter has finished, we use alternating rows to increment the counter from southeast to northwest then to copy of the bits from northwest to southeast unless the northwest bit just rolled over to northeast from 1 to 0. In the latter case, the diamond tile presents a strength-2 side with a label not found on any other diamond tiles, thus halting the counter. (The strength-2 side will be used in our next construction; here, any strength would suffice). There is a special diamond tile for the rightmost bit in the first increment row right the seed row. This diamond tile contains a strength-2 side to initiate the $u - v$ diagonal, thus filling in the rest of the diamond. Overall, the counter requires 17 tiles; the seed row requires $n - 1$ tiles; the two diagonals require 4 tiles and there are two blank tiles.*

We can do much better: by recursively iterating the above construction one can produce $N \times N$ diamonds with $N \geq \underbrace{2^{2^{2^{\cdots^2}}}}_{n \, times} \overset{def}{=} 2 * * n$. Define $log^* N$ as the least n such that $2 * * n \geq N$.

Theorem 4. $\mathcal{K}_{DA}^2(N) = O_{i.0.}(log^* N)$.

Proof. *Our proof is by induction. Let S^n refer to a diamond tile system containing fewer than $22n$ diamond tiles (including the u, v and blank diamond tiles) that uniquely produces an $N \times N$ full diamond such that*

- $N > 2 * * n$.
- *All binding domains on the northwest and northwest bottom are strength 1 or 0.*

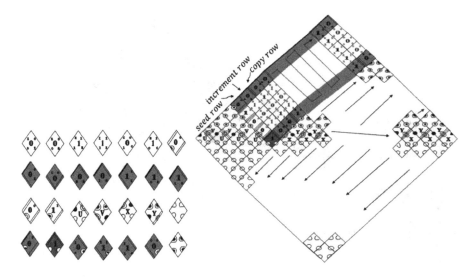

Fig. 5. Formation of an $N \times N$ diamond using $O(\log N)$ tiles.

- *All binding domains on the southwest have the strength-1 blank label.*
- *The binding domains on the northwest upper side conform to the pattern xy^*zb^*a where x is a strength-2 binding domain that occurs nowhere else and y, z, b and a are distinct strength-1 binding domains.*

We show that S^n exists for all n. The base case $n = 1$ is trivial. The inductive step is illustrated in Fig. 6. (In Fig. 6, given a set of tiles S^{n+1} that produce an $N \times N$ full diamond that satisfies recurrence, the addition of 22 new tiles results in S^{n+1} and produces a $(N + 2 \times 2^N) \times (N + 2 \times 2^N)$ full diamond. New side labels (with doubled symbols) prevent counter tiles from S^n from incorporating in the S^{n+1} counter). First, there are 5 diamond tiles that, initiated by x, produce an initial string of $0's$ for a new fixed-width counter and provide a strength-2 side for a new $\mathbf{u} - \mathbf{v}$ diagonal. Then there are 16 diamond tiles equivalent to the counter diamond tile in Theorem 3 but using new side labels; the counter counts to 2^N. The diagonal fills in the rest of the diamond, now with sides of length $N + 2 \times 2^N > 2^N > 2 * * (n + 1)$. Therefore S^n exist for all n and for those n,

$$22 \, log^* N \geq 22n \geq D^2_{DA}(N).$$

- *$log^* N$ is an exceedingly slowly growing function; the above construction shows that very large diamonds can be assembled with a very small number of diamond tiles. But we can do much better yet! By embedding the simulation of a Turing Machine in the growth of a diamond we show that:*

Theorem 5. $\mathcal{K}^2_{DA}(N) = O_{i.o.}(f(N))$ *for $f(N)$ any non-decreasing unbounded computable function.*

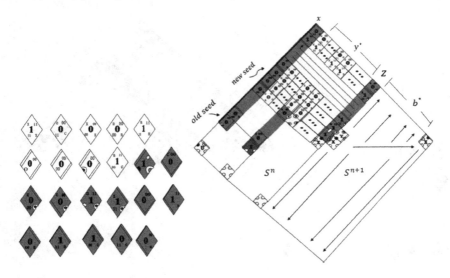

Fig. 6. Formation of an $N \times N$ diamond using $O_{i.o.}(log^* N)$

Proof. *Our proof relies on a self-assembly version of the Busy Beaver problem* [Ra1]. *Define:*

$$B_{DA}^T(n) = max\{N s.t. N, n) \in D^T\}.$$

To prove Theorem 5, we first show

$$B_{DA}^2(n) = \Omega(F(n)) \quad \text{for any computable function} \quad F(n). \tag{1}$$

Theorem 5 follows from (1) by contradiction: if false, then there exists a computable, non-decreasing, unbounded function $f(N)$ such that $\exists N_0$ s.t. $\forall N > N_0$, $\mathcal{K}_{DA}^2(N) \geq f(N)$.

Let $F(n) = max \{N$ s.t $N = 0$ or $f(N) \leq n\}$; this is a computational function. Note that $B_{DA}^2(n) \geq F(n)$ requires that $\exists (N, n) \in D^2$ s.t. $N \geq F(n)$ and therefore $f(N) > n$ and $\mathcal{K}_{DA}^2(N) \leq n$. For $N > N_0$ this contradicts $\mathcal{K}_{DA}^2(N) \geq f(N)$. Therefore, for all $n > f(N_0)$, $B_{DA}^2(n) < F(n)$, contradicting (1) and establishing Theorem 5.

Recall that $B_t(m) = \Omega(F'(m))$ for any computable function $F'(m)$ where:

$$B_t(m) = max\{t \text{ s.t. } m = qs \text{ and there exists a } q\text{-state, } s\text{-symbol}$$
$$\text{Turing machine that halts on a blank tape in } t \text{ steps}\}$$

Let M be a q-state, s-symbol Turing machine that halts on a blank tape in $B_t(m)$ steps, where $m = qs$. We will construct a diamond of size $N = 2B_t(m) + 3$ using $n = 12qs + 4s + 9$ diamond tiles by simulating M with tiles, similar to the construction of Robinson [Ro1]. Given any $n > 41$, we will use $s_n = 2$, $q_n = \lfloor \frac{n-17}{24} \rfloor$ and $m_n = q_n s_n$; our construction will need only $12q_n s_n + 4s_n + 9 < n$ diamond tiles. Then $B_{DA}^2(n) \geq 2B_t(m_n) + 3 = \Omega(F'(m_n))$. For any computable *function $F(n)$, we can find another computable function $F'(m)$ s.t. $\forall n$, $F'(m_n) >$ $F(n)$. Therefore, we arrive at (1). In Fig. 7, The Busy Beaver machine simulated here has three states ($q_0 = A$, $q_1 = B$, $q_2 = C$) and two symbols ($s_0 = 0$, $s_1 = 1$). Note that R denotes right, L denotes left and $4x$ indicates that four variations of a tile are used, one for each compass direction.*

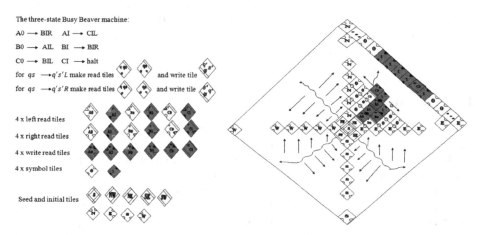

Fig. 7. Formation of an $N \times N$ diamond by growing four identical simulations of a given Turing machine.

We construct the diamond by growing four identical simulations of the Turing machine M, one from each side of a seed tile. Each simulation stays within one of the four regions bounded by the diagonals of the diamond; when M halts, the diamond is complete. We require 4 diamond tiles to create the four half-diagonals defining these boundaries between simulations. For each simulation we require 1 initial state that matches the seed diamond tile, s symbol diamond tiles, qs write diamond tiles and $2qs$ read diamond tiles giving a total of $3qs + s + 1$ diamond tiles per simulation. We describe these diamond tiles for the TM simulation to the northeast of the diamond seed tile. Recall that a diamond tile is a 4-tuple $(\sigma_{NW}, \sigma_{NE}, \sigma_{SE}, \sigma_{SW})$ representing the northwest, northeast, southeast and southwest binding domains. Binding domain strengths are 1 unless noted. Each of the four simulations has its own version of the side labels described, distinguished by superscripts (we omit the superscript ^{N} from the description of northwest facing simulation below).

The symbol diamond tile for symbol s is $(\sigma_s, \sigma_e, \sigma_s, \sigma_e)$ where σ_s is a binding domain representing the symbol s and σ_e is a binding domain indicating that the TM head is not present. For each state-symbol pair (q, s) the left read diamond tile $(\sigma_{q,s}, \sigma_e, \sigma_s, \sigma_q)$ and the right read tile $(\sigma_{q,s}, \sigma_q, \sigma_s, \sigma_e)$ represent the TM head in state q entering a tape cell (from the left or from the right) and reading the symbol s. The binding domain $\sigma_{q,s}$ have strength 2; this is necessary for the TM head to enter the next row of the simulation. The write diamond tiles, representing the action the TM head takes depend on the form of the state transition table entry. For each entry of the form $(q, s) \rightarrow (q', s', L)$ there is a write *diamond tile* $(\sigma_{s'}, \sigma_e, \sigma_{q,s}, \sigma_{q'})$; *for each entry of the form* $(q, s) \rightarrow (q', s', R)$ *there is a write diamond tile* $(\sigma_{s'}, \sigma_{q'}, \sigma_{q,s}, \sigma_e)$; *for each entry of the form* $(q, s) \rightarrow$ *halt there is a write diamond tile* $(\sigma_{halt}, \sigma_e, \sigma_{q,s}, \sigma_e)$.

To start the Turing Machine in state q_0 *reading the blank symbol* s_0, *the initial tile for the northeast simulation is* $NE = (\sigma_{q_0, s_0}, \sigma_e, \sigma_S, \sigma_e)$ *where* σ_S *is a strength-2 binding domain. The initial diamond tiles for all four simulations bind to the diamond seed tile* $S = (\sigma_S^{NW}, \sigma_S^{NE}, \sigma_S^{SE}, \sigma_S^{SW})$. *The four diagonal diamond tiles,* $N = (\sigma_{s_0}^N, \sigma_e^N, \sigma_e^W, \sigma_{s_0}^W)$, $E = (\sigma_{s_0}^N, \sigma_{s_0}^E, \sigma_e^E, \sigma_e^N)$, $S = (\sigma_e^E, \sigma_{s_0}^E, \sigma_{s_0}^S, \sigma_e^S)$ *and* $W = (\sigma_e^W, \sigma_e^S, \sigma_{s_0}^S, \sigma_{s_0}^W)$ *pad the tapes with extra cells containing the blank symbol* s_0 *and delimit the four simulations.*

4   Conclusion

This paper discussed the program-size complexity of self-assembled diamonds, where complexity is measured by the number of distinct diamond tile types involved. An alternative complexity measure is the minimum number of distinct side labels required uniquely to produce the object. The number of labels will be relevant in a physical system where the number of distinct binding interactions are limited due to imperfect specificity of binding. A main conclusion of this paper is that the program-size complexity of self-assembled objects (at $\mathcal{T} = 2$) looks remarkably similar to the usual program-size complexity with respect to Turing Machines.

References

[LL1] Lathrop, J.I., Lutz, J.H., Patitz, M.J., Summers, S.M.: Computability and complexity in self-assembly. Theory Comput. Syst. **48**(3), 617–647 (2011)

[Ra1] Rado, T.: On non-computable functions. Bell Syst. Tech. J. **41**(3), 877–884 (1962)

[RE1] Rothemund, P.W.K., Winfree, E.: The program-size complexity of self-assembled squares (extended abstract), Oregon, USA, pp. 1–37 (2000)

[Ro1] Robinson, R.M.: Undecidability and non periodicity of tilings of the plane. Inventions Math. **12**, 177–209 (1971)

[Se1] Seeman, N.C.: DNA nanotechnology: novel DNA constructions. Annu. Rev. Biophys. Biomol. Struct. **27**, 225–248 (1998)

[SE1] Soloveichik, D., Winfree, E.: Complexity of self-assembled shapes. SIAM J. Comput. **36**(6), 1544–1569 (2007)

[Su1] Summers, S.M.: Reducing tile complexity for the self-assembly of scaled shapes through temperature programming. Algorithmica **63**(1–2), 117–136 (2012)

[Wa1] Wang, H.: Proving theorems by pattern recognition, II. Bell Syst. Tech. J. **40**, 1–42 (1961)

[Wi1] Winfree, E.: Simulations of computing by self-assembly. In: Proceedings of the Fourth DIMACS Meeting on DNA Based Computers, pp. 1–27 (1998)

[Wi2] Winfree, E.: Algorithmic self-assembly of DNA. Ph.D. thesis, California Institute of Technology, June 1998

[WL1] Winfree, E., Liu, F., Wenzler, L.A., Seeman, N.C.: Design and self-assembly of two-dimensional DNA Crystals. Nature **394**, 539–544 (1998)

Exponential Spline Method for One Dimensional Nonlinear Benjamin-Bona-Mahony-Burgers Equation

A. S. V. Ravi Kanth$^{(\boxtimes)}$ and Sirswal Deepika

Department of Mathematics, National Institute of Technology, Kurukshetra,
Kurukshetra 136 119, Haryana, India
asvravikanth@yahoo.com

Abstract. In this paper, a numerical method based on exponential spline for solving one dimensional nonlinear Benjamin-Bona-Mahony-Burgers equation is presented. Stability analysis of the present method is analyzed by means of Von Neumann stability analysis and is proven to be unconditionally stable. Few numerical evidences are given to prove the validation of the proposed method.

Keywords: Benjamin-Bona-Mahony-Burgers equation
Exponential spline · Von Neumann technique

1 Introduction

Nonlinear partial differential equations are used in many real world applications [8]. To study such nonlinear partial differential equations, we analyze each individual nonlinear partial differential equation separately either by analytical methods or numerical methods. Benjamin-Bona-Mahony-Burgers (BBMB) equation is one among such nonlinear partial differential equations. Several analytic methods and numerical methods are developed for solving BBMB equation, for instance, Tanh Method [5], He's Variational Iteration Method [4], Homotopy Analysis Method [3], Adomian Decomposition Method [1], Lie Symmetry Method [6], Finite Difference Method [7], Radial Basis Function (RBF) Method [2] and Cubic B-spline Collocation Method [10].

In this article, we use the exponential spline method for solving one dimensional nonlinear BBMB equation

$$u_t - u_{xxt} - \zeta u_{xx} + \eta u_x + u u_x = 0, (x,t) \in [a,b] \times [0,T], \tag{1}$$

subject to the boundary conditions

$$u(a,t) = 0, u(b,t) = 0, t \in [0,T], \tag{2}$$

© Springer Nature Singapore Pte Ltd. 2018
D. Ghosh et al. (Eds.): ICMC 2018, CCIS 834, pp. 205–215, 2018.
https://doi.org/10.1007/978-981-13-0023-3_20

and the initial condition

$$u(x,0) = f(x), \ a \le x \le b, \tag{3}$$

represents the unidirectional propagation of long waves having small amplitudes in nonlinear dispersive media, $u(x,t)$ is the velocity of the fluid, x is the distance in the horizontal direction of propagation and ζ, η are the positive constants.

In comparison with the finite difference method, spline method provides the functional values and its derivative values between the mesh points. Functional values are only available at the chosen knots in the case of finite difference method. And in comparison with the finite element method, there is no need to evaluate the quadratures in spline method. The polynomial spline of lower order decreases the accuracy while the higher order leads the complexity of calculations, the use of exponential spline overcomes this problem because of its smoothing and handy nature.

The article is organised as follows: In Sect. 2, we first design the numerical scheme for the model problem (1)–(3). Then we discuss the stability estimates of the presented scheme in Sect. 3. In Sect. 4, we demonstrate the performance of the proposed numerical method by extensive numerical examples. Section 5 ends with conclusion.

2   Exponential Spline Method

For the positive integers N and M, let the partition of $[a,b] \times [0,T]$ be defined by $\Omega \ : \ \Omega_h \times \Omega_k$ where $\Omega_h \ = \ \{x_i \mid x_i = a + ih, 0 \le i \le N\}, \Omega_k = \{t_j \mid t_j = jk, 0 \le j \le M\}$ and $k = \frac{T}{M}, h = \frac{b-a}{N}$ are the temporal and spatial step size respectively.

Let $Z_i^j = Z(x_i, t_j)$ be an approximation to $u_i^j = u(x_i, t_j)$ obtained by the segment $P_i(x, t_j)$ of the mixed spline function passing through the points (x_i, Z_i^j) and (x_{i+1}, Z_{i+1}^j). Each segment can be written as [9]

$$P_i(x, t_j) = a_i(t_j)e^{\omega(x-x_i)} + b_i(t_j)e^{-\omega(x-x_i)}$$
$$+ c_i(t_j)(x - x_i) + d_i(t_j), \tag{4}$$

for each $i = 0, 1, \ldots, N - 1$. To obtain the coefficients $a_i(t_j), b_i(t_j), c_i(t_j)$ and $d_i(t_j)$ in terms of Z_i^j, Z_{i+1}^j, S_i^j and S_{i+1}^j, we define

$$\begin{cases} P_i(x_i, t_j) = Z_i^j, \\ P_i(x_{i+1}, t_j) = Z_{i+1}^j, \\ P_i^{(2)}(x_i, t_j) = S_i^j, \\ P_i^{(2)}(x_{i+1}, t_j) = S_{i+1}^j, \end{cases} \tag{5}$$

where

$$P_i^{(2)}(x, t) = \frac{\partial^2}{\partial x^2} P_i(x, t).$$

By using Eqs. (4) and (5), we obtain

$$
\begin{cases}
a_i = \frac{h^2(S_{i+1}^j - e^{-\theta}S_i^j)}{2\theta^2 \sinh(\theta)}, \\
b_i = \frac{h^2(e^{\theta}S_i^j - S_{i+1}^j)}{2\theta^2 \sinh(\theta)}, \\
c_i = \frac{(Z_{i+1}^j - Z_i^j)}{h} - \frac{h(S_{i+1}^j - S_i^j)}{\theta^2}, \\
d_i = Z_i^j - \frac{h^2 S_i^j}{\theta^2},
\end{cases}
\tag{6}
$$

where $a_i = a_i(t_j)$, $b_i = b_i(t_j)$, $c_i = c_i(t_j)$, $d_i = d_i(t_j)$ and $\theta = h\omega$.

Using the continuity condition of first derivative at $x = x_i$

$$
P_i^{(1)}(x_i, t_j) = P_{i-1}^{(1)}(x_i, t_j).
$$

Equations (4) and (6) yield the relation

$$
\alpha\, S_{i-1}^j + 2\beta\, S_i^j + \alpha\, S_{i+1}^j = \frac{1}{h^2}\left[Z_{i-1}^j - 2\,Z_i^j + Z_{i+1}^j\right],
$$
$$
i = 1, 2, \ldots, N-1, j \geq 0,
\tag{7}
$$

where

$$
\alpha = \frac{\sinh(\theta) - \theta}{\theta^2 \sinh(\theta)}, \beta = \frac{\theta \cosh(\theta) - \sinh(\theta)}{\theta^2 \sinh(\theta)}.
$$

Considering Eq. (7) at $(j+1)^{th}$ time level, we have

$$
\alpha\, S_{i-1}^{j+1} + 2\beta\, S_i^{j+1} + \alpha\, S_{i+1}^{j+1} = \frac{1}{h^2}\left[Z_{i-1}^{j+1} - 2\,Z_i^{j+1} + Z_{i+1}^{j+1}\right],
$$
$$
i = 1, 2, \ldots, N-1, j \geq 0.
\tag{8}
$$

The discretized representation of Eq. (1) at the grid point (x_i, t_j) is given by

$$
(Z_t)_i^j - (Z_{xxt})_i^j - \zeta(Z_{xx})_i^j + \eta(Z_x)_i^j + (ZZ_x)_i^j = 0.
\tag{9}
$$

To incorporate the proposed method, we first discretize the time derivative in the usual forward finite difference way and then applying the Crank-Nicolson scheme for Eq. (9), we get

$$
\frac{Z_i^{j+1} - Z_i^j}{k} - \frac{(Z_{xx})_i^{j+1} - (Z_{xx})_i^j}{k} - \zeta\frac{(Z_{xx})_i^{j+1} + (Z_{xx})_i^j}{2}
$$
$$
+ \eta\frac{(Z_x)_i^{j+1} + (Z_x)_i^j}{2} + \frac{(ZZ_x)_i^{j+1} + (ZZ_x)_i^j}{2} + O(k^2 + h^2) = 0.
\tag{10}
$$

Using the Taylor series, we have the following finite difference approximation

$$
(ZZ_x)_i^{j+1} = Z_i^{j+1}(Z_x)_i^j + Z_i^j(Z_x)_i^{j+1} - Z_i^j(Z_x)_i^j + O(k^2),
\tag{11}
$$

$$
(Z_x)_{i-1}^j = \frac{-Z_{i+1}^j + 4Z_i^j - 3Z_{i-1}^j}{2h} + O(h^2),
\tag{12}
$$

$$(Z_x)_i^j = \frac{Z_{i+1}^j - Z_{i-1}^j}{2h} + O(h^2), \tag{13}$$

$$(Z_x)_{i+1}^j = \frac{3Z_{i+1}^j - 4Z_i^j + Z_{i-1}^j}{2h} + O(h^2). \tag{14}$$

Using Eqs. (11)–(14) in Eq. (10) and rearranging the terms, we obtain

$$(\zeta k + 2)(Z_{xx})_i^{j+1} + (\zeta k - 2)(Z_{xx})_i^j = 2\left(Z_i^{j+1} - Z_i^j\right) + \frac{k}{2h}\left(\eta + Z_i^j\right)$$
$$\left(Z_{i+1}^{j+1} - Z_{i-1}^{j+1}\right) + \frac{k}{2h}\left(\eta + Z_i^{j+1}\right)\left(Z_{i+1}^j - Z_{i-1}^j\right) + O(k^3 + kh^2). \tag{15}$$

Multiplying Eqs. (7)–(8) by the factors $(\zeta k - 2)$ and $(\zeta k + 2)$, respectively and adding them, we get

$$(\zeta k + 2)\left[\alpha S_{i-1}^{j+1} + 2\beta S_i^{j+1} + \alpha S_{i+1}^{j+1}\right] + (\zeta k - 2)\left[\alpha S_{i-1}^j + 2\beta S_i^j + \alpha S_{i+1}^j\right]$$
$$= \frac{1}{h^2}(\zeta k + 2)\left[Z_{i-1}^{j+1} - 2Z_i^{j+1} + Z_{i+1}^{j+1}\right] + \frac{1}{h^2}(\zeta k - 2)\left[Z_{i-1}^j - 2Z_i^j + Z_{i+1}^j\right]. \tag{16}$$

Using the operator notations $E(Z(x,t)) = Z(x+h,t), D(Z(x,t)) = Z_x(x,t), I(Z(x,t)) = Z(x,t), E = e^{hD}$, we obtain

$$(\zeta k + 2)S_i^{j+1} + (\zeta k - 2)S_i^j = \frac{1}{h^2}\left(\frac{E^{-1} - 2I + E^{+1}}{\alpha E^{-1} + 2\beta I + \alpha E^{+1}}\right)$$
$$\left((\zeta k + 2)Z_i^{j+1} + (\zeta k - 2)Z_i^j\right). \tag{17}$$

Expanding them in powers of hD, we obtain

$$(\zeta k + 2)S_i^{j+1} + (\zeta k - 2)S_i^j = \frac{1}{2(\alpha + \beta)}\left((\zeta k + 2)(Z_{xx})_i^{j+1} + (\zeta k - 2)(Z_{xx})_i^j\right)$$
$$+ \frac{1 - 6\gamma}{24(\alpha + \beta)}h^2\left((\zeta k + 2)(Z_{xxxx})_i^{j+1} + (\zeta k - 2)(Z_{xxxx})_i^j\right) + O(h^4), \tag{18}$$

where $\gamma = \frac{\alpha}{\alpha+\beta}$.
Substituting Eq. (15) in Eq. (18), we get

$$(\zeta k + 2)S_i^{j+1} + (\zeta k - 2)S_i^j = \frac{1}{(\alpha + \beta)}\left(Z_i^{j+1} - Z_i^j\right) + \frac{k}{4h(\alpha + \beta)}\left(\eta + Z_i^j\right)$$
$$\left(Z_{i+1}^{j+1} - Z_{i-1}^{j+1}\right) + \frac{k}{4h(\alpha + \beta)}\left(\eta + Z_i^{j+1}\right)\left(Z_{i+1}^j - Z_{i-1}^j\right)$$
$$+ \frac{1 - 6\gamma}{24(\alpha + \beta)}h^2\left((\zeta k + 2)(Z_{xxxx})_i^{j+1} + (\zeta k - 2)(Z_{xxxx})_i^j\right) + O(k^3 + kh^2 + h^4). \tag{19}$$

In order to eliminate S_i^j from Eq. (16) substituting Eq. (19) in Eq. (16), we get

$$
\left[-\frac{(\zeta k + 2)}{h^2} + \frac{\alpha}{\alpha + \beta} - \frac{k\eta}{2h} - \frac{3k\alpha}{2h(\alpha + \beta)} Z_{i-1}^j - \frac{(\beta - 2\alpha)k}{2h(\alpha + \beta)} Z_i^j \right] Z_{i-1}^{j+1}
$$

$$
+ \left[\frac{2(\zeta k + 2)}{h^2} + \frac{2\beta}{(\alpha + \beta)} + \frac{k(\beta - 2\alpha)}{2h(\alpha + \beta)} Z_{i+1}^j - \frac{k(\beta - 2\alpha)}{2h(\alpha + \beta)} Z_{i-1}^j \right] Z_i^{j+1}
$$

$$
+ \left[-\frac{(\zeta k + 2)}{h^2} + \frac{\alpha}{\alpha + \beta} + \frac{k\eta}{2h} + \frac{3k\alpha}{2h(\alpha + \beta)} Z_{i+1}^j + \frac{(\beta - 2\alpha)k}{2h(\alpha + \beta)} Z_i^j \right] Z_{i+1}^{j+1}
$$

$$
= \left[\frac{\zeta k - 2}{h^2} + \frac{\alpha}{\alpha + \beta} + \frac{k\eta}{2h} \right] Z_{i-1}^j + \left[\frac{-2(\zeta k - 2)}{h^2} + \frac{2\beta}{\alpha + \beta} \right] Z_i^j
$$

$$
+ \left[\frac{\zeta k - 2}{h^2} + \frac{\alpha}{\alpha + \beta} - \frac{k\eta}{2h} \right] Z_{i+1}^j + T_i^j,
$$

$$
i = 1, 2, \ldots, N - 1, j \geq 0. \quad (20)
$$

where

$$
T_i^j = \frac{1 - 6\gamma}{24(\alpha + \beta)} h^2 \left\{ \alpha \left((\zeta k + 2)(Z_{xxxx})_{i-1}^{j+1} + (\zeta k - 2)(Z_{xxxx})_{i-1}^j \right) \right.
$$

$$
+ 2\beta \left((\zeta k + 2)(Z_{xxxx})_i^{j+1} + (\zeta k - 2)(Z_{xxxx})_i^j \right)
$$

$$
\left. + \alpha \left((\zeta k + 2)(Z_{xxxx})_{i+1}^{j+1} + (\zeta k - 2)(Z_{xxxx})_{i+1}^j \right) \right\} + O(k^3 + kh^2 + h^4).
$$

is the truncation error of the system (20). The system (20) contains $(N - 1)$ equations with $(N + 1)$ unknowns. To get a solution to this system, we need two additional equations. These equations are obtained from the boundary conditions (2), the discretized form for the boundary conditions are

$$
Z_0^j = 0, Z_N^j = 0, j \geq 0.
$$

Remark: It is clear from the system (20) that:

(i) For suitable arbitrary values of α, β, our scheme is of $O(k^2 + h^2)$.
(ii) For $\gamma = \frac{1}{6}$ i.e, $\beta = 5\alpha$, our scheme is of $O(k^3 + kh^2 + h^4)$.

3   Stability Analysis

The stability has been proven by using the Von Neumann technique. We replace the nonlinear term in Eq. (20) by a local constant (d^*) and the numerical solution can be expressed by means of a Fourier series

$$
Z_i^j = \xi^j \exp(n\phi ih), \quad (21)
$$

where $n = \sqrt{-1}$, ϕ is the wave number and ξ^j is the amplitude at the jth time level. Substituting Eq. (21) in Eq. (20), we obtain

$$\xi^{j+1}\left\{\left(-\frac{(\zeta k+2)}{h^2}+\frac{\alpha}{\alpha+\beta}-\frac{k(\eta+d^*)}{2h}\right)\exp(n(i-1)\sigma)+\left(\frac{2(\zeta k+2)}{h^2}+\frac{2\beta}{\alpha+\beta}\right)\right.$$

$$\left.\exp(ni\sigma)+\left(-\frac{(\zeta k+2)}{h^2}+\frac{\alpha}{\alpha+\beta}+\frac{k(\eta+d^*)}{2h}\right)\exp(n(i+1)\sigma)\right\}$$

$$=\xi^{j}\left\{\left(\frac{(\zeta k-2)}{h^2}+\frac{\alpha}{\alpha+\beta}+\frac{k\eta}{2h}\right)\exp(n(i-1)\sigma)+\left(\frac{-2(\zeta k-2)}{h^2}+\frac{2\beta}{\alpha+\beta}\right)\right.$$

$$\left.\exp(ni\sigma)+\left(\frac{(\zeta k-2)}{h^2}+\frac{\alpha}{\alpha+\beta}-\frac{k\eta}{2h}\right)\exp(n(i+1)\sigma)\right\},$$

$$(22)$$

where $\sigma = \phi h$. Using the Euler's formula in Eq. (22), we get

$$\xi = \frac{X_1+nY_1}{X_2+nY_2},$$

where

$$X_1 = \frac{-2(\zeta k-2)}{h^2}(1-\cos\theta)+\frac{2}{\alpha+\beta}(\alpha\cos\theta+\beta),$$

$$X_2 = \frac{2(\zeta k+2)}{h^2}(1-\cos\theta)+\frac{2}{\alpha+\beta}(\alpha\cos\theta+\beta),$$

$$Y_1 = \frac{-k\eta}{h}\sin\theta, Y_2 = \frac{k(\eta+d^*)(\alpha+\beta)}{h}\sin\theta.$$

For stability we need $|\xi| \le 1$, for this we require $A = X_1^2 + Y_1^2 - X_2^2 - Y_2^2 \le 0$. Since

$$A = \frac{-16\zeta k(1-\cos\theta)}{h^4}\left[2(1-\cos\theta)+\frac{h^2}{(\alpha+\beta)}(\alpha\cos\theta+\beta)\right] \qquad (23)$$

$$-\frac{k^2\sin^2\theta}{h^2}\left[(\eta+d^*)^2-\eta^2\right].$$

As $A \le 0$ provided $\alpha > 0$ and $\beta > 0$, so $|\xi| \le 1$ thus the proposed method is unconditionally stable.

4 Numerical Results

To show the applicability and efficiency of the proposed numerical scheme, we solve the following problems:

Example-1: Consider the one dimensional nonlinear homogenous BBMB equation

$$u_t - u_{xxt} - \zeta u_{xx} + \eta u_x + uu_x = 0,$$
$$(x,t) \in [a,b] \times [0,T], \qquad (24)$$

subject to the boundary conditions

$$u(a,t) = g_0(t) = 0, u(b,t) = g_1(t) = 0, t \in [0,T], \qquad (25)$$

and the initial condition

$$u(x,0) = f(x) = \exp(-x^2), \ a \le x \le b. \tag{26}$$

Table 1 shows the maximum absolute errors for various values of h at time levels $t = 5$ and $t = 10$ for $0 \le x \le 1$. It can be observe from the table that as the spatial mesh points are increasing, the error norm L_∞ is decreasing. In Fig. 1, the numerical solutions at the various time levels are presented for $-10 \le x \le 10$.

Table 1. Maximum absolute errors for Example 1 at $k = 1/20, \alpha = 1/6, \beta = 1/3, \zeta = 1$ and $\eta = 1$.

h	$t = 5$	$t = 10$
	L_∞	L_∞
1/4	1.5727×10^{-3}	1.7380×10^{-5}
1/8	9.7825×10^{-4}	1.4682×10^{-5}
1/16	5.4456×10^{-4}	9.1292×10^{-6}
1/32	2.8622×10^{-4}	5.0628×10^{-6}
1/64	1.4668×10^{-4}	2.6598×10^{-6}
1/128	7.4255×10^{-5}	1.3628×10^{-6}
1/256	3.7355×10^{-5}	6.8978×10^{-7}

Example-2: Consider the one dimensional nonlinear homogenous BBMB equation

$$u_t - u_{xxt} - \zeta u_{xx} + \eta u_x + u u_x = 0,$$
$$(x,t) \in [0,1] \times [0,T], \tag{27}$$

subject to the boundary conditions

$$u(0,t) = 0, u(1,t) = 0, t \in [0,T], \tag{28}$$

and the initial condition

$$u(x,0) = f(x) = \exp(-x)\sin(\pi x), \ 0 \le x \le 1. \tag{29}$$

The physical interpretation of Example 2 is plotted in Figs. 2 and 3. There is dispersive effects for various values of η and fixed ζ which is shown in Fig. 2. From Fig. 3, it is noticed that there is no dissipative effects for various values of ζ with fixed η.

Example-3: Consider the one dimensional nonlinear non-homogenous BBMB equation

$$u_t - u_{xxt} - \zeta u_{xx} + \eta u_x + u u_x = g(x,t),$$
$$(x,t) \in [0,\pi] \times [0,T], \tag{30}$$

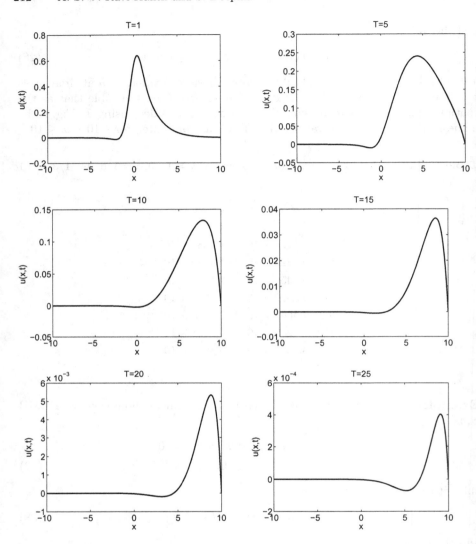

Fig. 1. Graphs of numerical solution for Example 1 with $h = 1/10, k = 1/100,$ $\alpha = 1/12, \beta = 5/12, \zeta = 1$ and $\eta = 1$.

subject to the boundary conditions

$$u(0, t) = 0, u(\pi, t) = 0, t \in [0, T], \tag{31}$$

and the initial condition

$$u(x, 0) = f(x) = \sin x, \ 0 \le x \le \pi, \tag{32}$$

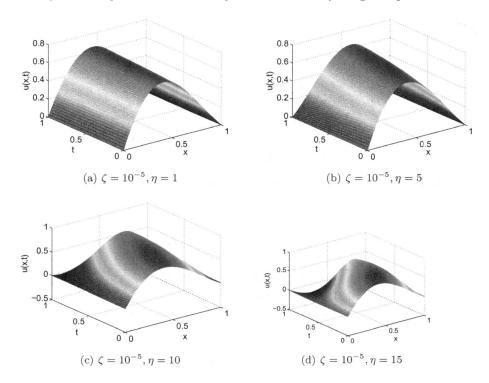

(a) $\zeta = 10^{-5}, \eta = 1$ (b) $\zeta = 10^{-5}, \eta = 5$

(c) $\zeta = 10^{-5}, \eta = 10$ (d) $\zeta = 10^{-5}, \eta = 15$

Fig. 2. Graphs of numerical solution for Example 2 with $h = 1/100, k = 1/100,$ $\alpha = 1/12, \beta = 5/12$ and $T = 1,$

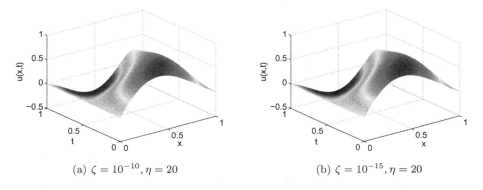

(a) $\zeta = 10^{-10}, \eta = 20$ (b) $\zeta = 10^{-15}, \eta = 20$

Fig. 3. Graphs of numerical solution for Example 2 with $h = 1/100, k = 1/100,$ $\alpha = 1/12, \beta = 5/12$ and $T = 1.$

where $g(x,t) = \exp(-t) \left[\cos x - \sin x + \frac{1}{2}\exp(-t)\sin(2x)\right].$ The exact solution of this problem is

$$u(x,t) = \exp(-t)\sin x. \tag{33}$$

Table 2. Errors for the present method for Example 3 with $h = \pi/20, \alpha = 1/12$, $\beta = 5/12, T = 1, \zeta = 1$ and $\eta = 1$.

k	RMS	L_2	L_∞
1/2	6.9000×10^{-3}	1.2230×10^{-2}	1.1385×10^{-2}
1/4	1.7238×10^{-3}	3.0554×10^{-3}	2.8992×10^{-3}
1/8	4.6532×10^{-4}	8.2477×10^{-4}	8.1760×10^{-4}
1/16	1.8894×10^{-4}	3.3488×10^{-4}	3.1616×10^{-4}
1/32	1.4823×10^{-4}	2.6273×10^{-4}	2.1109×10^{-4}

Table 3. Errors for the present method for Example 3 with $k = 1/20, \alpha = 1/12$, $\beta = 5/12, T = 1, \zeta = 1$ and $\eta = 1$.

h	RMS	L_2	L_∞
$\pi/4$	4.0726×10^{-3}	7.2185×10^{-3}	6.1837×10^{-3}
$\pi/8$	9.3034×10^{-4}	1.6489×10^{-3}	1.3856×10^{-3}
$\pi/16$	2.4215×10^{-4}	4.2920×10^{-4}	3.5166×10^{-4}
$\pi/32$	9.3192×10^{-5}	1.6517×10^{-4}	1.6462×10^{-4}
$\pi/64$	7.0796×10^{-5}	1.2548×10^{-4}	1.2254×10^{-4}

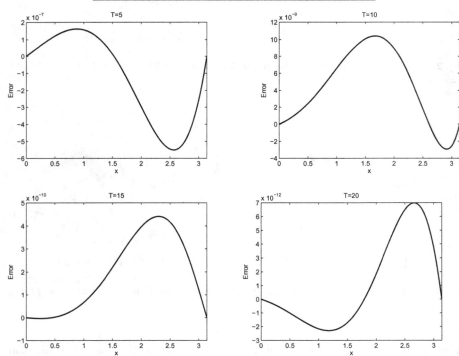

Fig. 4. Plots of errors for Example 3 taking $N = 200, k = 0.01, \alpha = 1/12, \beta = 5/12$, $\zeta = 1$ and $\eta = 1$.

Table 2 exhibits the RMS, L_2 and L_∞ error norms for the fixed value of $h = \pi/20$ and for the different values of k. The same error norms are computed for fixed $k = 1/20$ with different values of h which are reported in Table 3 and it can be noticed from both the tables that the errors are in better agreement with the exact solution. The error plots are given for different values of time in Fig. 4, the continuous decrement of error can be noticed from the figures.

5   Conclusion

Exponential spline method is used for the numerical solution of BBMB type equations. The method is proven to be unconditionally stable by using the Von Neumann stability process. The truncation error of the numerical scheme is also discussed. Numerical results establish the efficiency of the proposed method.

References

1. Al-Khaled, K., Momani, S., Alawneh, A.: Approximate wave solutions for generalized Benjamin-Bona-Mahony-Burgers equations. Appl. Math. Comput. **171**, 281–292 (2005)
2. Dehghana, M., Abbaszadeha, M., Mohebbib, A.: The numerical solution of nonlinear high dimensional generalized Benjamin-Bona-Mahony-Burgers equation via the meshless method of radial basis functions. Comput. Math. Appl. **68**, 212–237 (2014)
3. Fakhari, A., Domairry, G.: Ebrahimpour: approximate explicit solutions of nonlinear BBMB equations by homotopy analysis method and comparison with the exact solution. Phys. Lett. A **368**, 64–68 (2007)
4. Ganji, D.D., Babazadeh, H., Jalaei, M.H., Tashakkorian, H.: Application of He's variational iteration method for solving nonlinear BBMB equations and free vibration of systems. Acta Appl. Math. **106**, 359–367 (2009)
5. Gozukizil, O.F., Akcagil, S.: Exact solutions of Benjamin-Bona-Mahony-Burgers-type nonlinear pseudo-parabolic equations. Bound. Value Probl. **1**, 1–12 (2012)
6. Kumar, V., Gupta, R.K., Jiwari, R.: Painlevé analysis, Lie symmetries and exact solutions for variable coefficients Benjamin-Bona-Mahony-Burger (BBMB) equation. Commun. Theor. Phys. **60**(2), 175–182 (2013)
7. Omrani, K., Ayadi, M.: Finite difference discretization of the Benjamin-Bona-Mahony-Burgers equation. Numer. Methods Partial Differ. Equ. **24**(1), 239–248 (2007)
8. Stephenson, G.: Partial Differential Equations for Scientists and Engineers. Imperial College Press, Ames (1996)
9. Zahra, W.K., El Mhlawy, A.M.: Numerical solution of two-parameter singularly perturbed boundary value problems via exponential spline. J. King Saud Univ. Sci. **25**, 201–208 (2013)
10. Zarebnia, M., Parvaz, R.: On the numerical treatment and analysis of Benjamin-Bona-Mahony-Burgers equation. Appl. Math. Comput. **284**, 79–88 (2016)

A Fuzzy Regression Technique Through Same-Points in Fuzzy Geometry

Debdas Ghosh$^{(\boxtimes)}$ (iD), Ravi Raushan, and Gaurav Somani

Department of Mathematical Sciences,
Indian Institute of Technology (BHU) Varanasi,
Varanasi, Uttar Pradesh 221 005, India
{debdas.mat,gaurav.somani.mat14}@iitbhu.ac.in,
ravi.raushan.mat14@iitbhu.ac.in

Abstract. In this short article, a method to obtain a fuzzy regression curve for a set of imprecise locations is proposed. The given imprecise locations are presented by fuzzy points. The studied fuzzy regression curve is obtained with the help of a smooth regression technique for a set of precise locations. We observe the given imprecise points as a bunch of *same points* with varied membership values. For a set of same points, we obtain a smooth regression curve. The union of all these smooth regression curves, with different membership values, for the same points yields the proposed fuzzy regression curve. The method is demonstrated with a numerical example.

Keywords: Fuzzy points · Same points · Fuzzy curves
Smooth regression · Fuzzy regression

1 Introduction

Identification of a fuzzy regression curve to recognize the pattern of a given set of imprecise locations or data is an important topic of research in recent days. Although there is a plethora of regression methods to recognize the pattern of a set of crisp or precise data, the regression analysis for a set of imprecise data is not yet focused rigorously. After the introduction of fuzzy geometry [2–4], it is observed that a mathematical presentation of imprecise locations can be appropriately done by *fuzzy points*. In this article, we thus attempt to show how simply we can obtain a fuzzy regression curve for a set of fuzzy points.

It is shown in [2,5] that in the construction of a fuzzy line, the concept of *same points* is very useful. As fuzzy regression is closely associated with fuzzy line/curve fitting, in this article on fuzzy regression, we attempt to observe the given set of fuzzy points as a collection of same points with varied membership values. Hence, collectively, the problem under consideration can be observed as identifying a fuzzy regression curve that passes through the same points.

The fuzzy regression curve, in this article, is observed as a collection of conventional regression curves for each set of same points with different membership

© Springer Nature Singapore Pte Ltd. 2018
D. Ghosh et al. (Eds.): ICMC 2018, CCIS 834, pp. 216–224, 2018.
https://doi.org/10.1007/978-981-13-0023-3_21

values. Towards identifying an appropriate regression curve for a set of same points, we use the conventional smooth regression method given in [1]. We have chosen the smooth regression technique of [1] due to the facts that it is a simple heuristic approach and it is well-known by its applicability in diversified fields.

The delineation of the presented work is as follows. The next Sect. 2 gives some basic definitions and a brief presentation of a conventional smooth regression technique which is applied to obtain the proposed fuzzy regression technique. Section 3 gives the proposed method with a numerical illustration.

2   Preliminaries

In this section, at first, we give a brief sketch of the smooth regression technique presented in [1]. Then, a few definitions from fuzzy geometry is presented and those are used throughout the paper.

2.1   A Smooth Regression Technique

Consider n pairs of independent random variables $(X_1, Y_1), (X_2, Y_2), \ldots,$ (X_n, Y_n) with joint probability density function $f(x, y)$. The goal of regression is to discover a functional relationship between X and Y, given by

$$m(x) = E(Y \mid X) = \frac{\int y f(x, y) dy}{\int f(x, y) dy}. \tag{1}$$

A heuristic approach has been presented by Watson in [1] to obtain an estimate of $\hat{m}(x)$, using the sample of n variables that tends to $m(x)$ as $n \to \infty$ regardless of the nature of $f(x, y)$. An estimate of the *marginal probability density function* on X can be expressed in terms of sum of non-negative functions of the sequence $\delta_n(z)$, where $\delta_n(z)$ has a total area as unity and tends to the Dirac delta function as $n \to \infty$:

$$\hat{f}_1(x) = \frac{1}{n} \sum_{i=1}^{n} \delta_n(x - X_i). \tag{2}$$

At the continuous points, the estimate (2) converges to $f_1(x)$ as $n \to \infty$, provided $\delta_n(z) \to \delta(z)$. On applying a similar analogy, in two dimension, by substituting a joint density estimator $\hat{f}_n(x, y)$ in (1), we get the required estimator for $m(x)$ in finding a smoothed regression as

$$\hat{m}(x) = \frac{\sum_{i=1}^{n} Y_i \delta_n(x - X_i)}{\sum_{i=1}^{n} \delta_n(x - X_i)}.$$

In simple terms, it forms a weighted average of Y_i's corresponding to X_i's that are in nearby region of x and thus satisfying the domain of $\delta_n(x - X_i)$. It is clear that the weight of Y_i on X_i must decrease as one goes away from x and vary according to the sample in consideration.

This is a non-parametric regression where the degree of smoothing is dependent on function $\delta_n(z)$ parametrized by a smoothing parameter denoted by δ. The choice of this smoothing function may vary according to the sample in observation. Some of the commonly used choices are Triangular, Epanechnikov and Gaussian functions. The triangular smoothing is used in [1] for the purpose of fuzzy regression. It is given by the following function

$$\delta_n(z) = \begin{cases} \delta\left(1 - \delta|z|\right) & \text{for } |z| \leq 1/\delta \\ 0 & \text{for } |z| > 1/\delta. \end{cases}$$

The value of δ can be chosen according to the sample. However, the judgement is being based mainly on whether the associated estimate seems sensible and smoothed.

2.2   The Extension Principle and Same Points

• *The extension principle*

Let φ be a real-valued function of n variables $x_1, x_2, \ldots, x_n$. The extension principle extends this function to a fuzzy set $\widetilde{Y} = \widetilde{\varphi}(\widetilde{x}_1, \widetilde{x}_2, \ldots, \widetilde{x}_n)$ whose membership function is defined by

$$\mu(y|\widetilde{Y}) = \begin{cases} \displaystyle\sup_{y=\varphi(x_1, x_2, \ldots, x_n)} \min_{i=1,2,\ldots,n} \left(\mu(x_i|\widetilde{x}_i)\right) & \text{if } \varphi^{-1}(y) \neq \emptyset \\ 0 & \text{if } \varphi^{-1}(y) = \emptyset. \end{cases}$$

Definition 1 (α-cut of a fuzzy set). *For a fuzzy set $\widetilde{A}$ of $\mathbb{R}^n$ its α-cut, denoted $\widetilde{A}(\alpha)$, is defined by*

$$\widetilde{A}(\alpha) = \begin{cases} \{x : \mu(x|\widetilde{A}) \geq \alpha\} & \text{if } 0 < \alpha \leq 1 \\ closure\{x : \mu(x|\widetilde{A}) > 0\} & \text{if } \alpha = 0. \end{cases}$$

The set $\{x : \mu(x|\widetilde{A}) > 0\}$ is called the support of the fuzzy set $\widetilde{A}$. The 0-cut is named as base of the fuzzy set $\widetilde{A}$.

Definition 2 (Fuzzy point [3]). *A fuzzy point at (a_1, a_2) in $\mathbb{R}^2$, denoted $\widetilde{P}(a_1, a_2)$, is defined by the membership function*

(i) $\mu((x_1, x_2)|\widetilde{P}(a_1, a_2))$ is upper semi-continuous,
(ii) $\mu((x_1, x_2)|\widetilde{P}(a_1, a_2)) = 1$ if and only if $(x_1, x_2) = (a_1, a_2)$, and
(iii) $\widetilde{P}(a_1, a_2)(\alpha)$ is a compact and convex subset of $\mathbb{R}^2$, for all α in $[0, 1]$.

The notations $\widetilde{P}_1$, $\widetilde{P}_2$, $\widetilde{P}_3$, ... are usually used to represent fuzzy points.

Example 1. Let (a_1, b_1) be a point in $\mathbb{R}^2$. Consider the right elliptical cone with elliptical base $\{(x, y) : (\frac{x-a_1}{\ell_1})^2 + (\frac{x_2-a_2}{\ell_2})^2 \leq 1\}$ and vertex (a_1, a_2). This right elliptical cone can be taken as the membership function of a fuzzy point $\widetilde{P}(a_1, a_2)$ at (a_1, a_2). The mathematical form of $\mu(. | \widetilde{P}(a_1, a_2))$ is:

$$\mu((x_1, x_2) | \widetilde{P}(a_1, a_2)) = \begin{cases} 1 - \sqrt{\left(\frac{x_1-a_1}{\ell_1}\right)^2 + \left(\frac{x_2-a_2}{\ell_2}\right)^2} & \text{if } (\frac{x_1-a_1}{\ell_1})^2 + (\frac{x_2-a_2}{\ell_2})^2 \leq 1 \\ 0 & \text{elsewhere.} \end{cases}$$

Definition 3 (Same points for fuzzy numbers [3]). *Let x_1 and x_2 be two numbers in the supports of the continuous fuzzy numbers $\widetilde{a}_1$ and $\widetilde{a}_2$, respectively. The numbers x_1 and x_2 are said to be same points with respect to $\widetilde{a}_1$ and $\widetilde{a}_2$ if:*

(i) $\mu(x_1 | \widetilde{a}_1) = \mu(x_2 | \widetilde{a}_2)$, and
(ii) $x_1 \leq a_1$ and $x_2 \leq a_2$, or $x_1 \geq a_1$ and $x_2 \geq a_2$, where a_1 and b_1 are midpoints of $\widetilde{a}_1(1)$ and $\widetilde{a}_2(1)$, respectively.

Definition 4 (Same points for fuzzy points [3]). *Let (x_1, y_1) and (x_2, y_2) be two points on the supports of the continuous fuzzy points $\widetilde{P}(a_1, b_1)$ and $\widetilde{P}(a_2, b_2)$, respectively. Suppose that L_1 be the line joining (x_1, y_1) and (a_1, b_1); L_2 be the line joining (x_2, y_2) and (a_2, b_2).*

As $\widetilde{P}(a_1, b_1)$ is a fuzzy point, along the line L_1 there is a fuzzy number, $\widetilde{r}_1$ say, on the support of $\widetilde{P}(a_1, b_1)$. The membership function of this fuzzy number $\widetilde{r}_1$ can be written as: $\mu((x_1, y_1) | \widetilde{r}_1) = \mu((x_1, y_1) | \widetilde{P}(a_1, b_1))$ for (x_1, y_1) in L_1, and 0 otherwise.

Similarly, along the line L_2 there is a fuzzy number on the support of $\widetilde{P}(a_2, b_2)$. We let this fuzzy number be $\widetilde{r}_2$. The points (x_1, y_1) and (x_2, y_2) are said to be same points with respect to $\widetilde{P}(a_1, b_1)$ and $\widetilde{P}(a_2, b_2)$ if

(i) (x_1, y_1) and (x_2, y_2) are same points with respect to $\widetilde{r}_1$ and $\widetilde{r}_2$, and
(ii) L_1 and L_2 make the same angle with the line joining (a_1, b_1) and (a_2, b_2).

For instance, we consider the fuzzy points $\widetilde{P}_1(a_1, b_1)$ and $\widetilde{P}_2(a_2, b_2)$ in the Fig. 1. In the Fig. 1, the dotted circles are the boundaries of $\widetilde{P}_1$ and $\widetilde{P}_2$, respectively. The two dotted lines inside the circles make an identical angle θ with

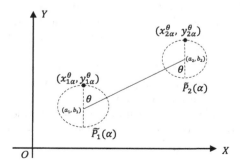

Fig. 1. Illustration of same points

the straight line joining (a_1, b_1) and (a_2, b_2). We note that the points $(x_{1\alpha}^\theta, y_{1\alpha}^\theta)$ and $(x_{2\alpha}^\theta, y_{2\alpha}^\theta)$ are with the same membership value and satisfies the restriction (ii) in the Definition 4. Thus, $(x_{1\alpha}^\theta, y_{1\alpha}^\theta)$ and $(x_{2\alpha}^\theta, y_{2\alpha}^\theta)$ is a same points with respect to $\tilde{P}_1$ and $\tilde{P}_2$.

3   Proposed Technique on Fuzzy Regression

The aim of this paper is to demonstrate a fuzzy regression technique for a given set of n fuzzy points $\tilde{P}_1(a_1, b_1), \tilde{P}_2(a_2, b_2), \ldots, \tilde{P}_n(a_n, b_n)$ using the concept of same points illustrated in the previous section. For the purpose of feasibility, we consider the membership function for each of the fuzzy points to be a right circular cone with vertex at (a_i, b_i). The membership function, for each $i = 1, 2, \ldots, n$, is given by

$$\mu\left((x, y)\Big|\tilde{P}_i(a_i, b_i)\right) = \begin{cases} 1 - \sqrt{(\frac{x-a_i}{R_i})^2 + (\frac{y-b_i}{R_i})^2} & (x-a_i)^2 + (y-b_i)^2 \le R_i^2 \\ 0 & \text{otherwise.} \end{cases}$$

Each R_i is a positive quantity.

We discretize the *base* of each fuzzy point into several sets of crisp points and form a set of points such that each set of crisp points have identical membership value. Further, for a given membership value α, the sets of crisp points with membership value α from different fuzzy points forms same points.

The rule for discretization is based on repetitive selection of different α's and θ's, where

(i) α is chosen from a set of discrete membership values in $[0, 1]$ and
(ii) θ, that represents the angle made by the lines through the cores of a fuzzy point and the x-axis, is chosen from a discrete set of values from $[0, 2\pi]$.

For each value of α we consider the α-cuts of the fuzzy points. Then we consider a θ value. We note that the line L_i, passing through the core and has an angle θ with the x-axis will intersect the boundary of the α-cut of the fuzzy point exactly at two points. The collection of the intersecting points that lie on the same half-space of the lines joining the consecutive cores of the fuzzy points will give us a set of same points. Thus, for a given α and θ values we will have two sets of same points. Varying α and θ across their possible values will yield the complete collection of same points.

The boundary of the α-cut for $P_i(a_i, b_i)$ is a circle around the core with radius $R_{i\alpha} = R_i(1-\alpha)$, i.e., the collection of points (x, y)'s so that $(x-a_i)^2 + (y-b_i)^2 = R_{i\alpha}^2$. There are, evidently, infinitely many points on this circle representing the boundary of the α-cut. We consider a finite number of collection of same points for the given fuzzy points. We represent a collection of same points by the core points $(a_1, b_1), (a_2, b_2), \ldots, (a_n, b_n)$.

Each set of same points $\{(x_{i\alpha}^\theta, y_{i\alpha}^\theta) | i = 1, 2, \cdots, n\}$ is parameterized by the value of α and θ as follows:

$$(x_{i\alpha}^\theta, y_{i\alpha}^\theta) = (a_i + R_{i\alpha}\cos\theta, \; b_i + R_{i\alpha}\sin\theta).$$

In order to have a fuzzy regression curve for the given set of fuzzy points, we construct a regression curve for a set of same points with the help of the regression method illustrated in Subsect. 2.1. Each set of same points with the membership value α corresponds to a unique regression curve. We denote the regression for the same points with a given α and θ value by $\mathcal{R}_\alpha^\theta$. We associate a membership value α to the regression curve $\mathcal{R}_\alpha^\theta$. The fuzzy regression curve $\widetilde{C}$, say, for the fuzzy points $\widetilde{P}_1, \widetilde{P}_2, \ldots, \widetilde{P}_n$ is defined by the union, through the sup-min composition of the extension principle, of all the regression curves $\mathcal{R}_\alpha^\theta$ with memebrship value α. That is,

$$\widetilde{C} = \bigvee_{\alpha \in [0,1], \; \theta \in [0, 2\pi]} \mathcal{R}_\alpha^\theta.$$

For a geometric visualization of the constructed fuzzy regression curve, we try to observe the membership values by an optical density. We put a grey level associated to the α-value to $\mathcal{R}_\alpha^\theta$. The higher membership value is represented darker region. On applying this process for different α-values, we obtain a fuzzy curve by the collection of smooth regression curves of discretized same points. It is important to note that the resultant group of overlapping curves corresponding to different membership values can be superimposed by the curve having the largest membership value according to the extension principle of fuzzy sets explained in the previous section. The shapes obtained from this method are shown in next section through an illustrative example.

4   Algorithm

Let A $= \{0.01, 0.02, \ldots 1\}$ be the set of discrete values of α.

Let $\Phi = \{10°, 20°, \ldots 360°\}$ be the set of discrete values of θ.

Smoothing parameter $\delta = 2, 5, 10, \cdots$.

Data: A sample of n fuzzy points, $\widetilde{P}_1(a_1, b_1), \widetilde{P}_2(a_2, b_2), \cdots, \widetilde{P}_n(a_n, b_n)$ with the radius of the circular support given by $R_1, R_2, \cdots, R_n$, respectively.

Task: Estimate a smooth regression fuzzy curve for given set of arbitrary fuzzy numbers $Q_1(\tilde{x}_1), \ldots, Q_m(\tilde{x}_m)$, $m \leq n$ with spread given by S_j.

Step 1: *Initialize*
Associate sample data into five sets each containing crisp data points for X_i, Y_i, x_j, R_i and S_j.

Step 2: *Discretize*
Given an α in A, compute $R_{i\alpha}$ and $R_{j\alpha}$ and iterate over each θ in Φ to obtain crisp points $\{(x_{i\alpha}^\theta, y_{i\alpha}^\theta)\}$ and $\{(x_{j\alpha}^\theta, y_{j\alpha}^\theta)\}$.

Step 3: for any $x_{j\alpha}^\theta$, iterate over all $x_{i\alpha}^\theta$ such that $x_{j\alpha}^\theta - x_{i\alpha}^\theta \leq \frac{1}{\delta}$ and compute the average on $y_{i\alpha}^\theta$.

Step 4: Repeat Step 3 for all j,

Step 5: Join the estimated $y_{j\alpha}^\theta$ by straight lines and assign it a grey-level inversely proportional to α.

Step 6: Repeat Step 2 for all α in A.

5   Illustrative Example

A sample of 100 points are taken where X is $N(0,1)$ and Y, given $X = x$, is $N\left(\frac{3x^2}{100}, 1\right)$ thus, $m(x) = \frac{3x^2}{100}$. In addition, fuzziness was introduced to each (X_i, Y_i) in the form of randomized radius or spread across the core, R_i from a uniform distribution on $[0, 0.125]$. Thereafter, a smooth estimate is depicted for eleven fuzzy points $\tilde{P}_i(a_i, b_i)$, chosen at intervals of length 1.0 from $[-5, 5]$ and randomized in the same manner as (X_i, Y_i). This experiment is done using the triangular function for different values of smoothing parameter $\delta = 2, 4, 6, \ldots$ as shown in Figs. 2, 3 and 4, respectively. The results in the Table 1 shows the predicted value of crisp y_k along with the lower and upper bounds induced by fuzziness (R_k) represented by l_k and u_k, respectively.

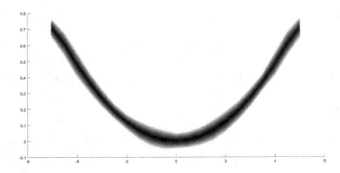

Fig. 2. Fuzzy regression curves with $\delta = 2$

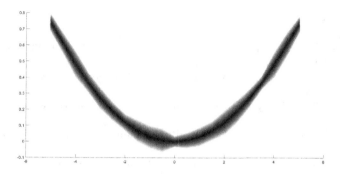

Fig. 3. Fuzzy regression curves with $\delta = 4$

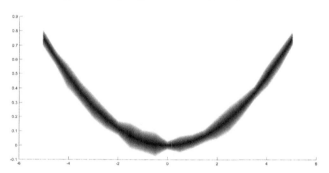

Fig. 4. Fuzzy regression curves with $\delta = 6$

Table 1. Illustration of smooth fuzzy regression using same points

Sample data			Predicted data		
k	$\tilde{P}_i(x_k, y_k)$	R_k	$(l_k, y_k, u_k)_{\delta=2}$	$(l_k, y_k, u_k)_{\delta=4}$	$(l_k, y_k, u_k)_{\delta=6}$
1	$(-5, 0.75)$	0.0706	$(0.6466, 0.7061, 0.7657)$	$(0.6786, 0.7352, 0.7917)$	$(0.6900, 0.745, 0.8)$
2	$(-4, 0.48)$	0.1100	$(0.4015, 0.4812, 0.5610)$	$(0.3989, 0.4802, 0.5614)$	$(0.3992, 0.48, 0.5609)$
3	$(-3, 0.27)$	0.0780	$(0.\ 2136, 0.2712, 0.3289)$	$(0.2011, 0.2702, 0.3392)$	$(0.1813, 0.27, 0.3588)$
4	$(-2, 0.12)$	0.0093	$(0.0717, 0.1212, 0.1708)$	$(0.0668, 0.1202, 0.1736)$	$(0.0684, 0.12, 0.1717)$
5	$(-1, 0.03)$	0.0293	$(-0.0325, 0.0312, 0.0950)$	$(-0.0415, 0.0302, 0.1018)$	$(-0.0496, 0.03, 0.1097)$
6	$(0, 0)$	0.1061	$(-0.0484, 0.0012, 0.0509)$	$(-0.0349, 0.0002, 0.0353)$	$(-0.0268, 0, 0.0269)$
7	$(1, 0.03)$	0.0197	$(-0.0267, 0.0312, 0.0892)$	$(-0.0297, 0.0302, 0.0901)$	$(-0.0214, 0.03, 0.0815)$
8	$(2, 0.12)$	0.0754	$(0.0534, 0.1212, 0.1891)$	$(0.0430, 0.1202, 0.1973)$	$(0.0440, 0.12, 0.1961)$
9	$(3, 0.27)$	0.0794	$(0.2111, 0.2712, 0.3313)$	$(0.2103, 0.2702, 0.3301)$	$(0.1934, 0.27, 0.3466)$
10	$(4, 0.48)$	0.0978	$(0.4066, 0.4812, 0.5559)$	$(0.4004, 0.4802, 0.5600)$	$(0.4049, 0.48, 0.5552)$
11	$(5, 0.75)$	0.0393	$(0.6376, 0.7061, 0.7746)$	$(0.6857, 0.7352, 0.7846)$	$(0.6948, 0.745, 0.7953)$

Acknowledgement. The first author gratefully acknowledges the financial support through *Early Career Research Award* (ECR/2015/000467), Science & Engineering Research Board, Government of India.

References

1. Geoffrey, S.W.: Smooth regression analysis, Sankhya, Series A (1961–2002), vol. 26, no. 4, pp. 359–372, December 1964
2. Chakraborty, D., Ghosh, D.: Analytical fuzzy plane geometry II. Fuzzy Sets Syst. **243**, 84–110 (2014)
3. Ghosh, D., Chakraborty, D.: Analytical fuzzy plane geometry I. Fuzzy Sets Syst. **209**, 66–83 (2012)
4. Ghosh, D., Chakraborty, D.: Analytical fuzzy plane geometry III. Fuzzy Sets Syst. **283**, 83–107 (2016)
5. Ghosh, D., Chakraborty, D.: On general form of fuzzy lines and its application in fuzzy line fitting. J. Intell. Fuzzy Syst. **29**, 659–671 (2015)

Bidirectional Associative Memory Neural Networks Involving Zones of No Activation/Dead Zones

V. Sree Hari Rao[1,2P(✉)] and P. Raja Sekhara Rao[3]

[1] Department of Mathematics and Statistics, Missouri University
of Science & Technology, Rolla, MO, USA
vshrao@researchfoundation.in, vshrao@gmail.com
[2] Foundation for Scientific Research and Technological Innovation (FSRTI),
Hyderabad 500 102, India
[3] Government Polytechnic, Kalidindi 521344, A.P., India
raoprs@gmail.com

Abstract. This article purports to present a systematically developed survey on the influence of zones of no activation/dead zones in bidirectional associative memory (BAM) neural networks. The modeling effort based on the concept of dead zones is capable of explaining very intricate phenomena concerned with the functioning of the human brain. Activation dynamic models presented in this article provide a platform for the development of artificial neural network models. Several questions of importance that arise in the modeling of these systems have been discussed in this article. More precisely, the influence of a dead zone on the global stability, which is associated with the recall of memories, is investigated and various easily verifiable sets of sufficient conditions are provided. Directions for further research related to the incorporation of various possible kinds of dead zones that occur naturally in biological or artificial systems are discussed.

1 Introduction

The basic building block of the nervous system is the neuron, the cell that communicates information to and from the various parts of the body. Networks in which neurons are connected amongst themselves are known as Neural Networks. We shall discuss briefly the functioning of biological neural systems (BNS for short). A neuron has a cell body with a nucleus called the soma, an axon that carries the signal away from the neuron and dendrites that receive the signals from other neurons (see Fig. 1).

It is known that in the mammals, the processing of information at the neuronal level is rather slow implying that the neuron is not very efficient, but it

This work is supported by the Foundation for Scientific Research and Technological Innovation (FSRTI) - A constituent division of Sri Vadrevu Seshagiri Rao Memorial Charitable Trust, Hyderabad, India.

© Springer Nature Singapore Pte Ltd. 2018
D. Ghosh et al. (Eds.): ICMC 2018, CCIS 834, pp. 225–242, 2018.
https://doi.org/10.1007/978-981-13-0023-3_22

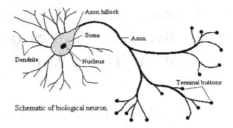

Fig. 1. A biological neuron.

is noticed that when these neurons are connected in a network, their efficiency increases. Hence, the study of networks of neurons is very significant. Networks are usually thought of as an aggregate of neurons spread across various layers and connected amongst themselves in the same layer and also with those in different layers (Fig. 2).

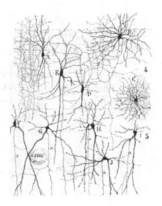

Fig. 2. A network of biological neurons.

Human brain is such a complex system that the mathematics available to day is not enough to describe its characteristics completely. Thus, activities of brain are segregated into various kinds and they are described by mathematical model equations. They may be termed as basic models of the brain. These models are used to develop artificial intelligent machines. These are called artificial neural systems. If human brain is regarded as an infinite dimensional space, an artificial neural system corresponds to a finite dimensional subspace of it. Efforts are going by the human brain to create its replica modifying the artificial neural systems (Fig. 3).

Study of networks of neurons has, thus, gained increasing prominence over the past decades and several networks of neurons such as spiking neural networks, cellular neural networks, bidirectional associative networks etc. are developed for

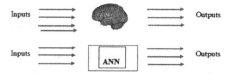

Fig. 3. Brain and an artificial neural system.

real world applications besides understanding of the brain. In particular, BAM networks are those, in which the flow of information will be both in forward and backward directions, with some stable behavior. In mathematical description, these networks are regard as dynamical systems. Two types of dynamics namely synaptic dynamics and activation dynamics are studied in general. Synaptic dynamics are concerned with the learning (training) aspects of the study while activation dynamics are associated with the recall of memory. Simply, synaptic dynamics correspond to evolution process and activation dynamics represent consolidation of evolved brain. Though evolution is important and is of primary concern, it is the activation dynamics that makes the system ready for application or tackle the issues - behavior is the concern here. It is the activation dynamic that verifies whether the learning is properly done or not. It is thus, both a testing tool and an application tool.

Knowing well that a brain is not simple we begin with a network of n number of neurons, n could be large, as desired. A simple BAM network, popularly known as Hopfield model, describing the activation dynamics of neurons in one single neuronal field may be expressed by the following system of equations.

$$x_i'(t) = -a_i x_i(t) + \sum_{j=1}^{n} b_{ij} f_j(x_j(t)) + I_i, \tag{1.1}$$

$i = 1, 2, \ldots, n.$

In Eq. (1.1), a_i for $i = 1, 2, \ldots, n$ represents the passive decay rates, b_{ij} represents the synaptic connection weight between the ith and the jth neurons, $f_j(x_j)$ for $j = 1, 2, \ldots, n$ represents the signal propagation functions and I_i for $i = 1, 2, \ldots, n$ represents the exogenous inputs to the ith neuron. This network is auto-associative in the sense that its topology is confined to a single neuronal field.

The generalization of the above Hopfield model for a network of neurons in two neuronal fields assumes the following form.

$$x_i'(t) = -a_i x_i(t) + \sum_{j=1}^{n} b_{ij} f_j(y_j(t)) + I_i,$$

$$y_i'(t) = -c_i y_i(t) + \sum_{j=1}^{n} d_{ij} f_j(x_j(t)) + J_i \tag{1.2}$$

for $i = 1, 2, \ldots, n$, and is studied by [13–15,21]. In (1.3), a_i, c_i are positive constants known as decay rates. I_i, J_i are exogenous inputs and b_{ij}, d_{ij} are the synaptic connection weights and all these are assumed to be real constants. These weight connections connect the ith neuron in one neuronal field to the jth neuron in another neuronal field. The functions f_i and g_i are the neuronal output response functions more commonly known as the signal functions. This network exhibits the hetero-associative property as it involves the dynamics of neurons in two neuronal fields, in which the synaptic connection matrices $B = (b_{ij})$ and $D = (d_{ij})$ are symmetric and satisfy $B^T = D$.

Further, these models donot take into account the time delays that occur in the transmission of information. Keeping these observations in view, the following system of equations is proposed in [22] to describe a BAM network model.

$$x'_i(t) = -a_i x_i(t) + \sum_{j=1}^{n} b_{ij} \int_{\infty}^{t} K_{ij}(t-s) f_j(\lambda_j, y_j(s)) ds + I_i$$

$$y'_i(t) = -c_i y_i(t) + \sum_{j=1}^{n} d_{ij} \int_{-\infty}^{t} L_{ij}(t-s) g_j(\mu_j, x_j(s)) ds + J_i \qquad (1.3)$$

for $i, j = 1, 2, \cdots, n$.

Other terms being explained above, the constants λ_i and μ_i in (1.3) represent the neuronal gains associated with the neuronal activations. The delay kernels K_{ij} and L_{ij} for $i = 1, 2, \cdots, n, j = 1, 2, \cdots, n$ are real valued non negative continuous functions defined on $[0, \infty)$.

The system (1.3) is more general in the sense that it allows both synaptic connection matrices to be non-symmetric and also need not satisfy the relation $B^T = D$. Further, signal response functions can be chosen from a more general class of functions, which need not necessarily satisfy a Lipschitz condition in order to establish the existence of unique equilibrium patterns to the given system. A set of sufficient conditions for the existence of equilibrium patterns to the system Eq. (1.3) and also various sets of sufficient conditions for the global asymptotic stability of the positive equilibrium are presented in [22–24].

Activation dynamics of a system are usually concluded by (i) global stability which is the eventual stabilization of activations of all processing elements (neurons) from any initial input and/or (ii) global convergence which is the eventual minimization of error between the desired and the computed processing elements.

Global stability guarantees that all inputs are mapped on to the same fixed point. The importance of global stability may be described as follows. Recall of memories is one of the processes by which the brain returns in some sense from a current state to another state in which it has been before. In neural network models, memory corresponds to a temporally stationery or non- stationery equilibrium and recall is modelled by the convergence of neuronal activations in the neuronal activation space, to the equilibrium.

We shall now come to the main aspect of this study. It is well known that in biological/behavioural systems, abnormal fluctuations, often drive the system to instability. For example, in human systems seizures, blood clots, sudden rising of blood pressure or blood sugar, in some cases lead to cerebral hemorrhage which may turn out to be fatal. This undesired behavior may be controlled by

(i) slowing down the activations with the help of drugs etc. or (ii) by removal of damaged portions by surgery or controlling by yoga and meditation. The first approach is a global one and the second way is a local approach. One of the most practical ways of controlling such behavior in a system is to create a dead zone or a zone of no activation. Often dead zones are created in biological systems, which exhibit over excitations due to some irregular/defective functioning of vital organs, mainly to avoid fatality. Often brain fails to recall an already stored memory - exhibits signs of instability. This may be due to the absence of enough information (missing or fading) during the process of recollection. Such symptoms are common in old age,dementia, in people with Alzhemier's disease, children with autism etc. Similarly, hyperactivity in children or overreactions in brain to certain inputs also represent a dead zone.

In analog communication network systems, a concept somewhat related to our dead zones, known as dead bands are introduced to annihilate noise, since the noise and the signal operate at different frequency levels. Another analogy to our dead zones may be observed in VLSI technology which primarily aims at designing a fault tolerant system [17]. The non functioning of one or more vital components of a VLSI system is generally viewed as a fault. The main problem here is to design a circuit which performs despite the occurrence of fault(s). Thus, in neural network models, fault tolerance may be viewed as a graceful degradation when the connections are damaged For fault tolerant neural network models both biological and artificial, we refer the readers to [1–4, 7–12, 18]. On the other hand, in [21, 26–28] dead zones are viewed as missing information in input and methods are proposed to obtain the same output as zone free system.

Our approach here is not to look at the dead zones as faults or missing information in the system but to answer the important question as how the presence of dead zones influences the stability behavior of the system. Henceforth, we use term Dead Zone to mean the creation of a region of no activation.

Mathematically a dead zone may be defined as

$$\phi(u(t)) = \begin{cases} \phi^*(u(t), \delta) \text{ for } u(t) > \delta, \\ \phi^{**}(u(t), \delta) \text{ for } u(t) < -\delta, \\ 0 \qquad \text{otherwise.} \end{cases} \qquad (1.4)$$

The following figure (Fig. 4) describes a dead zone in a given signal function.

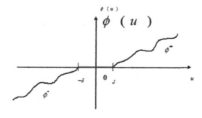

Fig. 4. A dead zone of given length.

A model incorporating a dead zone of activation for a BAM network may be described by the following system of equations:

$$u_i'(t) = -a_i\, u_i(t) + \sum_{j=1}^{n} b_{ij} \int_0^\infty K_{ij}(s)\, \psi_j\,(v_j(t-s))\, ds,$$

$$v_i'(t) = -c_i\, v_i(t) + \sum_{j=1}^{n} d_{ij} \int_0^\infty L_{ij}(s)\, \phi_j\,(u_j(t-s))\, ds. \tag{1.5}$$

for $i = 1, 2, \ldots, n$, in which the functions ϕ_i and ψ_i are defined as follows:

$$\phi_i(u_i(t)) = \begin{cases} g_i^*(u_i(t), h_i) & \text{for } u_i(t) > h_i, \\ g_i^{**}(u_i(t), h_i) & \text{for } u_i(t) < -h_i, \\ 0 & \text{otherwise.} \end{cases} \tag{1.6}$$

and

$$\psi_i(v_i(t)) = \begin{cases} f_i^*(v_i(t), k_i) & \text{for } v_i(t) > k_i, \\ f_i^{**}(v_i(t), k_i) & \text{for } v_i(t) < -k_i, \\ 0 & \text{otherwise.} \end{cases} \tag{1.7}$$

for each $i = 1, 2, \ldots, n$, h_i, k_i are positive numbers for each $i = 1, 2, \ldots, n$.

In view of the above, it is very important to study the influence of the presence of a dead zone in a system on the global asymptotic stability of the positive equilibrium pattern and our work in this paper is taken from [23, 24]. The concept of dead zone in communication systems and control systems is studied by [5, 6, 19, 20, 26–28]. The authors would understand the conceptual difference between the present study and those mentioned above as we go further.

The paper is organized as follows. For the present model (1.5), results on the existence of unique equilibrium patterns under fairly general circumstances are discussed in Sect. 2 of this paper. We present, in Sect. 3, different sets of sufficient conditions for the global asymptotic stability of the positive equilibrium pattern for the system (1.5) in the presence of a dead zone.

In Sect. 4, we present various kinds of dead zones that occur in artificial and natural systems to enthuse readers providing further scope of research in this area. A discussion concludes our study in the final section.

2   Equilibrium Patterns

The equilibria of system (1.5) are the solutions of the following algebraic system

$$a_i x_i^* = \sum_{j=1}^{n} b_{ij} f_j(\lambda_j, y_j^*) + I_i$$

$$c_i y_i^* = \sum_{j=1}^{n} d_{ij} g_j(\mu_j, x_j^*) + J_i \tag{2.1}$$

for $i = 1, 2, \cdots, n$.

We refer the readers to Theorem 2.1 of [22] which gives sufficient conditions for the existence of a unique positive equilibrium pattern.

The following initial conditions are assumed on the system (1.5).

$$x_i(s) = \tilde{p}_i(s), \quad y_i(s) = \tilde{q}_i(s) \quad \text{for} \quad s \in (-\infty, 0], i = 1, 2, \cdots n, \tag{2.2}$$

where $\tilde{p}_i, \tilde{q}_i$ are continuous, bounded functions on $(-\infty, 0]$.

We now rewrite system (1.5) as

$$X'(t) = F(X(t)) \tag{2.3}$$

in which

$$X(t) = (x_1(t), x_2(t), \cdots x_n(t)) \quad \text{and}$$

$$F(X(t)) = (\xi_1(t), \xi_2(t), \cdots, \xi_n(t), \eta_1(t), \eta_2(t), \cdots, \eta_n(t)) \quad \text{where}$$

$$\xi'_p(t) = -a_p \xi_p(t) + \sum_{j=1}^{n} b_{pj} \int_{-\infty}^{t} K_{pj}(t-s) f_j(\lambda_j, \eta_j(s)) ds + I_p$$

and

$$\eta'_q(t) = -c_q \eta_q(t) + \sum_{j=1}^{n} d_{qj} \int_{-\infty}^{t} L_{qj}(t-s) g_j(\mu_j, \xi_j(s)) ds + J_q,$$

for $p, q = 1, 2, \cdots, n$.

Now let S be an open subset of $\Re^{2n}$ and for any $\xi \in \Re^{2n}$ define $|\xi| = \sum_{i=1}^{2n} |\xi_i|$.
We now state our first result from [24].

Theorem 2.1. Let $F : S \to \Re^{2n}$ be continuous and satisfy the following condition: Corresponding to each point $\xi \in S$ and its neighbourhood U, there exist constants k_1 and k_2 which are non negative and satisfy $k_1 + k_2 \neq 0$, and functions h_j and ϕ_l for $j = 1, 2, \cdots, n$ and $l = 1, 2, \cdots, n, n+1, \cdots, 2n$ such that

$$|F(\xi) - F(\eta)| \leq k_1 |\xi - \eta| + k_2 \sum_{l=1}^{2n} |\phi_l(h_j(\xi)) - \phi_l(h_j(\eta))| \tag{2.4}$$

on U where each $h_j : R \to R$ is continuously differential function in ξ satisfying the relation

$$\sum_{i=1}^{n} \frac{\partial h_j(\xi)}{\partial \xi_i} F_i(\xi) \neq 0 \quad \text{on} \quad U \quad \text{and}$$

each $\phi_l : R \to R, l = 1, 2, \cdots, 2n$ is continuous and of bounded variation on bounded sub intervals. Then the initial value problem (2.3), (2.2) has a unique solution on any interval containing the initial functions (2.2).

The next result ensures that the system (1.5) has a unique equilibrium pattern. It is interesting to note that the signal functions f_j and g_j need not necessarily satisfy a Lipschitzian hypothesis here.

Theorem 2.2 ([24]). Assume that response functions f_j and $g_j, j = 1, 2, \cdots, n$ satisfy the hypotheses of Theorem 2.1. Then the system (2.3) admits a unique solution, yielding a unique equilibrium pattern for the system (1.5).

Henceforth, we tacitly assume that system (1.5) has a unique positive equilibrium (x^*, y^*) where $x^* = (x_1^*, x_2^*, \ldots, x_n^*)$ and $y^* = (y_1^*, y_2^*, \ldots, y_n^*)$.

Example 2.3. The system of equations

$$x'(t) = -a\, x(t) + \int_{-\infty}^{t} y^{\frac{5}{9}}(s)\, ds + I,$$

$$y'(t) = -c\, y(t) + \int_{-\infty}^{t} x^{\frac{5}{9}}(s)\, ds + J, \tag{2.5}$$

with $x(s) = 0 = y(s)$ for $s \in (-\infty, 0]$, has a unique solution by virtue of Theorem 2.1 although the functions $f(y) = y^{\frac{5}{3}}$ and $g(x) = x^{\frac{5}{3}}$ do not satisfy the Lipschitz conditions.

Remark 2.4. It is important to mention here that Theorem 3.4 of [23] follows from Theorem 2.2 for the choice of $k_1 = k_2 = k$. Theorem 3.4 of [23] works only for those Nonlinear signal functions that have a Lipschitz part where as Theorem 2.2 above applies to both Lipschitz and non Lipschtz functions or combinations of both. Moreover, Theorem 3.4 of [23] is limited to sublinearities in signal functions. The study in [24] allows us to super linear response functions also. For some commonly used response functions which are not Lipschitzian, we refer the readers to [16] in this context.

3   Global Stability for Dead Zone Model

In this Section, we present some important results of [23, 24] that analyzed the influence of a dead zone on the global asymptotic stability of the equilibrium pattern (x^*, y^*) of system (1.5).

By the following change of variables,

$$u_i(t) = x_i(t) - x_i^* \qquad v_i(t) = y_i(t) - y_i^*,$$

$$\phi_i(u_i) = g_i(\mu_i, x_i(t)) - g_i(\mu_i, x_i^*) = g_i(\mu_i, u_i(t) + x_i^*) - g_i(\mu_i, x_i^*)$$

$$\psi_i(v_i) = f_i(\lambda_i, y_i(t)) - f_i(\lambda_i, y_i^*) = f_i(\lambda_i, v_i(t) + y_i^*) - f_i(\lambda_i, y_i^*)$$

for each $i = 1, 2, \cdots, n$,

system (1.5) assumes the form,

$$u_i'(t) = -a_i u_i(t) + \sum_{j=1}^{n} b_{ij} \int_{-\infty}^{t} K_{ij}(t - s)\psi_j(v_j(s))ds$$

$$v_i'(t) = -c_i v_i(t) + \sum_{j=1}^{n} d_{ij} \int_{-\infty}^{t} L_{ij}(t - s)\phi_j(u_j(s))ds, \tag{3.1}$$

$i = 1, 2, \cdots, n.$

Observe that $(0,0)$ is the unique equilibrium pattern of (3.1).

We assume that the delay kernels K_{ij} and L_{ij} for $i, j = 1, 2, \ldots, n$ satisfy the following conditions.

$$\int_0^\infty K_{ij}(s)ds = 1 = \int_0^\infty L_{ij}(s)ds, \tag{3.2}$$

$$\int_0^\infty sK_{ij}(s)ds < \infty, \int_0^\infty sL_{ij}(s)ds < \infty \tag{3.3}$$

To incorporate a dead zone in the BAM network, we consider the model Eq. (3.1) with ϕ_j, ψ_j defined as follows:

$$\phi_j(u_j(t)) = \begin{cases} \phi_j^*(u_j(t), \delta_j) & \text{for} \quad u_j(t) > \delta_j \\ 0 & \text{for} \quad |u_j(t)| \leq \delta_j \\ \phi_j^{**}(u_j(t), \delta_j) & \text{for} \quad u_j(t) < -\delta_j \end{cases}$$

and

$$\psi_j(v_j(t)) = \begin{cases} \psi_j^*(v_j(t), \gamma_j) & \text{for} \quad v_j(t) > \gamma_j \\ 0 & \text{for} \quad |v_j(t)| \leq \gamma_j \\ \psi_j^{**}(v_j(t), \gamma_j) & \text{for} \quad v_j(t) < -\gamma_j, \end{cases} \tag{3.4}$$

where δ_j and γ_j are positive numbers for each $j = 1, 2, \cdots n$.

The forthcoming results of this section are from [23]. For the sake of an enthusiastic reader and to understand the techniques we use to prove some of the results. The following theorem describes a situation, in which the equilibrium of (3.1) is globally asymptotically stable in the presence of a dead zone.

Theorem 3.1. Assume that the hypotheses (3.2) and (3.3) are satisfied. Further assume that

$$a_i > \frac{1}{2}\sum_{j=1}^n |b_{ij}|, \quad c_i > \frac{1}{2}\sum_{j=1}^n |d_{ij}| \tag{3.5}$$

for $i = 1, 2, \ldots, n$ are satisfied. Then the equilibrium pattern of (3.1) is globally asymptotically stable provided the following inequalities hold:

For each $i = 1, 2, \ldots, n$,

$$\min\{A_i + m_i, A_i + \widetilde{m}_i\} > 0 \text{ and } \min\{B_i + n_i, B_i + \widetilde{n}_i\} > 0 \tag{3.6}$$

where for each $i = 1, 2, \ldots, n$

$$A_i = a_i - \frac{1}{2}\sum_{j=1}^n |b_{ij}|, \qquad B_i = c_i - \frac{1}{2}\sum_{j=1}^n |d_{ij}|,$$

$$m_i = \min_{u_i(t) > h_i}\left[a_i\left(\frac{g_i^*(u_i(t), h_i)}{u_i(t)}\right) - \left(\frac{g_i^*(u_i(t), h_i)}{u_i(t)}\right)^2 \sum_{j=1}^n\left(\frac{1}{2}|b_{ij}| + |d_{ji}|\right)\right],$$

$$\widetilde{m}_i = \min_{u_i(t) < -h_i}\left[a_i\left(\frac{g_i^{**}(u_i(t), h_i)}{u_i(t)}\right) - \left(\frac{g_i^{**}(u_i(t), h_i)}{u_i(t)}\right)^2 \sum_{j=1}^n\left(\frac{1}{2}|b_{ij}| + |d_{ji}|\right)\right],$$

$$n_i = \min_{v_i(t)>k_i} \left[c_i \left(\frac{f_i^*(v_i(t), k_i)}{v_i(t)} \right) - \left(\frac{f_i^*(v_i(t), k_i)}{v_i(t)} \right)^2 \sum_{j=1}^n (\frac{1}{2} |d_{ij}| + |b_{ji}|) \right],$$

$$\tilde{n}_i = \min_{v_i(t)<-k_i} \left[c_i \left(\frac{f_i^{**}(v_i(t), k_i)}{v_i(t)} \right) - \left(\frac{f_i^{**}(v_i(t), k_i)}{v_i(t)} \right)^2 \sum_{j=1}^n \left(\frac{1}{2} |d_{ij}| + |b_{ji}| \right) \right].$$

The following is a special case of Theorem 3.1.

Corollary 3.2. Assume that the conditions (3.2), (3.3) and (3.5) are satisfied. Further suppose that the following conditions hold:

$$(i) \quad xF(x, c) > 0 \text{ for } x \neq 0 \tag{3.7}$$

and any constant $c > 0$, where F in (3.11) is replaced by each of f_i^*, f_i^{**}, g_i^* and g_i^{**}.

$$(ii) \quad \begin{array}{l} \lim_{v_i(t)\to\infty} \frac{f_i^*(v_i(t),k_i)}{v_i(t)} = 0 = \lim_{v_i(t)\to-\infty} \frac{f_i^{**}(v_i(t),k_i)}{v_i(t)} \\ \lim_{u_i(t)\to\infty} \frac{g_i^*(u_i(t),h_i)}{u_i(t)} = 0 = \lim_{u_i(t)\to-\infty} \frac{g_i^{**}(u_i(t),h_i)}{u_i(t)} \end{array} \tag{3.8}$$

for each $i = 1, 2, \ldots, n$.

Then the equilibrium pattern of (3.1) is globally asymptotically stable.

We now illustrate our theorem in detail with the following example.

Example 3.3. We now consider the system (3.1) in which the functions ϕ_i and ψ_i assume the following form:

$$\phi_i(u_i) = \begin{cases} \beta_i(h_i)(u_i - h_i) & \text{for } u_i > h_i, \\ \beta_i(h_i)(u_i + h_i) & \text{for } u_i < -h_i, \\ 0 & \text{otherwise.} \end{cases} \tag{3.9}$$

and

$$\psi_i(v_i) = \begin{cases} \alpha_i(k_i)(v_i - k_i) & \text{for } v_i > k_i, \\ \alpha_i(k_i)(v_i + k_i) & \text{for } v_i < -k_i, \\ 0 & \text{otherwise.} \end{cases} \tag{3.10}$$

for $i = 1, 2, \ldots, n$, where the functions $\alpha_i(k_i), \beta_i(h_i)$ satisfy the following conditions:

$$\lim_{k_i \to 0} \alpha_i(k_i) = \frac{c_i}{\sum_{j=1}^n \left(\frac{1}{2}|d_{ij}| + |b_{ji}| \right)}, \quad \lim_{h_i \to 0} \beta_i(h_i) = \frac{a_i}{\sum_{j=1}^n \left(\frac{1}{2}|b_{ij}| + |d_{ji}| \right)} \tag{3.11}$$

for each $i = 1, 2, \ldots, n$.

Here, we have $f_i^*(v_i, k_i) = \alpha_i(k_i)(v_i - k_i), f_i^{**}(v_i, k_i) = \alpha_i(k_i)(v_i + k_i), g_i^*(u_i, h_i) = \beta_i(h_i)(u_i - h_i)$ and $g_i^{**}(u_i, h_i) = \beta_i(h_i)(u_i + h_i)$ for each $i = 1, 2, \ldots, n$. Then

$$m_i = \tilde{m}_i = a_i \, \beta_i(h_i) - \beta_i^2(h_i) \sum_{j=1}^n \left(\frac{1}{2}|b_{ij}| + |d_{ji}| \right)$$

and

$$n_i = \tilde{n}_i = c_i\,\alpha_i(k_i) - \alpha_i^2(k_i)\sum_{j=1}^{n}\left(\frac{1}{2}|d_{ij}| + |b_{ji}|\right)$$

for each $i = 1, 2, \ldots, n$.

Finally, we see that all the hypotheses of Theorem 3.1 are satisfied and accordingly the equilibrium, the origin of (3.1) is globally asymptotically stable, provided

$$A_i + m_i > 0 \text{ and } B_i + n_i > 0 \tag{3.12}$$

for each $i = 1, 2, \ldots, n$.

It should be noted that conditions (3.12) guarantee global stability when the dead zones are in a sense permanent. In this context, a natural question that needs to be examined is how best one could improve the neural gains so as to have a satisfactory (globally stable) performance of the network?

We now write the conditions (3.12) as

$$\beta_i < \frac{a_i + \sqrt{a_i^2 + 4\left(\sum_{j=1}^{n}\left(\frac{1}{2}|b_{ij}| + |d_{ji}|\right)\right)\left(a_i - \sum_{j=1}^{n}\frac{1}{2}|b_{ij}|\right)}}{2\left(\sum_{j=1}^{n}\left(\frac{1}{2}|b_{ij}| + |d_{ji}|\right)\right)}$$

$$\alpha_i < \frac{c_i + \sqrt{c_i^2 + 4\left(\sum_{j=1}^{n}\left(\frac{1}{2}|d_{ij}| + |b_{ji}|\right)\right)\left(c_i - \sum_{j=1}^{n}\frac{1}{2}|d_{ij}|\right)}}{2\left(\sum_{j=1}^{n}\left(\frac{1}{2}|d_{ij}| + |b_{ji}|\right)\right)} \tag{3.13}$$

for each $i = 1, 2, \ldots, n$.

The inequalities (3.13) clearly provide the upper bounds for α_i and β_i. That is, if the decay rates and the synaptic connection weights are fixed (in a certain sense) then, one can understand that α_i and β_i should not exceed the optimal bounds as in (3.13).

Further, we notice that if the width of the dead zones h_i and $k_i \to 0$ (which corresponds to the case of diminishing dead zones), then the Eq. (3.11) will provide the minimum values for the neural gains α_i and β_i respectively and the corresponding global stability conditions reduces to $A_i > 0$ and $B_i > 0$ for each $i = 1, 2, \ldots, n$.

In order to highlight further the influence of dead zones on the global stability of the equilibrium pattern, we consider the network described by the equations

$$\begin{aligned}\dot{x}_i &= -a_i\,x_i(t) + \sum_{j=1}^{2} b_{ij}\int_{-\infty}^{t} K_{ij}(t-s)\,\psi_j(y_j(s))\,ds, \\ \dot{y}_i &= -c_i\,y_i(t) + \sum_{j=1}^{2} d_{ij}\int_{-\infty}^{t} L_{ij}(t-s)\,\phi_j(x_j(s))\,ds.\end{aligned} \tag{3.14}$$

where i = 1, 2, in which we choose

$$a_1 = 0.5, \quad a_2 = 0.42, \quad c_1 = 0.51, \quad c_2 = 0.49,$$

$$[b_{ij}] = \begin{bmatrix} 1/6 & 1/4 \\ -1/4 & 1/4 \end{bmatrix}, \quad [d_{ij}] = \begin{bmatrix} 1/3 & 1/5 \\ 1/10 & 1/8 \end{bmatrix}.$$

With regard to this numerical example, we now compute the various estimates described in the relations (3.11)–(3.13).

Accordingly,

$$\begin{aligned} &\lim_{k_1 \to 0} \alpha_1 = 0.81, \quad \lim_{k_2 \to 0} \alpha_2 = 0.80, \\ &\lim_{h_1 \to 0} \beta_1 = 0.78, \quad \lim_{h_2 \to 0} \beta_2 = 0.73, \end{aligned} \quad (3.15)$$

which clearly provide the lower bounds. Again following the inequalities (3.13) and writing the lower bounds in (3.15) we see that the neural gains α_i and β_i for $i = 1, 2$ should satisfy the following inequalities:

$$\begin{aligned} 0.81 \leq \alpha_1(k_1) < 1.19, \quad 0.78 \leq \beta_1(h_1) < 1.17, \\ 0.80 \leq \alpha_2(k_2) < 2.94, \quad 0.73 \leq \beta_2(h_2) < 2.17. \end{aligned} \quad (3.16)$$

Thus, it follows that the equilibrium pattern described by (3.14) is globally asymptotically stable provided the neural gains satisfy the inequalities (3.16).

In our attempt to model the situation arising out of wild activations in a biological (or artificial) neural network, we began our analysis with the presumption that creation of a dead zone would help control wild activities (thus bringing a system otherwise tending to instability to normalcy). In this special situation, in which the activation functions ϕ_i and ψ_i assuming the form given in (3.9) and (3.10), we have suggested a mechanism to control the wildness by providing the upper bounds (inequalities (3.13)) and the lower bounds (Eq. (3.11)) for the neural gains which clearly depend on the widths of the zones.

In the case of human neurobiological systems modeled by the network equations involving dead zones as above, a clinical procedure may correspond either to administering a drug with doses depending upon the bounds on the neural gains or to a surgical procedure leading to the removal of a carefully selected tissue of the brain in such a manner that the functionality of other mechanisms of the body are not disturbed (or foregone).

Following results improve the applicability of our methods by expanding the space of parameters for the stability of equilibrium pattern using a useful inequality [24].

Theorem 3.5. Assume that the delay kernels satisfy (3.2) and (3.3) and the response functions satisfy the conditions

$$u_i \phi_i(u_i) > 0 \quad \text{for} \quad u_i \neq 0 \quad \text{and} \quad v_i \psi_i(v_i) > 0 \quad \text{for} \quad v_i \neq 0. \quad (3.17)$$

Then the equilibrium pattern $(0, 0)$ of (3.1) is globally asymptotically stable provided there exist constants η_1, η_2, η_3 and η_4, all positive, such that the following inequalities hold for $i = 1, 2, \cdots, n$,

$$\min\{A_i + m_i, A_i + \bar{m}_i\} > 0 \quad \text{and} \quad \min\{B_i + n_i, B_i + \bar{n}_i\} > 0,$$

where

$$A_i = a_i - \frac{1}{4\eta_1}\sum_{j=1}^{n}|b_{ij}|, \quad B_i = c_i - \frac{1}{4\eta_2}\sum_{j=1}^{n}|d_{ij}|,$$

$$m_i = \min_{u_i(t)>\delta_i}\left\{a_i\left(\frac{\phi_i^*(u_i(t))}{u_i(t)}\right) - \left[\frac{1}{4\eta_3}\sum_{j=1}^{m}|b_{ij}| + (\eta_2+\eta_4)\sum_{j=1}^{n}|d_{ji}|\right]\left(\frac{\phi_i^*(u_i(t))}{u_i(t)}\right)^2\right\},$$

$$n_i = \min_{v_i(t)>\gamma_i}\left\{c_i\left(\frac{\psi_i^*(v_i(t))}{v_i(t)}\right) - \left[\frac{1}{4\eta_4}\sum_{j=1}^{n}|d_{ij}| + (\eta_1+\eta_3)\sum_{j=1}^{n}|b_{ji}|\right]\left(\frac{\psi_i^*(v_i(t))}{v_i(t)}\right)^2\right\},$$

$$\bar{m}_i = \min_{u_i(t)<-\delta_i}\left\{a_i\left(\frac{\phi_i^{**}(u_i(t))}{u_i(t)}\right) - \left[\frac{1}{4\eta_3}\sum_{j=1}^{m}|b_{ij}| + (\eta_2+\eta_4)\sum_{j=1}^{n}|d_{ji}|\right]\left(\frac{\phi_i^{**}(u_i(t))}{u_i(t)}\right)^2\right\},$$

$$\bar{n}_i = \min_{v_i(t)<-\gamma_i}\left\{c_i\left(\frac{\psi_i^{**}(v_i(t))}{v_i(t)}\right) - \left[\frac{1}{4\eta_4}\sum_{j=1}^{n}|d_{ij}| + (\eta_1+\eta_3)\sum_{j=1}^{n}|b_{ji}|\right]\left(\frac{\psi_i^{**}(v_i(t))}{v_i(t)}\right)^2\right\}.$$

Remark 3.6. In Theorems 3.1 or 3.5, in view of the conditions (3.8), the quantities $m_i, \bar{m}_i, n_i$ and $\bar{n}_i$ may assume values in the extended real number system implying that the functions $\phi_i^*, \phi_i^{**}, \psi_i^*$ and ψ_i^{**} cannot be super linear in their arguments. In order to deal with situations when the response functions can be super linear, we use the following definition.

$$\phi_j^*(u_j(t)) = F_j^*(u_j(t)) \quad \text{for} \quad \delta_j < u_j(t) \le \Delta_j$$

$$\phi_j^*(u_j(t)) = F_j^*(\Delta_j) \quad \text{for} \quad u_j(t) > \Delta_j$$

$$\phi_j^{**}(u_j(t)) = F_j^{**}(u_j(t)) \quad \text{for} \quad -\Delta_j \le u_j(t) < -\delta_j$$

$$\phi_j^{**}(u_j(t)) = F_j^{**}(-\Delta_j) \quad \text{for} \quad u_j(t) < -\Delta_j$$

$$\psi_j^*(v_j(t)) = G_j^*(v_j(t)) \quad \text{for} \quad \gamma_j < v_j(t) \le \Gamma_j$$

$$\psi_j^*(v_j(t)) = G_j^*(\Gamma_j) \quad \text{for} \quad v_j(t) > \Gamma_j$$

$$\psi_j^{**}(v_j(t)) = G_j^{**}(v_j(t)) \quad \text{for} \quad -\Gamma_j \le v_j(t) < -\gamma_j$$

$$\psi_j^{**}(v_j(t)) = G_j^{**}(-\Gamma_j) \quad \text{for} \quad v_j(t) < -\Gamma_j$$

in which the functions $F_j^*, F_j^{**}, G_j^*, G_j^{**}$ can be super linear in their arguments.

Theorem 3.7. Assume that the delay kernels satisfy (3.2) and (3.3) and the response functions satisfy (3.4). Then the equilibrium pattern $(0,0)$ of (3.1) is globally asymptotically stable provided there exist positive constants η_1, η_2, η_3 and η_4 such that

$$\min\{A_i + m_i^*, A_i + \bar{m}_i^*\} > 0, \quad \min\{B_i + n_i^*, B_i + \bar{n}_i^*\} > 0,$$

for each $i = 1, 2, \cdots, n$ where

$$A_i = a_i - \frac{1}{4\eta_1}\sum_{j=1}^{n}|b_{ij}|, \quad B_i = c_i - \frac{1}{4\eta_2}\sum_{j=1}^{n}|d_{ij}|$$

and

$$m_i^* = \min_{u(t) > \delta_i} \left\{ a_i \left(\frac{F_i^*(u_i(t))}{u_i(t)} \right) - \left[\frac{1}{4\eta_3} \sum_{j=1}^{n} |b_{ij}| + (\eta_2 + \eta_4) \sum_{j=1}^{n} |d_{ji}| \right] \left(\frac{F_i^*(u_i(t))}{u_i(t)} \right)^2 \right\},$$

$$\bar{m}_i^* = \min_{u(t) < -\delta_i} \left\{ a_i \left(\frac{F_i^{**}(u_i(t))}{u_i(t)} \right) - \left[\frac{1}{4\eta_3} \sum_{j=1}^{n} |b_{ij}| + (\eta_2 + \eta_4) \sum_{j=1}^{n} |d_{ji}| \right] \left(\frac{F_i^{**}(u_i(t))}{u_i(t)} \right)^2 \right\},$$

$$n_i^* = \min_{u_i(t) > \gamma_i} \left\{ c_i \left(\frac{G_i^*(v_i(t))}{v_i(t)} \right) - \left[\frac{1}{4\eta_4} \sum_{j=1}^{n} |d_{ij}| + (\eta_1 + \eta_3) \sum_{j=1}^{n} |b_{ji}| \right] \left(\frac{G_i^*(v_i(t))}{v_i(t)} \right)^2 \right\},$$

$$\bar{n}_i^* = \min_{v_i(t) < -\gamma_i} \left\{ c_i \left(\frac{G_i^{**}(v_i(t))}{v_i(t)} \right) - \left[\frac{1}{4\eta_4} \sum_{j=1}^{n} |d_{ij}| + (\eta_1 + \eta_3) \sum_{j=1}^{n} |b_{ji}| \right] \left(\frac{G_i^{**}(v_i(t))}{v_i(t)} \right)^2 \right\}.$$

Remark 3.8. Theorems 3.5 and 3.7 improve significantly Theorem 3.1 in a number of ways. Theorem 3.1 follows from Theorem 3.5 for the choice of $\eta_1 = \eta_2 = \frac{1}{2}$, and clearly, Theorem 3.5 provides a larger region of stability. Further, Theorem 3.7 here is the super linear analogue of Theorem 3.1.

Now replacing the conditions (3.17) on the response functions ϕ_j and ψ_j by the assumption that there exist positive constants $\alpha_j(\lambda_j)$ and $\beta_j(\mu_j)$ such that

$$||\phi_j|| \leq \alpha_j |u_j|, \quad ||\psi_j|| \leq \beta_j |v_j|. \tag{3.18}$$

one may prove

Theorem 3.9. Assume that the delay kernels satisfy (3.2), (3.3) and the response functions satisfy (3.18). Then the equilibrium pattern $(0,0)$ of (3.1) is globally asymptotically stable provided there exist positive constants $\alpha_{i*}, \alpha_{i**}, \beta_{i*}, \beta_{i**}$ such that

$$\min \left\{ a_i - \alpha_{i*} \sum_{j=1}^{n} |d_{ji}|, \quad a_i - \alpha_{i**} \sum_{j=1}^{n} |d_{ji}| \right\} > 0$$

and

$$\min \left\{ c_i - \beta_{i*} \sum_{j=1}^{n} |b_{ji}|, \quad c_i - \beta_{i**} \sum_{j=1}^{n} |b_{ji}| \right\} > 0.$$

In the next result the condition (3.3) on the delay kernels is removed. However, this results in placing more restrictions on the parameters.

Theorem 3.10. Assume that the delay kernels satisfy (3.2) and the response functions satisfy (3.18). Then the equilibrium solution $(0,0)$ of (3.1) is globally asymptotically stable provided,

$$K_i = \min\{a_i, c_i\} > \max \left(\sum_{j=1}^{n} (\alpha_{j*}|d_{ij}| + \beta_{j*}|b_{ij}|), \sum_{j=1}^{n} (\alpha_{j**}|d_{ij}| + \beta_{j**}|b_{ij}|) \right),$$

for $i = 1, 2, \ldots, n$ where the positive constants $\alpha_{j*}, \alpha_{j**}, \beta_{j*}$ and β_{j**} are as in Theorem 3.9.

4   Future Work

We have seen that dead zones are considered in two ways. One as a missing information or a fault in the system and is regarded as a nonlinear, non smooth characteristic of the process [26–28]. On the positive sense, a dead zone is regarded as a control mechanism for either a collapsing or a turbulent system [23,24]. Understanding of system behavior and creation of appropriate zone of no activation is of utmost importance and is the first step in this area.

In both situations arising here, it is the preservation of stability that is under question. We have studied here the global stability of the system which is referred to as a physician's approach - control by drugs through any means - veins, oral or muscles. But the local stability is equally important as a surgeon's approach to remove the defective or infected parts or the practices of yoga or meditation focusing on problem locally. Thus, local stability analysis of (1.5) with (1.6) and (1.7) would be a useful topic for exploration.

Neurons are usually stimulated by external inputs which may be either fixed constants or time variants. Any gap or temporary absence of inputs may also influence the dynamics of the system considerable. Also a system with variable inputs is regarded as non autonomous in mathematical sense. For such systems inputs could be either discontinuous or impulsive. Thus, a dead zone in external inputs may introduce entirely new dynamics into the system. This is also an interesting area to explore along with the dead zones considered in state vector in this study. Further, non autonomous systems may not possess an equilibrium and hence, the stability of equilibrium as studied here makes little sense. Researchers may explore the situations such as the one in previous section (asymptotic equivalence). Also the study in [25] provides an innovative technique to select inputs to make the system approach pre-specified output values for such non autonomous systems. It would be interesting to see how the technique works out in case of dead zones.

In the present study we have considered the dead zone around an equilibrium pattern which may not always exist. A more natural way of setting a zone could be

$$\phi(u(t)) = \begin{cases} \phi^*(u(t), \delta) \text{ for } u(t) > \delta, \\ \phi^{**}(u(t), \delta) \text{ for } u(t) < -\delta, \\ \overline{\phi} \quad \text{otherwise.} \end{cases} \tag{4.1}$$

It is easy to see that (1.4) and (3.4) may be obtained from above definition letting $\overline{\phi} \equiv 0$ and $\overline{\phi} \equiv \phi(u^*)$ respectively.

Now consider the situation where

$$\lim_{u \to \delta^+} \phi^*(u(t), \delta) = \lim_{u \to -\delta^-} \phi^{**}(u(t), -\delta^-).$$

Then $\overline{\phi} \to 0$. This situation describes the presence of a **diminishing zone** in the system.

We shall now consider the case of multiple dead zones. Systems with missing information at many points during communication or supply of medicine at regular or appropriate time points during treatment are examples of such situations.

The following definition describes a function with two dead zones.

$$\phi(u(t)) = \begin{cases} \phi^*(u(t), \delta_1) \text{ for } u(t) < \delta_1, \\ 0 \text{ for } \delta_1 \le u(t) \le \delta_2, \\ \phi^{**}(u(t), \delta_2) \text{ for } \delta_2 < u(t) < \delta_3, \\ 0 \text{ for } \delta_3 \le u(t) \le \delta_4, \\ \phi^{***}(u(t), \delta_4) \text{ for } u(t) > \delta_4. \end{cases} \tag{4.2}$$

One may easily extend the results of Sect. 3 with same Lyapunov functionals and arguments for multiple dead zones. It would be interesting to see how the parameters withstand the pressure of dead zones to preserve the stability of the system.

We shall now consider the situation where the width of zone defined by δ is not a simple fixed constant but a variable one. This may be due to correction measures taken up to minimize the faults in the system or alternate methods or immunity developed during treatment the length of zone reduces. Similarly, if system suffers some more complications or develops side effects, the length of zone increases. Many practical systems suffer from such unwanted troubles. Thus, we need to consider the role of variable zones on the system dynamics. We propose the following.

An Oscillating Zone

Consider $\delta = \delta + f(\lambda)$, where the real parameter λ corresponds to an external factor that influences on the zone length. We have the following cases.

(i) $f(\lambda) = \frac{(-1)^\lambda}{\lambda}$, $\lambda \ne 0$. This produces oscillations in the zone which, however, eventually die out and hence, for large λ zone length tends to the given value δ.

(ii) $f(\lambda) = \sin \lambda\pi$. This produces standard oscillations and occasionally visits the original value δ at integer values of λ.

(iii) $f(\lambda) = e^{-\lambda} \sin \lambda\pi$ also produces an oscillatory dead zone but approaches δ for large values of λ and takes the same value δ for all integer values of λ as in above case.

One may similarly define expanding zones contrary to examples (i) and (iii) in which variations are contracting to original zone length to δ.

How about considering a zone of the type $\delta \equiv f(\delta, \lambda)$? Here also λ corresponds to an influencing factor on zone. This definition could represent a variety of zones including those described above.

5 Discussion

In this paper, we have considered a bidirectionalassociative memory neural network model described in [22,23] to study the activation dynamics of neurons in a pair of neuronal fields involving exogenous inputs, transmission delays and dead zones. A result on the existence and uniqueness of equilibrium patterns for the model (1.5) is presented that applies to not necessarily Lipschitz functions that could be sub linear or super linear. We have explained the concept of a dead zone introduced in [23] and provided several results on the global asymptotic stability of the equilibrium pattern in the presence of a dead zone. Several naturally possible types of dead zones are described for further exploration and development of theory to understand real world phenomena. These are left as open problems for interested readers.

Digital technology is the order of the day and almost every thing is trending towards digitization. In most of the real world situations, the data or information is available or measured in discrete time intervals only, though the phenomena are continuous time processes. Thus, discrete forms of continuous systems only are available for study in many cases. Then how far this discrete analogue reflects the original system is a basic question interest to deal with. We begin with a simple example. Consider the differential equation

$$y'' + 3y' + 2y = 0.$$

It is easy to see that solutions of this differential equation approach 0 as $t \to \infty$ where as the solutions of corresponding difference equation (its discrete version)

$$y(n + 2) + y(n + 1) = 0,$$

obtained by replacing y' with $\nabla y(n)$ and y'' with $y(n + 2) - 2y(n + 1) + y(n)$ do not preserve this property.

Thus, enough care should be taken while representing continuous dynamical systems by discrete systems. In this context, it is highly important to find suitable discretization schemes that preserve the dynamical behaviour of the continuous systems in the corresponding discretized systems and in addition, appropriate dead zones and analyze their influence via discrete analogues corresponding to the systems represented by continuous dynamics. We hope to pursue this challenge in our subsequent expositions.

References

1. Anderson, J.: A memory storage model utilizing spatial correlation functions. Kybernetik **5**, 113–119 (1968)
2. Anderson, J.: Two models for memory organization using interacting traces. Math. Biosci. **8**, 137–160 (1970)
3. Anderson, J.: A simple neural network generating an interactive memory. Math. Biosci. **14**, 197–220 (1972)
4. Anderson, J.: A theory for the recognition of items from short memorized lists. Psych. Rev. **80**, 417–438 (1973)
5. Chen, F.-C., Liu, C.-C.: Adaptively controlling nonlinear continuous-time systems using multilayer neural networks. IEEE Trans. Automat. Control **39**, 1306–1310 (1994)
6. Chen, F.-C., Khalil, H.K.: Adaptive control of nonlinear systems using neural networks - a dead zone approach. In: Proceedings of the 1991 American Control Conference, pp. 667–672 (1991)

7. Hopfield, J.J.: Neural networks and physical systems with emergent collective computational abilities. Proc. Nat. Acad. Sci. **79**, 2554–2558 (1982)
8. Hopfield, J.J.: Neurons with graded response have collective computational properties like those of two state neurons. Proc. Nat. Acad. Sci. **81**, 3088–3092 (1984)
9. Hopfield, J.J., Feinstein, D., Palmer, R.: Unlearning has a stabilizing effect in collective memories. Nature **304**, 158–159 (1983)
10. Kohonen, T.: Self-Organization and Associative Memory. Springer, New York (1988). https://doi.org/10.1007/978-3-642-88163-3
11. Kohonen, T.: Correlative associative memory. IEEE Trans. Comput. **C-21**, 353–359 (1972)
12. Kohonen, T.: Associative Memory - A System Theoretical Approach. Springer, New York (1977). https://doi.org/10.1007/978-3-642-96384-1
13. Kosko, B.: Adaptive bidirectional associative memories. Appl. Opt. **26**, 4947–4960 (1987)
14. Kosko, B.: Bidirectional associative memories. IEEE Trans. Syst. Man Cybern. SMC **18**, 49–60 (1988)
15. Kosko, B.: Feedback stability and unsupervised learning. In: Proceedings of the IEEE International Conference on Neural Networks, vol. I, pp. 141–152. IEEE, San Diego (1988)
16. Kosko, B.: Neural Networks and Fuzzy Systems - A Dynamical Systems Approach to Machine Intelligence. Prentice-Hall of India, New Delhi (1994)
17. Lala, P.K.: Fault Tolerant and Fault Testable Design. Prentice Hall International, Upper Saddle River (1985)
18. Lindsay, P., Norman, D.: Human Information Processing: An Introduction to Psychology. Academic Press, Orlando (1977)
19. Selmic, R.R., Lewis, F.L.: Dead zone compensation in motion control systems using neural networks. IEEE TAC **45**(4), 602–613 (2000)
20. Simpson, P.K.: Artificial Neural Systems - Foundations, Paradigms, Applications and Implementations. Pergamon Press, New York (1989)
21. Shyu, K.K., Liu, H.J., Hsu, K.C.: Design of large scale time delayed systems with dead zone input via variable structure control. Automatica **41**(7), 1239–1246 (2005)
22. Sree Hari Rao, V., Phaneendra, Bh.R.M., Prameela, V.: Global dynamics of bidirectional associative memory networks with transmission delays. Diff. Equ. Dyn. Syst. **4**, 453–471 (1996)
23. Sree Hari Rao, V., Phaneendra, Bh.R.M.: Global dynamics of bidirectional associative memory neural networks with transmission delays and dead zones. Neural Netw. **12**, 455–465 (1999)
24. Sree Hari Rao, V., Raja Sekhara Rao, P.: Stability of dead zone bidirectional associative memory neural networks involving time delays. Int. J. Neural Syst. **12**(1), 15–29 (2002)
25. Sree Hari Rao, V., Raja Sekhara Rao, P.: Time varying stimulations in simple neural networks and convergence to desired outputs. Diff. Equ. Dyn. Syst. **26**, 81–104 (2016). https://doi.org/10.1007/s12591-016-0312-z
26. Wang, X.S., Hong, H., Su, C.Y.: Robust adaptive control of a class of nonlinear systems with an unknown dead zones. Automatica **40**(3), 407–413 (2004)
27. Zhang, T., Sam Ge, S.: Adaptive neural network tracking control of MIMO nonlinear systems with unknown dead zones and control directions. IEEE Trans. Neural Netw. **20**, 483–497 (2009)
28. Zhou, J., Er, M.J., Veluvollu, K.C.: Adaptive output control of nonlinear time delayed systems with uncertain dead zone input. In: Proceedings of American Control Conference, Minneapolis, MN, 14–16 June 2006, pp. 5312–5316 (2006)

Pure Mathematics

Bohr's Inequality for Harmonic Mappings and Beyond

Anna Kayumova[1], Ilgiz R. Kayumov[1], and Saminathan Ponnusamy[2]([✉]) [ⓘ]

[1] Kazan Federal University, Kremlevskaya 18, 420 008 Kazan, Russia
anvas@inbox.ru, ikayumov@kpfu.ru
[2] Department of Mathematics, Indian Institute of Technology Madras,
Chennai 600036, India
samy@iitm.ac.in

Abstract. There has been a number of problems closely connected with the classical Bohr inequality for bounded analytic functions defined on the unit disk centered at the origin. Several extensions, generalizations and modifications of it are established by many researchers and they can be found in the literature, for example, in the multidimensional setting and in the case of the Dirichlet series, functional series, function spaces, etc. In this survey article, we mainly focus on the recent developments on this topic and in particular, we discuss new and sharp improvements on the classical Bohr inequality and on the Bohr inequality for harmonic functions.

Keywords: Bounded analytic functions · Univalent functions
Bohr radius · Rogosinski radius · Schwarz-Pick lemma · Subordination

Subject Classifications: Primary: 30A10, 30H05, 30C35; Secondary: 30C45

1 Introduction

A well-known inequality of H. Bohr–often referred to as the classical Bohr theorem, states the following [12]:

Theorem A. *If a power series $f(z) = \sum_{k=0}^{\infty} a_k z^k$ converges in the unit disk $\mathbb{D} := \{z \in \mathbb{C} : |z| < 1\}$ and its sum $f(z)$ has modulus less than 1 in $\mathbb{D}$, then*

$$M^f(r) := \sum_{n=1}^{\infty} |a_n| |z|^n \leq 1 - |a_0| = \operatorname{dist}(f(0), \partial\mathbb{D}) \tag{1}$$

for $|z| = r < 1/3$. Moreover, the constant $1/3$ cannot be improved. Here $\operatorname{dist}(f(0), \partial\mathbb{D})$ denotes the Euclidean distance between $f(0)$ and the boundary $\partial\mathbb{D}$ of the unit disk $\mathbb{D}$. Equality in (1) holds for constant functions only.

© Springer Nature Singapore Pte Ltd. 2018
D. Ghosh et al. (Eds.): ICMC 2018, CCIS 834, pp. 245–256, 2018.
https://doi.org/10.1007/978-981-13-0023-3_23

It is worth pointing out that Bohr's original article, compiled by G.H. Hardy from correspondence, indicates that Bohr initially obtained the inequality (1) for $r \leq 1/6$ which was quickly sharpened to $r \leq 1/3$ by M. Riesz, I. Schur and F. Wiener, independently. The number $1/3$ is called the classical Bohr radius for the class of analytic self-maps of the unit disk $\mathbb{D}$. Bohr's article contains both his own proof and the one of his colleagues. Bohr considered this problem while working on the absolute convergence problem for Dirichlet series but presently it has become an important area of interest in many different contexts. The proof of Theorem A is well-known and an elegant proof of this theorem may also be found in a recent article [23].

We remark that the existence of the number r, asserted by Theorem A, is independent of the coefficients of the power series. We describe this fact by saying that a Bohr phenomenon occurs in the class of analytic self-maps of the unit disk $\mathbb{D}$. In [6], the existence of a Bohr phenomenon in the space of holomorphic functions on complex manifold is proved. In this case, the proof uses the algebraic structure of holomorphic functions and the underlying fact seems to be nothing but the maximum principle.

Various extensions of Bohr's inequality have been proposed by different authors after Dixon [17] used it in the construction of a Banach algebra satisfying the non-unital von Neumann inequality and non-isomorphic to a subalgebra of $L(H)$, the algebra of bounded linear operators on the Hilbert space H. For example, Boas and Khavinson [11] and Aizenburg [4] studied Bohr's result for n-variable power series defined on polydisks in $\mathbb{C}^n$ and on ball or other domains. Bohr's phenomenon in a more general setting has created enormous interest on Bohr's inequality in variety of situations including function spaces point of view. See for example, [9,10,35], the recent survey on this topic by Abu-Muhanna et al. [7] and the references therein. See for instance, [1–3,5,7–9,11,35].

After the appearance of the survey [7], several new results have appeared on this topic. See for example [23–26,28,29,32].

Here is another well-known result equivalent to Theorem A for analytic functions with positive real part less than 1.

Theorem B. *If a power series $f(z) = \sum_{k=0}^{\infty} a_k z^k$ converges in the unit disk $\mathbb{D}$ and its sum f has its real part less than 1 in $\mathbb{D}$ such that $f(0) = a_0$ is positive, then (1) holds for $|z| < 1/3$ and the constant $1/3$ cannot be improved.*

One of the ways of proving Bohr's phenomenon in concrete cases is by establishing bounds on the coefficients to the functions in the hypothesis. The proofs of Theorems A and B are essentially same in the sense that in Theorem A one uses the inequality $|a_n| \leq 1 - |a_0|^2$ for all $n \geq 1$ whereas in Theorem B, the inequality $|a_n| \leq 2(1 - a_0)$ for all $n \geq 1$, is used. We see that Theorems A and B are special cases of Theorem C where the proof is simple and new.

The aim of this article is to present overview on the recent results as supplementary to the recent survey articles [7,23] and at the same time, we present also some new and improved Bohr's inequality for analytic functions.

2   Several Extensions of Bohr's Inequality

2.1   Bohr's Phenomenon for a Class of Subordinations

Many authors have discussed the Bohr radius and extended this notion to various settings which led to the introduction of Bohr's phenomenon in different but similar contexts. For example, using the "distance form" formulation of Bohr inequality stated as above, the notion of Bohr radius can be generalized to the class of functions f analytic in $\mathbb{D}$ which take values in a given simply connected domain Ω.

Definition 1. *Let f and g be analytic in $\mathbb{D}$. Then g is said to subordinate to f, written $g \prec f$ or $g(z) \prec f(z)$, iff there exists a function w analytic in $\mathbb{D}$ satisfying $w(0) = 0$, $|w(z)| < 1$ and $g(z) = f(w(z))$ for $z \in \mathbb{D}$.*

If f is univalent in $\mathbb{D}$, then $g \prec f$ if and only if $g(0) = f(0)$ and $g(\mathbb{D}) \subset f(\mathbb{D})$ (see [19, p. 190 and p. 253] and [34]). By the Schwarz lemma, it follows that

$$|g'(0)| = |f'(w(0))w'(0)| \leq |f'(0)|.$$

Now for a given f, let $S(f) = \{g : g \prec f\}$ and $\Omega = f(\mathbb{D})$. The family $S(f)$ is said to satisfy a Bohr phenomenon if there exists an r_f, $0 < r_f \leq 1$ such that whenever $g(z) = \sum_{n=0}^{\infty} b_n z^n \in S(f)$, then

$$\sum_{n=1}^{\infty} |b_n| r^n \leq \operatorname{dist}(f(0), \partial\Omega) \tag{2}$$

for $|z| = r < r_f$. Here $\operatorname{dist}(f(0), \partial\Omega)$ denotes the Euclidean distance between $f(0)$ and the boundary $\partial\Omega$ of the domain $\Omega = f(\mathbb{D})$.

We observe that if $f(z) = (a_0 - z)/(1 - \overline{a_0}z)$ with $|a_0| < 1$, $\Omega = \mathbb{D}$ and $g \in S(f)$, then $f(0) = a_0$ and $\operatorname{dist}(f(0), \partial\Omega) = 1 - |a_0| = 1 - |b_0|$ so that (2) holds with $r_f = 1/3$.

Here is a more general result which contains the proof of Theorems A and B.

Theorem C ([1]). *If f, g are analytic in $\mathbb{D}$ such that f is univalent and convex in $\mathbb{D}$ and $g \in S(f)$, then inequality (2) holds with $r_f = 1/3$. The sharpness of r_f is shown by the convex function $f(z) = z/(1 - z)$.*

Proof. Let $g(z) = \sum_{n=0}^{\infty} b_n z^n \prec f(z)$, where f is a univalent mapping of $\mathbb{D}$ onto a convex domain $\Omega = f(\mathbb{D})$. Then it is well known that (see, for instance, [19, p. 195, Theorem 6.4])

$$\frac{1}{2}|f'(0)| \leq \operatorname{dist}(f(0), \partial\Omega) \leq |f'(0)| \text{ and } |b_n| \leq |f'(0)|. \tag{3}$$

It follows that $|b_n| \leq 2\operatorname{dist}(f(0), \partial\Omega)$, and thus

$$\sum_{n=1}^{\infty} |b_n| r^n \leq \operatorname{dist}(f(0), \partial\Omega)\frac{2r}{1-r} \leq \operatorname{dist}(f(0), \partial\Omega)$$

provided $2r/(1 - r) \leq 1$, i.e., $r \leq 1/3$. When $f(z) = z/(1 - z)$, we obtain $\operatorname{dist}(f(0), \partial\Omega) = 1/2$ which gives sharpness.

An appropriate modification of (3) yields the following result obtained in [1, Theorem 1].

Theorem D. *If f, g are analytic in $\mathbb{D}$ such that f is univalent in $\mathbb{D}$ and $g \in S(f)$, then inequality (2) holds with $r_f = 3 - 2\sqrt{2} \approx 0.17157$. The sharpness of r_f is shown by the Koebe function $f(z) = z/(1 - z)^2$.*

Proof. In this case, instead of (3), we just need to use the following (see [19, p. 196] and [15]):

$$\frac{1}{4}|f'(0)| \leq \text{dist}(f(0), \partial\Omega) \leq |f'(0)| \text{ and } |b_n| \leq n|f'(0)|.$$

As a consequence of it, one quickly gets

$$\sum_{n=1}^{\infty} |b_n| r^n \leq |f'(0)| \sum_{n=1}^{\infty} n r^n \leq \text{dist}(f(0), \partial\Omega) \frac{4r}{(1-r)^2} \leq \text{dist}(f(0), \partial\Omega)$$

provided $4r \leq (1-r)^2$, that is, for $r \leq 3 - 2\sqrt{2}$. Sharpness part follows similarly.

There are still many cases where Theorem D could be improved. To state one such result, we need a result of Rogosinski [36, p. 64] which states that if $g(z) = \sum_{n=1}^{\infty} b_n z^n \prec f(z) = \sum_{n=1}^{\infty} a_n z^n$ in $\mathbb{D}$ and if, for $1 \leq k \leq n$, the numbers a_k are nonnegative, non-increasing, and convex, then $|b_k| \leq |a_1|$ for $k = 1, 2, \ldots, n$. Using this result and [22, Theorem 5], we have the following result.

Theorem 1. *If f, g are analytic in $\mathbb{D}$ such that $\text{Re } f'(z) > 0$ in $\mathbb{D}$ and $g \in S(f)$, then inequality (2) holds with $r_f = 1/5$.*

Proof. In this case, instead of (3), one just needs to use the following inequalities (see [22, Theorem 5] which uses the ideas of convex decreasing sequences)

$$\frac{1}{4}|f'(0)| \leq \text{dist}(f(0), \partial\Omega) \leq |f'(0)| \text{ and } |b_n| \leq |f'(0)|.$$

As a consequence of it, one quickly gets

$$\sum_{n=1}^{\infty} |b_n| r^n \leq |f'(0)| \sum_{n=1}^{\infty} r^n \leq \text{dist}(f(0), \partial\Omega) \frac{4r}{1-r} \leq \text{dist}(f(0), \partial\Omega)$$

provided $r \leq 1/5$.

Remark 1. Note that f satisfying the condition $\text{Re } f'(z) > 0$ in $\mathbb{D}$ is univalent in $\mathbb{D}$, but not necessarily convex in $\mathbb{D}$, and thus Bohr radius is expected to be bigger than the number $3 - 2\sqrt{2}$.

2.2   Improved Bohr's Inequality

Kayumov and Ponnusamy [25] improved the classical version of the Bohr theorem in four different formulations. Later in a survey article in [23], the authors have further improved couple of these results. We now recall them here.

Theorem E. *Suppose that $f(z) = \sum_{k=0}^{\infty} a_k z^k$ is analytic in $\mathbb{D}$, $|f(z)| \leq 1$ in $\mathbb{D}$ and S_r denotes the area of the Riemann surface of the function f^{-1} defined on the image of the subdisk $|z| < r$ under the mapping f. Then we have*

$$\sum_{k=0}^{\infty} |a_k| r^k + \frac{16}{9} \left(\frac{S_r}{\pi - S_r} \right) \leq 1 \ for \ r \leq \frac{1}{3} \tag{4}$$

and the number 16/9 cannot be improved. Furthermore,

$$|a_0|^2 + \sum_{k=1}^{\infty} |a_k| r^k + \frac{9}{8} \left(\frac{S_r}{\pi - S_r} \right) \leq 1 \ for \ r \leq \frac{1}{2} \tag{5}$$

and the number 9/8 cannot be improved.

It is worth pointing out that in [25], these two inequalities were proved by using the quantity S_r/π in place of $S_r/(\pi - S_r)$ in the inequalities (4) and (5). This observation shows that the inequalities (4) and (5) are indeed an improved versions of [25, Theorem 1]. In [25], the following results were also proved.

Theorem F. *Suppose that $f(z) = \sum_{k=0}^{\infty} a_k z^k$ is analytic in $\mathbb{D}$ and $|f(z)| \leq 1$ in $\mathbb{D}$. Then we have*

1. $|a_0| + \sum_{k=1}^{\infty} \left(|a_k| + \frac{1}{2} |a_k|^2 \right) r^k \leq 1 \ for \ r \leq \frac{1}{3}$, *and the constants 1/3 and 1/2 cannot be improved.*

2. $\sum_{k=0}^{\infty} |a_k| r^k + |f(z) - a_0|^2 \leq 1 \ for \ r \leq \frac{1}{3}$, *and the constant 1/3 cannot be improved.*

3. $|f(z)|^2 + \sum_{k=1}^{\infty} |a_k|^2 r^{2k} \leq 1 \ for \ r \leq \sqrt{\frac{11}{27}}$, *and the constant $\sqrt{11/27}$ cannot be improved.*

2.3   Bohr inequality for p-Symmetric Analytic Functions

Ali et al. [8], considered the problem of determining Bohr radius for symmetric functions and suggested to determine the Bohr radius for the class of odd functions f satisfying $|f(z)| \leq 1$ for all $z \in \mathbb{D}$. Indeed, motivated by the work of Ali et al. [8], Kayumov and Ponnusamy [24] considered the following general problem.

Problem 1 ([24]). Given $p \in \mathbb{N}$ and $0 \le m \le p$, determine the Bohr radius for the class of functions $f(z) = z^m \sum_{k=0}^{\infty} a_{pk} z^{pk}$ analytic in $\mathbb{D}$ and $|f(z)| \le 1$ in $\mathbb{D}$.

As a solution to this problem, the following result was established in [24].

Theorem G. *Let $p \in \mathbb{N}$ and $0 \le m \le p$, $f(z) = z^m \sum_{k=0}^{\infty} a_{pk} z^{pk}$ be analytic in $\mathbb{D}$ and $|f(z)| \le 1$ in $\mathbb{D}$. Then*

$$M^f(r) \le 1 \text{ for } r \le r_{p,m},$$

where $r_{p,m}$ is the maximal positive root of the equation

$$-6r^{p-m} + r^{2(p-m)} + 8r^{2p} + 1 = 0.$$

The number $r_{p,m}$ is sharp. Moreover, in the case $m \ge 1$ there exists an extremal function of the form $z^m(z^p - a)/(1 - az^p)$, where

$$a = \left(1 - \frac{\sqrt{1 - r_{p,m}^{2p}}}{\sqrt{2}}\right) \frac{1}{r_{p,m}^p}.$$

Several choices of r and m provide Bohr radii for gap series of different types. For example, for the case $p = m, 2m, 3m$, the Bohr radii give

$$r_{m,m} = 1/\sqrt[2m]{2}, \quad r_{2m,m} = \sqrt[m]{r_2}, \text{ and } r_{3m,m} = \sqrt[2m]{\frac{7 + \sqrt{17}}{16}},$$

respectively, where $r_2 = 0.789991\ldots$ is given by

$$r_2 = \frac{1}{4}\sqrt{\frac{B-2}{6}} + \frac{1}{2}\sqrt{\frac{1}{3}\sqrt[3]{\frac{6}{B-2}} - \frac{B}{24} - \frac{1}{6}},$$

with

$$B = (3601 - 192\sqrt{327})^{\frac{1}{3}} + (3601 + 192\sqrt{327})^{\frac{1}{3}}.$$

It is worth pointing out from the last case that $r_{3,1}$ gives the value $(\sqrt{7 + \sqrt{17}})/4$. The result for $m = 0$ gives

Corollary 1. *Let $p \ge 1$. If $f(z) = \sum_{k=0}^{\infty} a_{pk} z^{pk}$ is analytic in $\mathbb{D}$, and $|f(z)| \le 1$ in $\mathbb{D}$, then $M^f(r) \le 1$ for $0 \le r \le r_{p,0} = 1/\sqrt[3]{3}$. The radius $r_{p,0} = 1/\sqrt[3]{3}$ is best possible.*

For the case $a_0 = 0$, it was pointed out that the number $r_{p,0} = 1/\sqrt[3]{3}$ in Corollary 1 can be evidently replaced by $r_{p,0} = 1/\sqrt[2p]{2}$ which is the Bohr radius in this case. Moreover, the radius $r = 1/\sqrt[2p]{2}$ in this case is best possible as demonstrated by the function

$$\varphi_\alpha(z) = z^p \left(\frac{\alpha - z^p}{1 - \alpha z^p}\right)$$

with $\alpha = 1/\sqrt[2p]{2}$.

It is now appropriate to recall the following problem proposed in [24]. The problem remains open.

Problem 2. Find the Bohr radius for the class of odd functions f satisfying $0 < |f(z)| \le 1$ for all $0 < |z| < 1$.

2.4   Powered Bohr Inequality

In 2000, Djakov and Ramanujan [18] investigated the Bohr phenomenon from different point of view. For the class $\mathcal{B}$ of analytic self-maps f of the unit disk $\mathbb{D}$ and a fixed $p > 0$, we consider the powered Bohr sum $M_p^f(r)$ defined by

$$M_p^f(r) = \sum_{k=0}^{\infty} |a_k|^p r^k.$$

Observe that for $p = 1$, $M_p^f(r)$ reduces to the classical Bohr sum defined in (1) by $M^f(r)$. The best possible constant ρ_p for which

$$M_p^f(r) \leq 1 \text{ for all } r \leq \rho_p$$

is called the (powered) Bohr radius for the family $\mathcal{B}$. To recall some known results, let us introduce

$$M_p(r) := \sup_{f \in \mathcal{B}} M_p^f(r)$$

and

$$r_p := \sup \left\{ r : a^p + \frac{r(1-a^2)^p}{1-ra^p} \leq 1, \ 0 \leq a < 1 \right\} = \inf_{a \in [0,1)} \frac{1 - a^p}{a^p(1 - a^p) + (1 - a^2)^p}.$$

Theorem H ([18, Theorem 3]). *For each $p \in (1,2)$ and $f(z) = \sum_{k=0}^{\infty} a_k z^k$ belongs to $\mathcal{B}$, we have $M_p^f(r) \leq 1$ for $r \leq T_p$, where*

$$m_p \leq T_p \leq r_p$$

where r_p is as above and

$$m_p := \frac{p}{\left(2^{1/(2-p)} + p^{1/(2-p)} \right)^{2-p}}.$$

Djakov and Ramanujan [18] posed the following problem about the Bohr radius for $M_p^f(r)$.

Problem 3 ([18, Question 1, p. 71]). What is the exact value of the (powered) Bohr radius ρ_p, $p \in (1,2)$? Is it true that $\rho_p = r_p$?

Using the recent approach from [24,26], this problem is solved affirmatively in the following form.

Theorem 2 (Kayumov and Ponnusamy [27]). *If $f(z) = \sum_{k=0}^{\infty} a_k z^k$ belongs to $\mathcal{B}$ and $p \leq 2$, then*

$$M_p(r) = \max_{a \in [0,1]} \left[a^p + \frac{r(1-a^2)^p}{1-ra^p} \right], \quad 0 \leq r \leq 2^{p/2-1},$$

and

$$M_p(r) < \left(\frac{1}{1 - r^{2/(2-p)}} \right)^{1-p/2}, \quad 2^{p/2-1} < r < 1.$$

In particular, the following result of Bombieri [13] (see also Bombieri and Bourgain [14]) is obtained as a special case.

Corollary 2. *If $p \in (1, 2)$, then $M_p(r) = 1$ for $r \leq r_p$. Also, we have the sharp estimate:*

$$M_1(r) = \frac{1}{r}(3 - \sqrt{8(1 - r^2)}) \text{ for } r \in \left[\frac{1}{3}, \frac{1}{\sqrt{2}}\right].$$

2.5 Improved Bohr's Inequality for Locally Univalent Harmonic Mappings

A complex-valued function $f = u + iv$ defined on $\mathbb{D}$ is harmonic if u and v are real-harmonic in $\mathbb{D}$. Every harmonic function f admits the canonical representation $f = h + \overline{g}$, where h and g are analytic in $\mathbb{D}$ such that $g(0) = 0 = f(0)$. A locally univalent harmonic function f in $\mathbb{D}$ is said to be sense-preserving if the Jacobian $J_f(z)$, $J_f(z) = |h'(z)|^2 - |g'(z)|^2$, is positive in $\mathbb{D}$ or equivalently, its dilatation $\omega_f(z) = g'(z)/h'(z)$ satisfies the inequality $|\omega_f(z)| < 1$ for $z \in \mathbb{D}$ (see [16, 20, 31, 33]). Properties of harmonic mappings have been investigated extensively, especially after the appearance of the pioneering work of Clunie and Sheil-Small [16] in 1984.

A sense-preserving homeomorphism f from the unit disk $\mathbb{D}$ onto Ω', contained in the Sobolev class $W_{loc}^{1,2}(\mathbb{D})$, is said to be a K-*quasiconformal mapping* if, for $z \in \mathbb{D}$,

$$\frac{|f_z| + |f_{\bar{z}}|}{|f_z| - |f_{\bar{z}}|} = \frac{1 + |\omega_f(z)|}{1 - |\omega_f(z)|} \leq K, \text{ i.e., } |\omega_f(z)| \leq k = \frac{K - 1}{K + 1},$$

where $K \geq 1$ so that $k \in [0, 1)$ (cf. [30, 37]).

As a harmonic extension of the classical Bohr theorem, the following results were established in [29].

Theorem I. *Suppose that $f(z) = h(z) + \overline{g(z)} = \sum_{n=0}^{\infty} a_n z^n + \overline{\sum_{n=1}^{\infty} b_n z^n}$ is a sense-preserving K-quasiconformal harmonic mapping of the disk $\mathbb{D}$, where h is a bounded function in $\mathbb{D}$. Then we have*

1. $\displaystyle\sum_{n=0}^{\infty} |a_n| r^n + \sum_{n=1}^{\infty} |b_n| r^n \leq \|h\|_\infty$ *for* $r \leq \dfrac{K+1}{5K+1}$. *The constant $(K+1)/$ $(5K+1)$ is sharp.*

2. $\displaystyle|a_0|^2 + \sum_{n=1}^{\infty} (|a_n| + |b_n|) r^n \leq \|h\|_\infty$ *for* $r \leq \dfrac{K+1}{3K+1}$. *The constant $(K+1)/$ $(3K+1)$ is sharp.*

Theorem J. *Suppose that either $f = h + g$ or $f = h + \overline{g}$, where $h(z) = \sum_{n=1}^{\infty} a_n z^n$ and $g(z) = \sum_{n=1}^{\infty} b_n z^n$ are bounded analytic functions in $\mathbb{D}$. Then*

$$\sum_{n=1}^{\infty} (|a_n| + |b_n|) r^n \leq \max\{\|h\|_\infty, \|g\|_\infty\} \text{ for } r \leq \sqrt{\frac{7}{32}}.$$

This number $\sqrt{7/32}$ is sharp.

As in the symmetric case of analytic functions (see [8,24,26]), we have the following analog result for harmonic functions.

Theorem K. *Let $p \geq 2$. Suppose that $f(z) = h(z) + \overline{g(z)} = \sum_{n=0}^{\infty} a_n z^{pn+1} + \overline{\sum_{n=0}^{\infty} b_n z^{pn+1}}$ is a harmonic p–symmetric function in $\mathbb{D}$, where h and g are bounded functions in $\mathbb{D}$. Then*

$$\sum_{n=0}^{\infty} (|a_n| + |b_n|) r^{pn+1} \leq \max\{||h||_{\infty}, ||g||_{\infty}\} \ for \ r \leq \frac{1}{2}.$$

The number $1/2$ is sharp.

Moreover, Kayumov and Ponnusamy [27] investigated the powered Bohr radius also for sense-preserving harmonic mappings defined on the unit disk $\mathbb{D}$. In addition to a number of several new results, the authors proved the following results.

Theorem 3 ([27]). *Suppose that $f(z) = h(z) + \overline{g(z)} = \sum_{k=0}^{\infty} a_k z^k + \overline{\sum_{k=1}^{\infty} b_k z^k}$ is a harmonic mapping of the disk $\mathbb{D}$, where h is a bounded function in $\mathbb{D}$ and $|g'(z)| \leq |h'(z)|$ for $z \in \mathbb{D}$ (the later condition obviously holds if f is sense-preserving). If $p \in [0,2]$ then the following sharp inequality holds:*

$$|a_0|^p + \sum_{k=1}^{\infty} (|a_k|^p + |b_k|^p) r^k \leq ||h||_{\infty} \max_{a \in [0,1]} \left[a^p + \frac{2r(1-a^2)^p}{1-ra^p} \right]$$

for $r \leq (2^{1/(p-2)} + 1)^{p/2-1}$. In the case $p > 2$, we have

$$|a_0|^p + \sum_{k=1}^{\infty} (|a_k|^p + |b_k|^p) r^k \leq ||h||_{\infty} \max\{1, 2r\}.$$

Corollary 3. *Suppose that $f(z) = h(z) + \overline{g(z)} = \sum_{k=0}^{\infty} a_k z^k + \overline{\sum_{k=1}^{\infty} b_k z^k}$ is a sense-preserving harmonic mapping of the disk $\mathbb{D}$, where h is a bounded function in $\mathbb{D}$. Then the following sharp inequalities holds:*

1. $|a_0| + \sum_{k=1}^{\infty} (|a_k| + |b_k|) r^k \leq \dfrac{||h||_{\infty}}{r} (5 - 2\sqrt{6}\sqrt{1-r^2})$ *for* $\dfrac{1}{5} \leq r \leq \sqrt{\dfrac{2}{3}}$,

2. $|a_0| + \sum_{k=1}^{\infty} (|a_k| + |b_k|) r^k \leq ||h||_{\infty}$ *for $r \leq \dfrac{1}{5}$.*

For further results on Bohr radius for quasiconformal harmonic mappings, we refer to [29] where there are a couple of conjectures and Bohr radius for the space of harmonic Bloch functions. In the same spirit, as with the analytic case, Evdoridis et al. [21] improve the results of [25] for locally univalent harmonic mappings. We now recall them here.

Theorem L. *Suppose that $f(z) = h(z) + \overline{g(z)} = \sum_{k=0}^{\infty} a_k z^k + \overline{\sum_{k=1}^{\infty} b_k z^k}$ is a harmonic mapping of the disk $\mathbb{D}$, where h is a bounded function in $\mathbb{D}$ such that $|h(z)| < 1$ and $|g'(z)| \le |h'(z)|$ for $z \in \mathbb{D}$. If S_r, as in Theorem E, denotes the area of the image of the subdisk $|z| < r$ under the mapping f, then*

$$H_1(r) := |a_0| + \sum_{k=1}^{\infty} (|a_k| + |b_k|) r^k + \frac{108}{25} \left(\frac{S_r}{\pi} \right) \le 1 \text{ for } r \le \frac{1}{5},$$

and the constants $1/5$ and $c = 108/25$ cannot be improved. Moreover,

$$H_2(r) := |a_0|^2 + \sum_{k=1}^{\infty} (|a_k| + |b_k|) r^k + \frac{4}{3} \left(\frac{S_r}{\pi} \right) \le 1 \text{ for } r \le \frac{1}{3},$$

and the constants $1/3$ and $4/3$ cannot be improved.

Theorem M. *Suppose that $f(z) = h(z) + \overline{g(z)} = \sum_{k=0}^{\infty} a_k z^k + \overline{\sum_{k=1}^{\infty} b_k z^k}$ is a harmonic mapping in $\mathbb{D}$, with $\|h\|_{\infty} = 1$ and $|g'(z)| \le |h'(z)|$ for $z \in \mathbb{D}$. Then*

$$L(r) := |a_0| + \sum_{k=1}^{\infty} (|a_k| + |b_k|) r^k + \frac{3}{8} \sum_{k=1}^{\infty} (|a_k|^2 + |b_k|^2) r^k \le 1 \text{ for } r \le \frac{1}{5}.$$

The constants $3/8$ and $1/5$ cannot be improved.

Theorem N. *Suppose that $f(z) = h(z) + \overline{g(z)} = \sum_{k=0}^{\infty} a_k z^k + \overline{\sum_{k=1}^{\infty} b_k z^k}$ is a harmonic mapping in $\mathbb{D}$, where $\|h\|_{\infty} = 1$ and $|g'(z)| \le |h'(z)|$ for $z \in \mathbb{D}$. Then*

$$N(r) := |a_0| + \sum_{k=1}^{\infty} (|a_k| + |b_k|) r^k + |h(z) - a_0|^2 \le 1 \text{ for } r \le 1/5.$$

The constant $1/5$ is best possible.

Theorem O. *Suppose that $f(z) = h(z) + \overline{g(z)} = \sum_{k=0}^{\infty} a_k z^k + \overline{\sum_{k=1}^{\infty} b_k z^k}$ is a harmonic mapping of the disk $\mathbb{D}$, where h is a bounded function in $\mathbb{D}$ such that $\|h\|_{\infty} = 1$ and $|g'(z)| \le |h'(z)|$ for $z \in \mathbb{D}$. Then,*

$$|h(z)|^2 + \sum_{k=1}^{\infty} (|a_k|^2 + |b_k|^2) r^{2k} \le 1 \text{ for } r \le r_0,$$

where $r_0 = \sqrt{5/(9 + 4\sqrt{5})} \approx 0.527864$ is the unique positive root of the equation

$$(10 + 6r^2)^{3/2} + 144r^2 - 80 = 0$$

in the interval $(0, 1/\sqrt{2})$. The number r_0 is the best possible.

In the case of harmonic mappings, more flexible approach was suggested in a recent paper [26]. This approach is expected to be more efficient as demonstrated in a number results proved in [26].

Acknowledgements. The research of the first and the second authors were supported by Russian foundation for basic research, Proj. 17-01-00282. The work of the third author is supported by Mathematical Research Impact Centric Support of DST, India (MTR/2017/000367). The third author is currently at Indian Statistical Institute (ISI), Chennai Centre, Chennai, India.

References

1. Abu-Muhanna, Y.: Bohr phenomenon in subordination and bounded harmonic classes. Complex Var. Elliptic Equ. **55**(11), 1071–1078 (2010)
2. Abu-Muhanna, Y., Ali, R.M.: Bohr phenomenon for analytic functions into the exterior of a compact convex body. J. Math. Anal. Appl. **379**(2), 512–517 (2011)
3. Abu-Muhanna, Y., Ali, R.M., Ng, Z.C., Hasni, S.F.M.: Bohr radius for subordinating families of analytic functions and bounded harmonic mappings. J. Math. Anal. Appl. **420**(1), 124–136 (2014)
4. Aizenberg, L.: Multidimensional analogues of Bohr's theorem on power series. Proc. Am. Math. Soc. **128**(4), 1147–1155 (2000)
5. Aizenberg, L.: Generalization of results about the Bohr radius for power series. Stud. Math. **180**, 161–168 (2007)
6. Aizenberg, L., Aytuna, A., Djakov, P.: Generalization of a theorem of Bohr for bases in spaces of holomorphic functions of several complex variables. J. Math. Anal. Appl. **258**(2), 429–447 (2001)
7. Abu-Muhanna, Y., Ali, R.M., Ponnusamy, S.: On the Bohr inequality. In: Govil, N.K., Mohapatra, R., Qazi, M.A., Schmeisser, G. (eds.) Progress in Approximation Theory and Applicable Complex Analysis. SOIA, vol. 117, pp. 269–300. Springer, Cham (2017). https://doi.org/10.1007/978-3-319-49242-1_13
8. Ali, R.M., Barnard, R.W., Solynin, A.Y.: A note on the Bohr's phenomenon for power series. J. Math. Anal. Appl. **420**(1), 154–167 (2017)
9. Bénéteau, C., Dahlner, A., Khavinson, D.: Remarks on the Bohr phenomenon. Comput. Methods Funct. Theory **4**(1), 1–19 (2004)
10. Boas, H.P.: Majorant series. Math. Soc. **37**(2), 321–337 (2000)
11. Boas, H.P., Khavinson, D.: Bohr's power series theorem in several variables. Proc. Am. Math. Soc. **125**(10), 2975–2979 (1997)
12. Bohr, H.: A theorem concerning power series. Proc. Lond. Math. Soc. **13**(2), 1–5 (1914)
13. Bombieri, E.: Sopra un teorema di H. Bohr e G. Ricci sulle funzioni maggioranti delle serie di potenze. Boll. Unione Mat. Ital. **17**, 276–282 (1962)
14. Bombieri, E., Bourgain, J.: A remark on Bohr's inequality. Int. Math. Res. Not. **80**, 4307–4330 (2004)
15. de Branges, L.: A proof of the Bieberbach conjecture. Acta Math. **154**, 137–152 (1985)
16. Clunie, J.G., Sheil-Small, T.: Harmonic univalent functions. Ann. Acad. Sci. Fenn. Ser. A I **9**, 3–25 (1984)
17. Dixon, P.G.: Banach algebras satisfying the non-unital von Neumann inequality. Bull. Lond. Math. Soc. **27**(4), 359–362 (1995)
18. Djakov, P.B., Ramanujan, M.S.: A remark on Bohrs theorem and its generalizations. J. Anal. **8**, 65–77 (2000)
19. Duren, P.L.: Univalent Functions. Springer, New York (1983)
20. Duren, P.: Harmonic Mappings in the Plane. Cambridge University Press, Cambridge (2004)

21. Evdoridis, S., Ponnusamy, S., Rasila, A.: Improved Bohr's inequality for locally univalent harmonic mappings, 14 p. (2017). https://arxiv.org/pdf/1709.08944.pdf
22. Hallenbeck, D.J.: Convex hulls and extreme points of some families of univalent functions. Trans. Am. Math. Soc. **192**, 285–292 (1974)
23. Ismagilov, A., Kayumova, A., Kayumov, I.R., Ponnusamy, S.: Bohr type inequalities in certain classes of analytic functions. Math. Sci. (New York) (English) (2019)
24. Kayumov, I.R., Ponnusamy, S.: Bohr inequality for odd analytic functions. Comput. Methods Funct. Theory **17**, 679–688 (2017)
25. Kayumov, I.R., Ponnusamy, S.: Improved version of Bohr's inequality. Comptes Rendus Mathematique **356**, 272–277 (2018)
26. Kayumov, I.R., Ponnusamy, S.: Bohr's inequalities for the analytic functions with lacunary series and harmonic functions. J. Math. Anal. Appl. (2018, to appear)
27. Kayumov, I.R., Ponnusamy, S.: On a powered Bohr inequality (2018)
28. Kayumov, I.R., Ponnusamy, S.: Bohr-Rogosinski radius for analytic functions (2017)
29. Kayumov, I.R., Ponnusamy, S., Shakirov, N.: Bohr radius for locally univalent harmonic mappings, Math. Nachr., 12 (2017). https://doi.org/10.1002/mana.201700068
30. Lehto, O., Virtanen, K.I.: Quasiconformal Mappings in the Plane. Springer, Heidelberg (1973)
31. Lewy, H.: On the non-vanishing of the Jacobian in certain one-to-one mappings. Bull. Am. Math. Soc. **42**, 689–692 (1936)
32. Liu, G., Ponnusamy, S.: On Harmonic ν-Bloch and ν-Bloch-type mappings (2017). https://arxiv.org/abs/1707.01570
33. Ponnusamy, S., Rasila, A.: Planar harmonic and quasiregular mappings. In: Topics in Modern Function Theory. CMFT, RMS-Lecture Notes Series No. 19, pp. 267–333 (2013)
34. Pommerenke, C.: Boundary Behaviour of Conformal Maps. Springer, New York (1992). https://doi.org/10.1007/978-3-662-02770-7
35. Popescu, G.: Multivariable Bohr inequalities. Trans. Am. Math. Soc. **359**(11), 5283–5317 (2007)
36. Rogosinski, W.: On the coefficients of subordinate functions. Proc. Lond. Math. Soc. **48**(2), 48–82 (1943)
37. Vuorinen, M.: Conformal Geometry and Quasiregular Mappings. LNM, vol. 1319. Springer, Heidelberg (1988). https://doi.org/10.1007/BFb0077904

Application of the Fractional Differential Transform Method to the First Kind Abel Integral Equation

Subhabrata Mondal[1] and B. N. Mandal[2(⊠)]

[1] Department of Applied Mathematics, University of Calcutta,
92, A.P.C. Road, Kolkata 700009, India
[2] Physics and Applied Mathematics Unit, Indian Statistical Institute,
203, B.T Road, Kolkata 700108, India
bnm2006@rediffmail.com

Abstract. The fractional differential transform method is employed here for solving first kind Abel integral equation. Abel integral equation occurs in the mathematical modeling of several models in physics, astrophysics, solid mechanics and applied sciences. An analytic technique for solving Abel integral equation of first kind by the proposed method is introduced here. Also illustrative examples with exact solutions are considered to show the validity and applicability of the proposed method. Numerical results reveal that the proposed method works well and has good accuracy. The method introduces a promising tool for solving many linear and nonlinear fractional integral equation.

Keywords: Abel integral equation · Differential transform method
Fractional differential transform method

1 Introduction

Abel integral equation is one of the most important integral equations that is derived directly from a problem of mechanics. Abel integral equation was derived by Abel in the year 1826 when he was generalizing and solving the Tautochrone problem. It involves finding the total time required for a particle to fall along a given smooth curve in the vertical plane. Here we consider the Abel integral equations of first kind given by

$$\int_0^x \frac{\phi(t)}{(x-t)^\alpha}\, dt = f(x),\ \ x > 0\ (f(0) = 0) \tag{1.1}$$

where α is a real constant such that $0 < \alpha < 1$, $f(x)$ is a known function and $\phi(x)$ is an unknown function to be determined. The Eq. (1.1) is a particular case of a linear Volterra integral equation of the first kind, where the order of the fractional integral Eq. (1.1) is $(1 - \alpha)$.

© Springer Nature Singapore Pte Ltd. 2018
D. Ghosh et al. (Eds.): ICMC 2018, CCIS 834, pp. 257–267, 2018.
https://doi.org/10.1007/978-981-13-0023-3_24

Mandal et al. (1996) solved a system of generalized Abel integral equations by using fractional calculus. Yousefi (2006) obtained the numerical solution of Abel integral equation by using Legendre wavelets. Liu and Tao (2007) used mechanical quadrature methods for solving first kind Abel integral equations. Derili and Sohrabi (2008) obtained the numerical solution of Abel integral equations by using orthogonal functions. De et al. (2009) reinvestigated the water wave scattering problem involving two submerged plane thin vertical barriers by an approach leading to the problem of solving a system of Abel integral equations. Alipour and Rostamy (2011) used Bernstein polynomials to solve Abel integral equations. Kumar et al. (2015) obtained an analytical solution of Abel integral equation arising in astrophysics using homotopy perturbation transform method.

In this paper, we employ a new analytical technique, namely fractional differential transform method from fractional calculus, to solve first kind Abel integral equations. This semi-analytical numerical technique formalizes fractional power series expansion in a manner that differential transform method formalizes Taylor series expansion. The main aim of this paper is to present analytical and numerical solution of Abel integral equation by using new mathematical tool like fractional differential transform method where we have used the result

$$\int_0^x \frac{t^{\frac{k}{\beta}}}{(x-t)^\alpha} dt = \frac{\Gamma(1+\frac{k}{\beta})\Gamma(1-\alpha)}{\Gamma(2+\frac{k}{\beta}-\alpha)} x^{1+\frac{k}{\beta}-\alpha}, \ 0 < \alpha < 1. \tag{1.2}$$

2   Analysis of the Differential Transform Method (DTM)

The basic definitions and fundamental operations of the one-dimensional differential transform method (DTM) and its applicability for various kinds of differential equations, integral equations are given by Odibat (2008). For convenience, we present here a review of the DTM. The differential transform of the kth derivative of a function $f(x)$ in one variable is

$$F(k) = \frac{1}{k!} \left[\frac{d^k f(x)}{dx^k} \right]_{x=x_0} \tag{2.1}$$

where $f(x)$ is the original function and $F(k)$ is the transformed function. The differential inverse transform of $F(k)$ is defined as

$$f(x) = \sum_{k=0}^{\infty} F(k)(x - x_0)^k. \tag{2.2}$$

Eq. (2.2) implies that the concept of differential transform is derived from the Taylor series expansion.

In real applications, the function $f(x)$ is expressed by a finite series and Eq. (2.2) can be written as

$$f(x) = \sum_{k=0}^{n} F(k)(x - x_0)^k. \tag{2.3}$$

3   The Fractional Differential Transform Method (FDTM)

Arikoglu and Ozkol (2007) introduced the fractional differential transform method for solving fractional differential equations. For convenience, in this section a short review of the FDTM is presented with some preliminary concepts and definitions.

The fractional differential transform of the kth derivative of the analytic function $f(x)$ is defined as

$$
F(k) = \begin{cases} 0 & \frac{k}{\beta} \notin \mathbb{Z}^+ \\ \frac{1}{(\frac{k}{\beta})!} \left(\frac{d^{\frac{k}{\beta}}}{dx^{\frac{k}{\beta}}} f(x) \right)_{x=x_0} , & \frac{k}{\beta} \in \mathbb{Z}^+ \end{cases} \tag{3.1}
$$

where β is the order of the fraction.

There are several approaches to the generalization of the notion of differentiation to fractional orders. The fractional differentiation in Riemann-Liouville sense is defined by

$$
D_{x_0}^q f(x) = \frac{1}{\Gamma(m-q)} \frac{d^m}{dx^m} \left[\int_{x_0}^x \frac{f(t)}{(x-t)^{1+q-m}} dt \right] \tag{3.2}
$$

for $m - 1 \leq q < m$, $m \in \mathbb{Z}^+$, $x > x_0$ and q is the order of the corresponding fractional equation.

Concerning the practical applications encountered in various branches of science, the fractional initial conditions are frequently not available, and it may not be clear what their physical meaning is. Therefore, the definition in Eq. (3.2) should be modified to deal with integer ordered initial conditions in Caputo sense (1967) as follows:

$$
\begin{aligned}
D_{*x_0}^q f(x) &= D_{x_0}^q \left[f(x) - \sum_{k=0}^{m-1} \frac{1}{k!} (x-x_0)^k f^{(k)}(x_0) \right] \\
&= \frac{1}{\Gamma(m-q)} \frac{d^m}{dx^m} \left[\int_{x_0}^x \frac{f(t) - \sum_{k=0}^{m-1} \frac{1}{k!}(t-x_0)^k f^{(k)}(x_0)}{(x-t)^{1+q-m}} dt \right] \tag{3.3}
\end{aligned}
$$

for $k = 0, 1, 2, \ldots (\beta q - 1)$.

Thus the fractional differential inverse transform of $F(k)$ is defined as

$$
f(x) = \sum_{k=0}^{\infty} F(k)(x - x_0)^{\frac{k}{\beta}}, \tag{3.4}
$$

which implies that the concept of fractional differential transform is derived from fractional power series expansion. In practical application, the function $f(x)$ can be approximated by the finite series

$$f(x) \equiv \sum_{k=0}^{n} F(k)(x - x_0)^{\frac{k}{\beta}}. \tag{3.5}$$

The fundamental operations of fractional differential transform method are listed in Table 1 below.

Table 1. Operations of fractional differential transform

Original function	Transformed function
$f(x) = g(x) \pm h(x)$	$F(k) = G(k) \pm H(k)$
$f(x) = g(x)h(x)$	$F(k) = \sum_{l=0}^{k} G(l)H(k - l)$
$f(x) = cg(x)$	$F(k) = cG(k)$
$f(x) = x^p$	$F(k) = \delta(k - \beta p) = \begin{cases} 1 & if\ k = \beta p \\ 0 & if\ k \neq \beta p \end{cases}$
$f(x) = D_{*x_0}^q g(x)$	$F(k) = \dfrac{\Gamma(q+1+\frac{k}{\beta})}{\Gamma(1+\frac{k}{\beta})}G(k + \beta q)$
$f(x) = \int_{x_0}^{x} g(t)dt$	$F(k) = \frac{\beta G(k-\beta)}{k}$, $k \geq \beta$
$f(x) = g(x) \int_{x_0}^{x} h(t)dt$	$F(k) = \beta \sum_{k_1=\beta}^{k} \frac{H(k_1-\beta)}{k_1} G(k - k_1)$, $k \geq \beta$

4   Analytical Solution of Abel Integral Equation

To illustrate the basic idea of the FDTM for solution of singular integral equation of Abel type, we consider the following Abel integral equation of first kind as

$$\int_{0}^{x} \frac{\phi(t)}{\sqrt{x - t}}dt = f(x), \ f(0) = 0 \tag{4.1}$$

whose exact solution is

$$\phi(x) = \frac{1}{\pi} \frac{d}{dx} \left[\int_{0}^{x} \frac{f(t)}{\sqrt{x - t}}dt \right] \tag{4.2}$$

It is easy to see that the order of fraction of the fractional integral Eq. (4.1) is 2.

Let

$$\phi(x) = \sum_{k=0}^{\infty} \Phi(k)x^{\frac{k}{2}}. \tag{4.3}$$

Then

$$\int_{0}^{x} \frac{\phi(t)}{\sqrt{x - t}}dt = \sum_{k=0}^{\infty} \Phi(k) \int_{0}^{x} \frac{t^{\frac{k}{2}}}{\sqrt{x - t}}dt. \tag{4.4}$$

Let

$$f(x) = \sum_{k=0}^{\infty} F(k)x^{\frac{k}{2}} = \sum_{k=1}^{\infty} F(k)x^{\frac{k}{2}} \quad [F(0) = 0] \tag{4.5}$$

By using Eq. (4.4), Eq. (4.5) and the result (1.2) in Eq. (4.1), we obtain

$$\sum_{k=1}^{\infty} F(k)x^{\frac{k}{2}} = \sum_{k=0}^{\infty} \Phi(k)\frac{\sqrt{\pi}\Gamma(1+\frac{k}{2})}{\Gamma(\frac{k}{2}+\frac{3}{2})}x^{\frac{k+1}{2}}$$

$$= \sum_{k=1}^{\infty} \Phi(k-1)\frac{\sqrt{\pi}\Gamma(\frac{k+1}{2})}{\Gamma(\frac{k}{2}+1)}x^{\frac{k}{2}}.$$

Therefore,

$$F(k) = \sqrt{\pi}\,\Phi(k-1)\frac{\Gamma(\frac{k+1}{2})}{\Gamma(\frac{k}{2}+1)}$$

and hence

$$\Phi(k) = \frac{1}{\sqrt{\pi}}\,F(k+1)\frac{\Gamma(\frac{k+1}{2}+1)}{\Gamma(\frac{k}{2}+1)}. \tag{4.6}$$

Using Eq. (4.6) in Eq. (4.3) we get

$$\phi(x) = \sum_{k=0}^{\infty} \frac{1}{\sqrt{\pi}}\,F(k+1)\frac{\Gamma(\frac{k+1}{2}+1)}{\Gamma(\frac{k}{2}+1)}x^{\frac{k}{2}}$$

$$= \frac{1}{\sqrt{\pi}}\sum_{k=0}^{\infty} F(k+1)\frac{\Gamma(\frac{k+1}{2}+1)}{\Gamma(\frac{k}{2}+2)}(\frac{k}{2}+1)\,x^{\frac{k}{2}}$$

$$= \frac{1}{\pi}\sum_{k=1}^{\infty} F(k)\frac{\sqrt{\pi}\Gamma(\frac{k}{2}+1)}{\Gamma(\frac{k}{2}+\frac{3}{2})}\left(\frac{k}{2}+\frac{1}{2}\right)x^{(\frac{k}{2}+\frac{1}{2})-1}$$

$$= \frac{1}{\pi}\sum_{k=1}^{\infty} F(k)\frac{\sqrt{\pi}\Gamma(\frac{k}{2}+1)}{\Gamma(\frac{k}{2}+\frac{3}{2})}\frac{d}{dx}\left[x^{\frac{k}{2}+\frac{1}{2}}\right]$$

$$= \frac{1}{\pi}\sum_{k=1}^{\infty} F(k)\frac{d}{dx}\left[\frac{\sqrt{\pi}\Gamma(\frac{k}{2}+1)}{\Gamma(\frac{k}{2}+\frac{3}{2})}x^{\frac{k}{2}+\frac{1}{2}}\right]. \tag{4.7}$$

Using the result (1.2) we get from Eq. (4.7),

$$\phi(x) = \frac{1}{\pi}\frac{d}{dx}\left[\int_{0}^{x}\frac{\sum_{k=0}^{\infty}F(k)\,t^{\frac{k}{2}}}{\sqrt{x-t}}\,dt\right] \quad [F(0)=0] \tag{4.8}$$

Now, using Eq. (4.5) we obtain from Eq. (4.8) the solution of the integral Eq. (4.1) as given by Eq. (4.2).

5 Numerical Solution of Abel Integral Equation

To obtain the numerical solution of Eq. (1.1) we need to prove and use the following Theorem.

Theorem 1. If $\int\limits_{0}^{x} \frac{\phi(t)}{(x-t)^\alpha}\, dt = f(x)$, $(0 < \alpha < 1)$ then the fractional differential transform of the integral equation is

$$\sum_{k=0}^{N} \Phi(k)\, \frac{\Gamma(1-\alpha)\Gamma(1+\frac{k}{\beta})}{\Gamma(2+\frac{k}{\beta}-\alpha)}\, \delta[\,\bar{k} - (1 + \frac{k}{\beta} - \alpha)\,\beta\,] \;=\; F(\bar{k}),\ N \to \infty$$

where $\Phi(k)$ is the fractional differential transform of $\phi(x)$ and β is the order of the fraction.

Proof. Here β is the order of the fraction of the fractional integral equation

$$\int\limits_{0}^{x} \frac{\phi(t)}{(x-t)^\alpha}\, dt = f(x),\ (f(0) = 0) \tag{5.1}$$

Using fractional power series about origin and result (1.2) we obtain the L.H.S. of Eq. (5.1) as

$$\sum_{k=0}^{\infty} \Phi(k) \int\limits_{0}^{x} \frac{t^{\frac{k}{\beta}}}{(x-t)^\alpha}\, dt$$

$$= \sum_{k=0}^{\infty} \Phi(k) \left[\frac{\Gamma(1-\alpha)\Gamma(1+\frac{k}{\beta})}{\Gamma(2+\frac{k}{\beta}-\alpha)}\, x^{1+\frac{k}{\beta}-\alpha} \right]. \tag{5.2}$$

Let $F(\bar{k})$ be the fractional differential transform of $f(x)$. Then by using the fundamental operations in Table 1, we obtain from Eqs. (5.1) and (5.2)

$$\sum_{k=0}^{\infty} \Phi(k)\, \frac{\Gamma(1-\alpha)\Gamma(1+\frac{k}{\beta})}{\Gamma(2+\frac{k}{\beta}-\alpha)}\, \delta[\,\bar{k} - (1 + \frac{k}{\beta} - \alpha)\,\beta\,] \;=\; F(\bar{k}). \tag{5.3}$$

6 Illustrative Examples

In order to illustrate the advantages and the accuracy of the fractional differential transform method for solving the Abel integral equation, we have applied the method to solve some examples.

Example 1. Consider the following Abel integral equation of the first kind

$$\int\limits_{0}^{x} \frac{\phi(t)}{\sqrt{x-t}}\, dt = x\,, \tag{6.1}$$

with exact solution $\phi(x) = \frac{2}{\pi}\sqrt{x}$.

Here $f(x) = x$ and $\alpha = \frac{1}{2}$. We see that the order of fraction of Eq. (6.1) is $\beta = 2$.

Using the fundamental operations of Table 1, we obtain the fractional differential transform of $f(x) = x$ as $F(\bar{k}) = \delta(\bar{k} - 2)$.

Therefore,

$$F(\bar{k}) = \begin{cases} 1 & \bar{k} = 2 \\ 0, & \text{otherwise.} \end{cases}$$

By using Theorem 1, the fractional differential transform of Eq. (6.1) becomes

$$\sum_{k=0}^{\infty} \Phi(k) \frac{\Gamma(\frac{1}{2})\Gamma(1 + \frac{k}{2})}{\Gamma(\frac{3}{2} + \frac{k}{2})} \delta[\bar{k} - (1 + k)] = \delta(\bar{k} - 2)$$

so that $\quad \Phi(\bar{k} - 1) \dfrac{\Gamma(\frac{1}{2})\Gamma(1 + \frac{k-1}{2})}{\Gamma(\frac{3}{2} + \frac{k-1}{2})} = \delta(\bar{k} - 2).$

Putting $\bar{k} = 1, 2, 3, \ldots$ we obtain

$$\Phi(k) = \begin{cases} \frac{2}{\pi} & k = 1 \\ 0, & \text{otherwise.} \end{cases}$$

Now, we have from inverse fractional differential transform

$$\phi(x) = \sum_{k=0}^{\infty} \Phi(k)\, x^{\frac{k}{2}}$$

$$= \Phi(1)\, x^{\frac{1}{2}}$$

$$= \frac{2}{\pi}\, \sqrt{x}.$$

Thus we get the solution of Eq. (6.1) as $\phi(x) = \frac{2}{\pi}\sqrt{x}$, which is identical to the exact solution.

Example 2. As the second example consider the following Abel integral equation of the first kind

$$\int_0^x \frac{\phi(t)}{\sqrt{x - t}}\, dt = \frac{2}{105}\sqrt{x}(105 - 56x^2 + 48x^3), \tag{6.2}$$

with exact solution $\phi(x) = x^3 - x^2 + 1$.

Here $f(x) = \frac{2}{105}\sqrt{x}(105 - 56x^2 + 48x^3)$ and $\alpha = \frac{1}{2}$. We see that the order of fraction of Eq. (6.2) is $\beta = 2$.

Using the fundamental operations of Table 1, we obtain the fractional differential transform of $f(x) = \frac{2}{105}\sqrt{x}(105 - 56x^2 + 48x^3)$ as

$$F(\bar{k}) = 2\delta(\bar{k} - 1) - \frac{16}{15}\delta(\bar{k} - 5) + \frac{32}{35}\delta(\bar{k} - 7).$$

Therefore,

$$F(\bar{k}) = \begin{cases} 2 & \bar{k} = 1 \\ -\frac{16}{15} & \bar{k} = 5 \\ \frac{32}{35} & \bar{k} = 7 \\ 0, & \text{otherwise.} \end{cases}$$

By using Theorem 1, the fractional differential transform of Eq. (6.2) becomes

$$\sum_{k=0}^{\infty} \Phi(k) \frac{\Gamma(\frac{1}{2})\Gamma(1+\frac{k}{2})}{\Gamma(\frac{3}{2}+\frac{k}{2})} \delta[\bar{k} - (1+k)] = 2\delta(\bar{k} - 1) - \frac{16}{15}\delta(\bar{k} - 5) + \frac{32}{35}\delta(\bar{k} - 7)$$

so that $\quad \Phi(\bar{k} - 1) \dfrac{\Gamma(\frac{1}{2})\Gamma(1+\frac{k-1}{2})}{\Gamma(\frac{3}{2}+\frac{k-1}{2})} = 2\delta(\bar{k} - 1) - \dfrac{16}{15}\delta(\bar{k} - 5) + \dfrac{32}{35}\delta(\bar{k} - 7).$

Putting $\bar{k} = 1, 2, 3, \ldots$ we obtain

$$\Phi(k) = \begin{cases} 1 & k = 0, 6 \\ -1 & k = 4 \\ 0, & \text{otherwise.} \end{cases}$$

Now, we have from inverse fractional differential transform

$$\phi(x) = \sum_{k=0}^{\infty} \Phi(k) \, x^{\frac{k}{2}}$$
$$= \Phi(0) + \Phi(4) \, x^2 + \Phi(6) \, x^3$$
$$= 1 - x^2 + x^3.$$

Thus we get the solution of Eq. (6.2) as $\phi(x) = 1 - x^2 + x^3$, which is identical to the exact solution.

Example 3. Consider the following Abel integral equation of the first kind with another form

$$\int_0^x \frac{\phi(t)}{(x-t)^{\frac{1}{3}}} \, dt = \frac{2}{3\sqrt{3}} \, \pi x, \tag{6.3}$$

with exact solution $\phi(x) = x^{\frac{1}{3}}$.

Here $f(x) = \frac{2}{3\sqrt{3}} \pi x$ and $\alpha = \frac{1}{3}$. We see that the order of fraction of Eq. (6.3) is $\beta = 3$.

Using the fundamental operations of Table 1, we obtain the fractional differential transform of $f(x) = \frac{2}{3\sqrt{3}} \pi x$ as

$$F(\bar{k}) = \frac{2}{3\sqrt{3}} \pi \, \delta(\bar{k} - 3).$$

Therefore,

$$F(\bar{k}) = \begin{cases} \frac{2}{3\sqrt{3}} \pi & \bar{k} = 3 \\ 0, & \text{otherwise.} \end{cases}$$

By using Theorem 1, the fractional differential transform of Eq. (6.3) becomes

$$\sum_{k=0}^{\infty} \Phi(k) \frac{\Gamma(\frac{2}{3})\Gamma(1+\frac{k}{3})}{\Gamma(\frac{5}{3}+\frac{k}{3})} \delta[\bar{k}-(2+k)] = \frac{2}{3\sqrt{3}} \pi \delta(\bar{k}-3)$$

so that $\quad \Phi(\bar{k}-2) \dfrac{\Gamma(\frac{2}{3})\Gamma(1+\frac{\bar{k}-2}{3})}{\Gamma(\frac{5}{3}+\frac{\bar{k}-2}{3})} = \dfrac{2}{3\sqrt{3}} \pi \delta(\bar{k}-3).$

Putting $\bar{k} = 2, 3, \ldots$ we obtain

$$\Phi(k) = \begin{cases} 1 & k = 1 \\ 0, & otherwise. \end{cases}$$

Now, we have from inverse fractional differential transform

$$\phi(x) = \sum_{k=0}^{\infty} \Phi(k) \, x^{\frac{k}{3}}$$
$$= \Phi(1) \, x^{\frac{1}{3}}$$
$$= x^{\frac{1}{3}}.$$

Thus we get the solution of Eq. (6.3) as $\phi(x) = x^{\frac{1}{3}}$, which is identical to the exact solution.

Example 4. As the last example consider the following Abel integral equation of the first kind

$$\int_0^x \frac{\phi(t)}{(x-t)^{\frac{1}{2}}} \, dt = e^x - 1, \tag{6.4}$$

with exact solution $\phi(x) = \frac{e^x erf(\sqrt{x})}{\sqrt{\pi}}$.

Here $f(x) = e^x - 1$ and $\alpha = \frac{1}{2}$. We see that the order of fraction of Eq. (6.4) is $\beta = 2$.

Using the fundamental operations of Table 1, we obtain the fractional differential transform of $f(x) = e^x - 1$ as

$$F(\bar{k}) = \delta(\bar{k}-2) + \frac{1}{2!}\delta(\bar{k}-4) + \frac{1}{3!}\delta(\bar{k}-6) + \ldots$$

Therefore,

$$F(\bar{k}) = \begin{cases} 1 & \bar{k} = 2 \\ \frac{1}{2!} & \bar{k} = 4 \\ \frac{1}{3!} & \bar{k} = 6 \\ \frac{1}{4!} & \bar{k} = 8 \\ \ldots & \text{and so on} \end{cases}$$

By using Theorem 1, the fractional differential transform of Eq. (6.4) becomes

$$\sum_{k=0}^{\infty} \varPhi(k) \frac{\Gamma(\frac{1}{2})\Gamma(1+\frac{k}{2})}{\Gamma(\frac{3}{2}+\frac{k}{2})} \, \delta \, [\bar{k} - (1+k)] \; = \; F(\bar{k})$$

so that $\quad \varPhi(\bar{k} - 1) \dfrac{\Gamma(\frac{1}{2})\Gamma(1+\frac{\bar{k}-1}{2})}{\Gamma(\frac{3}{2}+\frac{\bar{k}-1}{2})} \; = \; F(\bar{k}).$

Putting $\bar{k} \; = \; 1, 2, 3, \ldots$ we obtain

$$\varPhi(k) = \begin{cases} \frac{2}{\pi} & k = 1 \\ \frac{4}{3\pi} & k = 3 \\ \frac{8}{15\pi} & k = 5. \\ \ldots & \text{and so on} \end{cases}$$

Now, we have from inverse fractional differential transform

$$\phi(x) = \sum_{k=0}^{\infty} \varPhi(k) \, x^{\frac{k}{2}}$$

$$= \varPhi(1) \, x^{\frac{1}{2}} + \varPhi(3) \, x^{\frac{3}{2}} + \varPhi(5) \, x^{\frac{5}{2}} + \ldots$$

$$= \frac{2}{\pi} \, x^{\frac{1}{2}} + \frac{4}{3\pi} \, x^{\frac{3}{2}} + \frac{8}{15\pi} \, x^{\frac{5}{2}} + \ldots$$

$$= \frac{e^x erf(\sqrt{x})}{\sqrt{\pi}}.$$

Thus we get the solution of Eq. (6.4) as $\phi(x) \; = \; \frac{e^x erf(\sqrt{x})}{\sqrt{\pi}}$, which is identical to the exact solution.

7   Conclusion

In this paper, a new fractional differential transform method is employed to obtain a quick and accurate solution of the singular integral equation of Abel type. The method provides the solutions in terms of convergent series with easily computable components in a direct way. The efficiency and reliability of the proposed technique has been demonstrated by considering several examples with known exact solution. From the results, it is seen that the solutions are identical to the exact solutions for all considered examples which show that the FDTM is a reliable tool for the solution of singular integral equation of Abel type.

References

Alipour, M., Rostamy, D.: Bernstein polynomials for solving Abel's integral equation. J. Math. Comput. Sci. **3**, 403–412 (2011)

Arikoglu, A., Ozkol, I.: Solution of fractional differential equations by using differential transform method. Chaos Solitons Fractals **34**, 1473–1481 (2007)

Caputo, M.: Linear models of dissipation whose Q is almost frequency independent. Part II. J. R. Austron. Soc. **13**, 529–539 (1967)

De, S., Mandal, B.N., Chakrabarti, A.: Water wave scattering by two submerged plane vertical barriers - Abel integral equations approach. J. Eng. Math. **65**, 75–87 (2009)

Derili, H., Sohrabi, S.: Numerical solution of singular integral equations using orthogonal functions. Math. Sci. (QJMS) **3**, 261–272 (2008)

Kumar, S., Kumar, A., Kumar, D., Singh, J., Singh, A.: Analytical solution of Abel integral equation arising in astrophysics via Laplace transform. J. Egypt. Math. Soc. **23**, 102–107 (2015)

Liu, Y., Tao, L.: Mechanical quadrature methods and their extrapolation for solving first kind Abel integral equations. J. Comput. Appl. Math. **201**, 300–313 (2007)

Mandal, N., Chakrabarti, A., Mandal, B.N.: Solution of a system of generalized Abel integral equations using fractional calculus. Appl. Math. Lett. **9**, 1–4 (1996)

Odibat, Z.M.: Differential transform method for solving Volterra integral equation with separable kernels. Math. Comput. Model. **48**, 1144–1149 (2008)

Yousefi, S.A.: Numerical solution of Abel's integral equation by using Legendre wavelets. Appl. Math. Comp. **175**, 574–580 (2006)

On the Relationship Between L-fuzzy Closure Spaces and L-fuzzy Rough Sets

Vijay K. Yadav[1(✉)], Swati Yadav[2], and S. P. Tiwari[2]

[1] Department of Mathematics, School of Mathematics, Statistics and Computational Sciences, Central University of Rajasthan, NH-8, Bandarsindari, Ajmer 305817, Rajasthan, India
vkymaths@gmail.com
[2] Department of Applied Mathematics, Indian Institute of Technology (Indian School of Mines), Dhanbad 826004, India
yswatimaths@gmail.com, sptiwarimaths@gmail.com

Abstract. This work is towards the establishment of bijective correspondence between the family of all L-fuzzy reflexive/tolerance approximation spaces and the family of all quasi-discrete L-fuzzy closure spaces satisfying a certain condition.

Keywords: L-fuzzy closure space
L-fuzzy reflexive approximation space
L-fuzzy tolerance approximation space

1 Introduction

Rough sets, firstly introduced by Pawlak [11] has been advanced notably with worthy of attention due to its widespread applications in both mathematics and computer sciences for the study of intelligent systems having insufficient, imprecise, uncertain and incomplete information. The partition or equivalence (indiscernibility) relations were the fundamental and abstract tools of the rough set theory introduced by Pawlak. Researchers have made several generalizations of rough sets using an arbitrary relation in place of an equivalence relation (cf., [4,7,21,22]). Dubois and Prade [3], proposed fuzzy version of rough sets in which fuzzy relations play a key roll instead of crisp relations. The fuzzy rough sets and their relationship with fuzzy topological spaces were described in detail by several authors (e.g., cf., [2,6,10,12–14,16,17,19,20]). Moreover, in [6,10,17], the set of all L-fuzzy preorder approximation spaces together with the set of all saturated L-fuzzy topological spaces were center of interest, and it was shown that under a certain extra condition there exists a bijective correspondence between them. The silence on such relationship between the set of other generalized approximation spaces (such as L-fuzzy reflexive approximation space and L-fuzzy tolerance approximation spaces) and the set of some L-fuzzy topological structures, in the cited work, attract our attention and lead us an

© Springer Nature Singapore Pte Ltd. 2018
D. Ghosh et al. (Eds.): ICMC 2018, CCIS 834, pp. 268–277, 2018.
https://doi.org/10.1007/978-981-13-0023-3_25

attempt to establish such relationships by using the concept of L-fuzzy closure spaces. Finally, we have established the similar result for the set of all, L-fuzzy preorder approximation spaces and L-fuzzy closure spaces, respectively.

2 Preliminaries

We begin by recalling the following concept of a residuated lattice from [1].

Definition 1. *An algebra $L = (L, \wedge, \vee, *, \rightarrow, 0, 1)$ define a **residuated lattice**, if $(L, \wedge, \vee, 0, 1)$ is a lattice having 0 and 1 as least and greatest element, respectively, $(L, *, 1)$ is a commutative monoid having unit 1, and $*$ and $\rightarrow$ form an adjoint pair, i.e., $\forall\ x, y, z \in L,\ x * y \leq z \Leftrightarrow x \leq y \rightarrow z$. Also, L is said to be a **complete residuated lattice** if lattice $(L, \vee, \wedge, 0, 1)$ is complete.*

Definition 2. *The **precomplement** on L is a map $\rightharpoondown\colon L \longrightarrow L$ such that $\rightharpoondown x = x \rightarrow 0, \forall x \in L$.*

Throughout, L denotes the complete residuated lattice. For a nonempty set X, L^X denote the collection of all L-fuzzy sets in X, for $\alpha \in L$, $\bar{\alpha}$ denotes the constant L-fuzzy set.

Definition 3. *A complete residuated lattice L is called **regular** if $\rightharpoondown (\rightharpoondown a) = a$, $\forall a \in L$.*

The basic properties of a complete regular residuated lattice, which we use in subsequent sections are listed in following proposition.

Proposition 1. *For all $a, b, a_i \in L$, $i \in J$ an index set, we have*

(i) $a * b =\rightharpoondown (a \rightarrow (\rightharpoondown b))$,
(ii) $a \rightarrow b =\rightharpoondown (a * (\rightharpoondown b))$,
(iii) $\rightharpoondown (\wedge\{a_i\}) = \vee\{\rightharpoondown a_i\}$,
(iv) $\rightharpoondown (\vee\{a_i\}) = \wedge\{\rightharpoondown a_i\}$.

Definition 4 [5]. *Let X be a nonempty set, then L-fuzzy relation on X is a map $R : X \times X \rightarrow L$.*
*For, properties of an L-fuzzy relation we refer to [5,10,15]. However, for completeness we emphasize from [10,15] that an L-fuzzy reflexive and L-fuzzy symmetric relation R is known as L-**fuzzy tolerance relation** and, if R is L-fuzzy reflexive as well as L-fuzzy transitive then it is called L-**fuzzy preorder**.*

Definition 5 [6,10,15,17]. *Let R be an L-fuzzy relation on a nonempty set X, then an L-**fuzzy approximation space** is a pair (X, R), which is further known as L-**fuzzy reflexive/tolerance/preorder approximation space**, respectively, according as underlying L-fuzzy relation R is an reflexive, tolerance or preorder.*

Throughout, set of all L-fuzzy approximation space over a nonempty set X is denoted by Ω.

Definition 6 [10,15,17]. *Consider an $(X, R) \in \Omega$ and $A \in L^X$. The **lower approximation** $\underline{apr}_R(A)$ of A and the **upper approximation** $\overline{apr}_R(A)$ of A in (X, R) are respectively defined as follows:*

$$\underline{apr}_R(A)(x) = \wedge\{R(x, y) \to A(y) : y \in X\}, and$$
$$\overline{apr}_R(A)(x) = \vee\{R(x, y) * A(y) : y \in X\}.$$

*For an $(X, R) \in \Omega$ and $A \in L^X$, we called the pair $(\underline{apr}_R(A), \overline{apr}_R(A))$ an **L-fuzzy rough set**.*

Proposition 2 [17]. *Consider an $(X, R) \in \Omega$, where L is regular as well, then for all $A \in L^X$,*

(i) $\underline{apr}_R(A) = \to \overline{apr}_R(\to A)$, and
(ii) $\overline{apr}_R(A) = \to \underline{apr}_R(\to A)$.

Proposition 3 [6,15,17]. *Consider an $(X, R) \in \Omega$, then $\forall A_i \in L^X, i \in J$ and $\alpha \in L$,*

(i) $\overline{apr}_R(\vee\{A_i : i \in J\}) = \vee\overline{apr}_R\{A_i : i \in J\}$,
(ii) $\underline{apr}_R(\wedge\{A_i : i \in J\}) = \wedge\underline{apr}_R\{A_i : i \in J\}$, and
(iii) $\overline{apr}_R(A * \bar{\alpha}) = \overline{apr}_R(A) * \bar{\alpha}$.

Proposition 4 [17]. *Consider an $(X, R) \in \Omega$, which is reflexive and $A \in L^X$, then*

(i) $\underline{apr}_R(A) \leq A$, and
(ii) $A \leq \overline{apr}_R(A)$.

Proposition 5 [17]. *Consider an $(X, R) \in \Omega$ and $A \in L^X$, then R is an L-fuzzy transitive relation on X iff $\overline{apr}_R(\overline{apr}_R(A)) \leq \overline{apr}_R(A)$.*

Proposition 6. *Let $(X, R), (X, S) \in \Omega$, then $R \leq S$ iff $\overline{apr}_R(A) \leq \overline{apr}_S(A)$, $\forall A \in L^X$.*

Proof. Let $\overline{apr}_R(A) \leq \overline{apr}_S(A), \forall A \in L^X$, i.e., $\vee\{R(x, y) * A(y)\} \leq \vee\{S(x, y) * A(y)\}, \forall A \in L^X$. Thus $R \leq S, \forall x, y \in X$.
Conversely, let $R \leq S$ and $x \in X$. Then $\overline{apr}_R(A)(x) = \vee\{R(x, y) * A(y) : y \in X\} \leq \vee\{S(x, y) * A(y) : y \in X\} = \overline{apr}_S(A)(x)$. Thus $\overline{apr}_R(A) \leq \overline{apr}_S(A)$.

The L-fuzzy topological concepts, we use here, are fairly standard and based on [8].

Definition 7. *An **L-fuzzy topology** τ over a nonempty set X is a subset of L^X closed under arbitrary suprema and finite infima and which contains all constant L-fuzzy sets.*

The pair (X, τ) is called an *L-bffuzzy topological space*. As usual, the member of τ are called *L-fuzzy τ-open sets*.

Definition 8. *A* **Kuratowski** *L-fuzzy* **closure operator** *over a nonempty set* X *is a map* $k : L^X \to L^X$, *whit property that* $\forall A, \in L^X$ *and* $\forall \alpha \in L$,

(i) $k(\bar{\alpha}) = \bar{\alpha}$,
(ii) $A \leq k(A)$,
(iii) $k(A \vee B) = k(A) \vee k(B)$, *and*
(iv) $k(k(A)) = k(A)$.

Proposition 7 [6]. *Consider an* $(X, R) \in \Omega$, *where* R *be an L-fuzzy reflexive relation, then* $\tau_R = \{A \in L^X : \underline{apr}_R(A) = A\}$ *is an L-fuzzy topology.*
One can easily verify that τ_R *is a saturated*[1] *L-fuzzy topology over* X.

Proposition 8 [17]. *Let* k *be as defined in Definition* 8, *then* $\exists$ *an L-fuzzy preorder* S_k *over* X *for which* $\overline{apr}_{S_k}(A) = k(A)$ *iff (i)* $\forall i \in J$ *an indexed set* $k(\vee\{A_i\}) = \vee\{k(A_i)\}$, $\forall A_i \in L^X$ *and (ii)* $k(A * \bar{\alpha}) = k(A) * \bar{\alpha}$, $\forall A \in L^X$, $\forall \alpha \in L$.

The concept of fuzzy closure spaces was proposed in (cf., [9]). Further, the concepts of subspace of a fuzzy closure space, sum of a family of pairwise disjoint fuzzy closure spaces and product of a family of fuzzy closure spaces were studied in [18]. Now, we introduce here the following concept of an L-fuzzy closure space as a generalization of the concept of a fuzzy closure space studied in [9,18].

Definition 9. *An L-fuzzy closure space over a nonempty set* X *is a pair* (X, c), *where the map* $c : L^X \to L^X$ *is such that* $\forall A, B \in L^X$ *and* $\forall \alpha \in L$,

(i) $c(\bar{\alpha}) = \bar{\alpha}$,
(ii) $A \leq c(A)$, *and*
(iii) $c(A \vee B) = c(A) \vee c(B)$.

Definition 10. *An L-fuzzy closure space* (X, c) *is called*

(i) **quasi-discrete** *if* $c\{\vee\{A_i : i \in J\}\} = \vee\{c(A_i) : i \in J\}$, $\forall A_i \in L^X$,
(ii) **symmetric** *if* $c(1_y)(x) = c(1_x)(y)$, $\forall x, y \in X$, *and*
(iii) $A \in L^X$ *is called* L-**fuzzy closed** *if* $c(A) = A$.

Proposition 9. *Let* (X, c) *be as in* 9, *then*

(i) for $A, B \in L^X$ *if* $A \leq B$ *then* $c(A) \leq c(B)$,
(ii) $c\{\wedge\{A_i : i \in J\}\} \leq \wedge\{c(A_i) : i \in J\}$, $\forall A_i \in L^X, i \in J$.

Proof. Follows obviously.

Proposition 10. *Consider L-fuzzy closure space* (X, c), $H \in L^X$ *and* $\bar{c} : L^X \to L^X$ *be a map such that* $\bar{c}(H) = \wedge\{K \in L^X : H \leq K \text{ and } c(K) = K\}$. *Then* $\bar{c}$ *is a Kuratowski L-fuzzy closure operator on* X.

---

[1] In the sense that arbitrary infimum of L-fuzzy τ_R-open sets is also, an L-fuzzy τ_R-open.

Proof. Obviously $\forall \alpha \in L, c(\bar{\alpha}) = \bar{\alpha}$ and $\forall H \in L^X, H \leq \bar{c}(H)$. Now, let $H, K \in X$. Then $\bar{c}(H \vee K) = \wedge\{G \in L^X : (H \vee K) \leq G$ and $c(G) = G\}$. Thus $\bar{c}(H \vee K) = \wedge\{G \in L^X : H \leq G, K \leq G$ and $c(G) = G\} = \{\wedge\{G \in L^X : H \leq G$ and $c(G) = G\} \vee \{\wedge(G \in L^X : K \leq G$ and $c(G) = G)\}\} = \bar{c}(H) \vee \bar{c}(K)$. Finally, $\bar{c}(\bar{c}(H)) = \bar{c}\{\{\wedge\{K \in L^X : H \leq K$ and $c(K) = K\}\} \leq \wedge\{\bar{c}(K) : H \leq K, c(K) = K\} = \wedge\{\wedge\{G : K \leq G, c(G) = G\} : H \leq K, c(K) = K\} = \wedge\{G : H \leq G, c(G) = G\} = \bar{c}(H)$.

Thus $\bar{c}$ induces an L-fuzzy topology, say, $\tau_{\bar{c}}$ and is given by $\tau_{\bar{c}} = \{H \in L^X : \bar{c}(\rightarrow H) = \rightarrow H\}$.

Proposition 11. *Let (X, c) be an L-fuzzy closure space. Then $\forall H \in L^X$,*

(i) $c(\bar{c}(H)) = \bar{c}(H)$, *i.e.,* $\bar{c}(H)$ *is L-fuzzy closed.*
(ii) $c(H) \leq \bar{c}(H)$,
(iii) $c(H) = H$ *iff* $\bar{c}(H) = H$.

Proof. (i) Let $H \in L^X$. Then from Proposition 9, $c(\bar{c}(H)) = c(\wedge\{K : H \leq K$ and $c(K) = K\}) \leq \wedge\{c(K) : H \leq K$ and $c(K) = K\} = \wedge\{K : H \leq K$ and $c(K) = K\} = \bar{c}(H)$.
(ii) $H \leq \bar{c}(H) \Rightarrow c(H) \leq c(\bar{c}(H)) = \bar{c}(H)$.
(iii) Let $c(H) = H, \forall H \in L^X$. Then H is L-fuzzy closed. Therefore $\bar{c}(H) \leq H$ (cf., Proposition 10). This together with $H \leq \bar{c}(H)$ shows that $\bar{c}(H) = H$. Conversely, let $\bar{c}(H) = H$. Then from (ii), $H \leq c(H) \leq \bar{c}(H) = H$. Thus $\bar{c}(H) = H$, whereby $c(H) = H$.

Proposition 12. *Let (X, c) be an L-fuzzy closure space. Then $\forall H \in L^X$, $c(H) = \bar{c}(H)$ iff $c(c(H)) = c(H)$.*

Proof. Let $c(H) = \bar{c}(H)$, $H \in L^X$. Then $c(c(H)) = c(\bar{c}(H)) = \bar{c}(H) = c(H)$. Conversely, let $c(c(H)) = c(H)$. Then $c(H)$ is L-fuzzy closed. Hence from Proposition 11 (iii), $c(H) = \bar{c}(H)$.

Proposition 13. *Let (X, c) be a quasi-discrete L-fuzzy closure space. Then the L-fuzzy topology $\tau_{\bar{c}}$ on X is a saturated L-fuzzy topology.*

Proof. Follows from Definition 10 and Propositions 10 and 12.

3 L-fuzzy Closure Spaces and L-fuzzy Approximation Spaces

The existence of a bijective correspondence between the set of all L-fuzzy reflexive approximation spaces and the set of all quasi-discrete L-fuzzy closure spaces under a certain extra condition is established here. The similar relationship between the set of all L-fuzzy tolerance approximation spaces and the set of all symmetric quasi-discrete L-fuzzy closure spaces satisfying a certain extra condition is also demonstrated.
We begin with the following.

Proposition 14. *Consider an $(X, R) \in \Omega$, where R is L-fuzzy reflexive relation then $(X, \overline{apr}_R)$ is a quasi-discrete L-fuzzy closure space such that $\overline{apr}_R(A * \bar{\alpha}) = \overline{apr}_R(A) * \bar{\alpha}$, $\forall A \in L^X$ and $\forall \alpha \in L$.*

Proof. Follows from Propositions 3 and 4.

Definition 11. *For $y \in X$ and $\alpha \in L$, the L-fuzzy subset $1_y * \bar{\alpha}$ of X is called an L-**fuzzy point** in X, and is denoted as y_α.*

Proposition 15. *Let (X, c) be a quasi-discrete L-fuzzy closure space such that $c(A * \bar{\alpha}) = c(A) * \bar{\alpha}$, $\forall A \in L^X$ and $\forall \alpha \in L$. Then $\exists$ a L-fuzzy reflexive relation R_c over X which is unique and satisfy $\overline{apr}_{R_c}(A) = c(A)$, $\forall A \in L^X$.*

Proof. Let (X, c) be a quasi-discrete L-fuzzy closure space such that $c(A * \bar{\alpha}) = c(A) * \bar{\alpha}$, $\forall A \in L^X$ and $\forall \alpha \in L$. Also, let $R_c(x, t) = c(1_t)(x)$, $\forall x, t \in X$. Then R_c is an L-fuzzy relation on X such that $1 = 1_x(x) \leq c(1_x)(x)$. Thus $c(1_x)(x) = 1$, whereby R_c is an L-fuzzy reflexive relation over X. Now, let $A \in L^X$, $\alpha \in L$ and $x \in X$. Then

$$
\begin{aligned}
\overline{apr}_{R_c}(A)(x) &= \overline{apr}_{R_c}(\vee\{t_\alpha : t \in X\})(x), \text{ where } \alpha = A(t) \\
&= \vee\{\vee\{R_c(x, r) * t_\alpha(r) : r \in X\} : t \in X\} \\
&= \vee\{\vee\{R_c(x, r) * t_\alpha(r) : r \in X, r \neq t\}, \\
&\qquad \vee\{R_c(x, t) * t_\alpha(r) : r \in X, \, r = t\} : t \in X\} \\
&= \vee\{0 \vee (R_c(x, t) * \alpha) : t \in X\} \\
&= \vee\{R_c(x, t) * \alpha : t \in X\} \\
&= \vee\{c(1_t)(x) * \alpha : t \in X\} \\
&= \vee\{c\{(1_t) * \bar{\alpha}\}(x) : t \in X\} \\
&= c\{\vee\{1_t * \bar{\alpha} : t \in X\}(x)\} \\
&= c(A).
\end{aligned}
$$

Hence $\overline{apr}_{R_c}(A) = c(A)$. To show the uniqueness of L-fuzzy relation R_c, let R' be another L-fuzzy reflexive relation on X such that $\overline{apr}_{R'}(A) = c(A)$, $\forall A \in L^X$. Then $R_c(x, t) = c(1_t)(x) = \overline{apr}_{R'}(1_t)(x) = \vee\{R'(x, r) * 1_t(r) : r \in X\} = R'(x, t)$. Thus $R_c = R'$. Hence the L-fuzzy relation R_c on X is unique.

Now, Propositions 14 and 15 lead us to the following.

Proposition 16. *Let $\mathcal{F}$ be the set of all L-fuzzy reflexive approximation spaces and $\mathcal{T}$ be the set of all quasi-discrete L-fuzzy closure spaces satisfying $c(A * \bar{\alpha}) = c(A) * \bar{\alpha}$, $\forall A \in L^X$ and $\forall \alpha \in L$. Then there exists a bijective correspondence between $\mathcal{F}$ and $\mathcal{T}$.*

Remark 1. In [6], it has been pointed out that for $A \in L^X$, $\underline{apr}_R(A)$ and $\overline{apr}_R(A)$ are not dual to each other. Therefore $\tau_{R_c} \neq \tau_{\bar{c}}$. The next proposition says that the equality holds if L is regular.

Proposition 17. *Let L be regular and (X, c) be a quasi-discrete satisfying $c(A * \bar{\alpha}) = c(A) * \bar{\alpha}, \forall A \in L^X, \forall \alpha \in L$. Then $\tau_{R_c} = \tau_{\bar{c}}$, where R_c is an L-fuzzy reflexive relation on X induced by c.*

Proof. Let $A \in \tau_{\bar{c}}$. Then $\bar{c}(\rightarrow A) =\rightarrow A$. As from Proposition 11, $c(A) \leq \bar{c}(A)$, $\forall A \in L^X$, $c(\rightarrow A) \leq \bar{c}(\rightarrow A)$, or that $A \leq \rightarrow c(\rightarrow A)$.

$$
\begin{aligned}
\text{Now,} \quad \rightarrow c(\rightarrow A) &= \rightarrow \overline{apr}_{R_c}(\rightarrow A) \\
&= \rightarrow \{\vee\{R_c(w,t) * (\rightarrow A(t))\} : t \in X\} \\
&= \rightarrow \{\vee\{\rightarrow \{R_c(w,t) \rightarrow (\rightarrow (\rightarrow A(t)))\}\} : t \in X\} \\
&= \rightarrow \{\vee\{\rightarrow \{R_c(w,t) \rightarrow A(t)\}\} : t \in X\} \\
&= \wedge\{\rightarrow\rightarrow \{R_c(w,t) \rightarrow A(t)\} : t \in X\} \\
&= \wedge\{R_c(w,t) \rightarrow A(t)\} \\
&= apr_{R_c}(A).
\end{aligned}
$$

Thus $A \leq apr_{R_c}(A)$. Also, $apr_{R_c} \leq A$, whereby $apr_{R_c} = A$. Hence $\tau_{\bar{c}} \leq \tau_{R_c}$.

Conversely, let $A \in \tau_{R_c}$. Then $apr_{R_c}(A) = A$, or that $\wedge\{R_c(w,t) \rightarrow A(t) : t \in X\} = A$, i.e., $\wedge\{\rightarrow \{R_c(w,t) * (\rightarrow A(t)) : t \in X\}\} = A$, or that $\rightarrow \{\vee\{R_c(w,t) * (\rightarrow A(t)) : t \in X\}\} = A$, i.e., $\vee\{R_c(w,t) * (\rightarrow A(t)) : t \in X\} =\rightarrow A$, or that $\overline{apr}_{R_c}(\rightarrow A) =\rightarrow A$, whereby $c(\rightarrow A) =\rightarrow A$. Thus from Proposition 11, $\bar{c}(\rightarrow A) =\rightarrow A$, whereby $A \in \tau_{\bar{c}}$, or that $\tau_{R_c} \leq \tau_{\bar{c}}$. Hence $\tau_{R_c} = \tau_{\bar{c}}$.

For a given quasi-discrete L-fuzzy closure space (X, c) satisfying $c(A * \bar{\alpha}) = c(A) * \bar{\alpha}, \forall A \in L^X, \forall \alpha \in L$ and its associated Kuratowski L-fuzzy closure operator $\bar{c}$, $(X, \bar{c})$ is obviously a quasi-discrete L-fuzzy closure space such that $\bar{c}(A * \bar{\alpha}) = \bar{c}(A) * \bar{\alpha}, \forall A \in L^X, \forall \alpha \in L$. Hence from Proposition 15, there exists an L-fuzzy reflexive relation, say, $S_{\bar{c}}$ on X, given by $S_{\bar{c}}(w,t) = \bar{c}(1_t)(w), \forall w, t \in X$.

Before stating next, we introduce the following.

Definition 12. *Let R and T be two L-fuzzy relations on X. Then T is called L-**fuzzy transitive closure** of R if T is the smallest L-fuzzy transitive relation containing R.*

Now, we have the following.

Proposition 18. *Let (X, c) be a quasi-discrete L-fuzzy closure space such that $c(H * \bar{\alpha}) = c(H) * \bar{\alpha}, \forall H \in L^X, \forall \alpha \in L$ and $\bar{c}$ be the associated Kuratowski L-fuzzy closure operator. Then the L-fuzzy relation $S_{\bar{c}}$ is L-fuzzy transitive closure of L-fuzzy relation R_c.*

Proof. Let $S_{\bar{c}} = \bar{c}(1_y)(x), \forall x, y \in X$. Transitivity of $S_{\bar{c}}$ follows from Propositions 5 and 15. Also, $R_c \leq S_{\bar{c}}$ follows from Proposition 11. To show the relation $S_{\bar{c}}$ is an L-fuzzy transitive closure of L-fuzzy relation R_c, it only remains to show that $S_{\bar{c}}$ is the smallest L-fuzzy reflexive and transitive relation containing R_c. For this, let T be another L-fuzzy reflexive and transitive relation on X such that $R_c \leq T$. Then from the reflexivity of T, $(X, \overline{apr}_T)$ is quasi-discrete L-fuzzy closure space. Now, from transitivity of T and Proposition 12 followed by Proposition 10, we have $\overline{apr}_T(H) = \wedge\{K \in L^X : H \leq K, \overline{apr}_T(K) = K\}, \forall H \in L^X$. Also, $S_{\bar{c}}$ being L-fuzzy reflexive and L-fuzzy transitive relation associated with Kuratowski L-fuzzy closure operator $\bar{c}$, from Proposition 8 $\overline{apr}_{S_{\bar{c}}}(H) = \bar{c}(H), \forall H \in L^X$ and $\bar{c}$ being Kuratowski L-fuzzy closure operator associated with quasi-discrete L-fuzzy closure space $(X, c), \forall H \in L^X$, it follows from Proposition 15 that $\bar{c}(H) = \wedge\{K \in L^X : H \leq K, c(K) = K\} = \wedge\{K \in L^X : H \leq K, \overline{apr}_{R_c}(K) = K\}$. Thus from Proposition 6, $\overline{apr}_{S_{\bar{c}}}(H) = \wedge\{K \in L^X : H \leq K, \overline{apr}_{R_c}(K) = K\} \leq \wedge\{K \in L^X : H \leq K, \overline{apr}_T(K) = K\} = \overline{apr}_T(H)$, whereby $\overline{apr}_{S_{\bar{c}}}(H) \leq \overline{apr}_T(H)$, showing that $S_{\bar{c}} \leq T$.

Now, we show that there is a bijective correspondence between the set of all L-fuzzy tolerance approximation spaces and the set of all symmetric quasi-discrete L-fuzzy closure spaces satisfying an extra condition.

Proposition 19. *Let (X, R) be an L-fuzzy tolerance approximation space. Then $(X, \overline{apr}_R)$ is a symmetric quasi-discrete L-fuzzy closure space such that $\overline{apr}_R(H * \bar{\alpha}) = \overline{apr}_R(H) * \bar{\alpha}, \forall H \in L^X$ and $\forall \alpha \in L$.*

Proof. From Propositions 3 and 4 it follows that $(X, \overline{apr}_R)$ is an L-fuzzy closure space and quasi-discrete. Now, $\forall x, y \in X, \overline{apr}_R(1_y)(x) = \vee\{R(x, t) * 1_y(t) : t \in X\} = \vee\{R(y, t) * 1_x(t) : t \in X\} = \overline{apr}_R(1_x)(y)$, showing that $(X, \overline{apr}_R)$ is symmetric. Also, for all $H \in L^X$ and $\alpha \in L, \overline{apr}_R(H * \bar{\alpha}) = \overline{apr}_R(H) * \bar{\alpha}$ follows from Proposition 3.

Proposition 20. *Let (X, c) be a symmetric quasi-discrete L-fuzzy closure space such that $c(H * \bar{\alpha}) = c(H) * \bar{\alpha}, \forall H \in L^X$ and $\forall \alpha \in L$. Then $\exists$ a L-fuzzy tolerance relation R_c over X which is unique and satisfy $\overline{apr}_{R_c}(H) = c(H), \forall H \in L^X$.*

Proof. Let (X, c) be a quasi-discrete and such that $c(H * \bar{\alpha}) = c(H) * \bar{\alpha}, \forall H \in L^X, \forall \alpha \in L$. Let L-fuzzy relation R_c on X be such that $R_c(x, y) = c(1_y)(x), \forall x, y \in X$. Then $1 = 1_x(x) \leq c(1_x)(x)$. Thus $c(1_x)(x) = 1$. Hence R_c is an L-fuzzy reflexive relation on X. Also, (X, c) being an L-fuzzy symmetric closure space, the L-fuzzy relation R_c is symmetric and $\overline{apr}_{R_c}(H) = c(H)$ (cf., Proposition 15). To show the uniqueness of L-fuzzy relation R_c, let R' be another L-fuzzy tolerance relation on X such that $\overline{apr}_{R'}(H) = c(H), \forall H \in L^X$. Then $R_c(x, y) = c(1_y)(x) = \overline{apr}_{R'}(1_y)(x) = \vee\{R'(x, t) * 1_y(t) : t \in X\} = R'(x, y)$. Thus $R_c = R'$. Hence the L-fuzzy relation R_c on X is unique.

Proposition 21. *Let $\mathcal{F}$ be the set of all L-fuzzy tolerance approximation spaces and $\mathcal{T}$ be the set of all symmetric quasi-discrete L-fuzzy closure spaces satisfying $c(H * \bar{\alpha}) = c(H) * \bar{\alpha}, \forall H \in L^X$ and $\forall \alpha \in L$, then $\exists$ a bijective correspondence between $\mathcal{F}$ and $\mathcal{T}$.*

Proof. Follows from Propositions 19 and 20.

Proposition 22. *Let (X, c) be a symmetric quasi-discrete L-fuzzy closure space such that $c(H * \bar{\alpha}) = c(H) * \bar{\alpha}$, $\forall H \in L^X$, $\forall \alpha \in L$ and $\bar{c}$ be the associated Kuratowski L-fuzzy closure operator. Then the L-fuzzy relation $S_{\bar{c}}$ is an L-fuzzy transitive closure of L-fuzzy relation R_c.*

Proof. Similar to that of Proposition 18.

Proposition 23. *Let (X, R) be an L-fuzzy preorder approximation space. Then $(X, \overline{apr}_R)$ is a quasi-discrete L-fuzzy closure space such that (i) $\overline{apr}_R(\overline{apr}_R(H)) = H$ and (ii) $\overline{apr}_R(H * \bar{\alpha}) = \overline{apr}_R(H) * \bar{\alpha}$, $\forall H \in L^X$, $\forall \alpha \in L$.*

Proof. Follows from Propositions 5 and 14.

Proposition 24. *Let (X, c) be a quasi-discrete L-fuzzy closure space such that (i) $c(c(H)) = c(H)$ and (ii) $c(H * \bar{\alpha}) = c(H) * \bar{\alpha}$, $\forall H \in L^X$, $\forall \alpha \in L$. Then there exists an unique L-fuzzy preorder R_c on X such that $\overline{apr}_{R_c}(H) = c(H)$, $\forall H \in L^X$.*

Proof. Follows from Propositions 8, 12 and 15.

Finally, the following is an equivalent characterization of the result regarding the bijective correspondence between the set of all L-fuzzy preorder approximation spaces and the set of all saturated L-fuzzy topological spaces observed in [6,10,17].

Proposition 25. *Let $\mathcal{F}$ be the set of all L-fuzzy preorder approximation spaces and $\mathcal{T}$ be the set of all quasi-discrete L-fuzzy closure spaces satisfying (i) $c(c(H)) = c(H)$ and (ii) $c(H * \bar{\alpha}) = c(H) * \bar{\alpha}$, $\forall H \in L^X$, $\forall \alpha \in L$. Then there exists a bijective correspondence between $\mathcal{F}$ and $\mathcal{T}$.*

Proof. Follows from Propositions 23 and 24.

4   Conclusion

The present paper established an association between L-fuzzy rough sets and L-fuzzy closure spaces. In literature, the bijective correspondence between the set of all L-fuzzy preorder approximation spaces and the set of all L-fuzzy topological spaces of certain type is well known (cf., [6,10]). But the work done in this paper shows that actual theory for such bijective correspondence begins from the notion of L-fuzzy closure spaces. In future we will try to associate L-fuzzy approximation spaces and L-fuzzy topological spaces in categorical point of view.

References

1. Blount, K., Tsinakis, C.: The structure of residuated lattices. Int. J. Algebra Comput. **13**, 437–461 (2003)
2. Boixader, D., Jacas, J., Recasens, J.: Upper and lower approximations of fuzzy sets. Int. J. Gen. Syst. **29**, 555–568 (2000)
3. Dubois, D., Prade, H.: Rough fuzzy set and fuzzy rough set. Int. J. Gen. Syst. **17**, 191–209 (1990)
4. Gautam, V., Yadav, V.K., Singh, A.K., Tiwari, S.P.: On the topological structure of rough soft sets. In: RSKT 2014, LNAI, vol. 8818, pp. 39-48 (2014)
5. Goguen, J.A.: L-fuzzy sets. J. Math. Anal. Appl. **18**, 145–174 (1967)
6. Hao, J., Li, Q.: The relationship between L-fuzzy rough set and L-topology. Fuzzy Sets Syst. **178**, 74–83 (2011)
7. Kondo, M.: On the structure of generalized rough sets. Inf. Sci. **176**, 586–600 (2006)
8. Lowen, R.: Fuzzy topological space and fuzzy compactness. J. Math. Anal. Appl. **56**, 621–633 (1976)
9. Mashhour, A.S., Ghanim, M.H.: Fuzzy closure spaces. J. Math. Anal. Appl. **106**, 154–170 (1985)
10. Ma, Z.M., Hu, B.Q.: Topological and lattice structures of L-fuzzy rough sets determined by lower and upper sets. Inf. Sci. **218**, 194–204 (2013)
11. Pawlak, Z.: Rough sets. Int. J. Comput. Inf. Sci. **11**, 341–356 (1982)
12. Qin, K., Pei, Z.: On the topological properties of fuzzy rough sets. Fuzzy Sets Syst. **151**, 601–613 (2005)
13. Qin, K., Yang, J., Pei, Z.: Generalized rough sets based on reflexive and transitive relations. Inf. Sci. **178**, 4138–4141 (2008)
14. Ramadan, A.A., Elkordy, E.H., El-Dardery, M.: L-fuzzy approximation space and L-fuzzy topological spaces. Iran. J. Fuzzy Syst. **13**(1), 115–129 (2016)
15. Radzikowska, A.M., Kerre, E.E.: Fuzzy rough sets based on residuated lattices. In: Peters, J.F., Skowron, A., Dubois, D., Grzymała-Busse, J.W., Inuiguchi, M., Polkowski, L. (eds.) Transactions on Rough Sets II. LNCS, vol. 3135, pp. 278–296. Springer, Heidelberg (2004). https://doi.org/10.1007/978-3-540-27778-1_14
16. Sharan, S., Tiwari, S.P., Yadav, V.K.: Interval type-2 fuzzy rough sets and interval type-2 fuzzy closure spaces. Iran. J. Fuzzy Syst. **12**, 127–135 (2015)
17. She, Y.H., Wang, G.J.: An axiomatic approach of fuzzy rough sets based on residuated lattices. Comput. Math. Appl. **58**, 189–201 (2009)
18. Srivastava, R., Srivastava, M.: On T_0- and T_1-fuzzy closure spaces. Fuzzy sets Syst. **109**, 263–269 (2000)
19. Tiwari, S.P., Sharan, S., Yadav, V.K.: Fuzzy closure spaces vs. fuzzy rough sets. Fuzzy Inf. Eng. **6**, 93–100 (2014)
20. Wu, W.-Z.: A study on relationship between fuzzy rough approximation operators and fuzzy topological spaces. In: Wang, L., Jin, Y. (eds.) FSKD 2005. LNCS (LNAI), vol. 3613, pp. 167–174. Springer, Heidelberg (2005). https://doi.org/10.1007/11539506_21
21. Yao, Y.Y.: Constructive and algebraic methods of the theory of rough sets. Inf. Sci. **109**, 21–47 (1998)
22. Zhu, W.: Generalized rough sets based on relations. Inf. Sci. **177**(22), 4997–5011 (2007)

Fixed Point Results for (ϕ, ψ)-Weak Contraction in Fuzzy Metric Spaces

Vandana Tiwari$^{(\boxtimes)}$ and Tanmoy Som

Department of Mathematical Sciences, Indian Institute of Technology (BHU),
Varanasi 221005, India
{vandanatiwari.rs.apm12,tsom.apm}@itbhu.ac.in

Abstract. In the present work, a fixed point result for generalized weakly contractive mapping in fuzzy metric space has been established. An example is cited to illustrate the obtained result.

Keywords: Weak contraction · Fuzzy metric · Fixed points

1 Introduction and Preliminaries

The concept of fuzzy metric spaces have been introduced in different ways by many authors. Among which, KM-fuzzy metric space, introduced by Kramosil and Michalek [2] and GV-fuzzy metric space, introduced by George and Veeramani [3], are two most widely used fuzzy metric spaces. KM-fuzzy metric space is similar to generalized Menger space [4]. George and Veeramani imposed a strong condition on the definition of Kramosil and Michalek for topological reasons. Several fixed point results in these fuzzy metric spaces can be found in [5,7,8,10,11].

Alber et al. extended the concept of Banach contraction to the weak contraction and established a fixed point result in Hilbert space [1]. There after B.E. Rhoades investigated this result in metric space [6]. Fixed point problem for weak contraction mapping have been investigated by many authors [12–15,17–24]. In [9] Dutta et al. extended the results of Rhoades. Motivated by the works of [9,16,25], in the present work, a fixed point result in fuzzy metric space, introduced by George and Veeramani, is obtained and an example is added in the support of main result.

Definition 1.1 [4]. *A continuous t-norm $*$ is a binary operation on $[0,1]$, which satisfies the following conditions:*

(i) $*$ is associative and commutative,
(ii) $x * 1 = x$, for all $x \in [0,1]$,
(iii) $x * y \le u * v$, whenever $x \le u$ and $y \le v$, for all $x, y, u, v \in [0,1]$,
(iv) $*$ is continuous.

For example: (a) The minimum t-norm, $*_M$, defined by $x *_M y = \min\{x, y\}$; (b) The product t-norm, $*_P$, defined by $x *_P y = x.y$, are two basic t-norms.

© Springer Nature Singapore Pte Ltd. 2018
D. Ghosh et al. (Eds.): ICMC 2018, CCIS 834, pp. 278–285, 2018.
https://doi.org/10.1007/978-981-13-0023-3_26

Definition 1.2 [3]. *The triplet $(X, M, *)$ is called fuzzy metric space if X is a non-empty set, $*$ is continuous t-norm and M is a fuzzy set on $X^2 \times (0, \infty)$ satisfying the following conditions:*

(i) $M(x, y, t) > 0$,
(ii) $M(x, y, t) = 1$ *if and only if* $x = y$,
(iii) $M(x, y, t) = M(y, x, t)$,
(iv) $M(x, z, t + s) \geq M(x, y, t) * M(y, z, s)$,
(v) $M(x, y, .) : (0, \infty) \to [0, 1]$ *is continuous,*

for all $t, s \in (0, \infty)$ and $x, y, z \in X$.

In this paper, we use the notion of fuzzy metric space introduced by George and Veeramani.

Definition 1.3 [3]. *Let $(X, M, *)$ be a fuzzy metric space. Then*

(i) *A sequence $\{x_n\} \subseteq X$ is said to converge to a point $x \in X$ if $\lim\limits_{n \to \infty} M(x_n, x, t) = 1$, for all $t > 0$.*
(ii) *A sequence $\{x_n\} \subseteq X$ is called a Cauchy sequence if for each $0 < \varepsilon < 1$ and $t > 0$, there exists an $N \in \mathbb{N}$ such that $M(x_n, x_m, t) > 1 - \varepsilon$, for each $m, n \geq N$.*
(iii) *A fuzzy metric space is called complete if every Cauchy sequence in this space is convergent.*

Lemma 1.1 [5]. *Let $(X, M, *)$ be a fuzzy metric space. Then $(X, M, .)$ is non-decreasing for all $x, y \in X$.*

Lemma 1.2 [25]. *If $*$ is a continuous t-norm, and $\{\alpha_n\}$, $\{\beta_n\}$ and $\{\gamma_n\}$ are sequences such that $\alpha_n \to \alpha, \gamma_n \to \gamma$ as $n \to \infty$, then $\overline{\lim}\limits_{k \to \infty} (\alpha_k * \beta_k * \gamma_k) = \alpha * \overline{\lim}\limits_{k \to \infty} \beta_k * \gamma$ and $\underline{\lim}\limits_{k \to \infty} (\alpha_k * \beta_k * \gamma_k) = \alpha * \underline{\lim}\limits_{k \to \infty} \beta_k * \gamma$.*

Lemma 1.3 [25]. *Let $\{f(k, .) : (0, \infty) \to (0, 1], k = 0, 1, 2, ...\}$ be a sequence of functions such that $f(k, .)$ is continuous and monotone increasing for each $k \geq 0$. Then $\overline{\lim}\limits_{k \to \infty} f(k, t)$ is a left continuous function in t and $\underline{\lim}\limits_{k \to \infty} f(k, t)$ is a right continuous function in t.*

2   Main Results

Theorem 2.1. *Let $(X, M, *)$ be a complete fuzzy metric space with an arbitrary continuous t-norm $'*'$ and let $T : X \to X$ be a self mapping satisfying the following condition:*

$$\psi(M(Tx, Ty, t)) \leq \psi(\min(M(x, y, t), M(x, Tx, t), M(y, Ty, t))) \qquad (2.1)$$
$$-\phi(\min(M(x, y, t), M(y, Ty, t))),$$

where $\psi, \phi : (0, 1] \to [0, \infty)$ are two functions such that:

(i) ψ *is continuous and monotone decreasing function with* $\psi(t) = 0$ *if and only if* $t = 1$,

(ii) ϕ *is lower semi continuous function with* $\phi(t) = 0$ *if and only if* $t = 1$.

Then T has a unique fixed point.

Proof: Let $x_0 \in X$. We define the sequence $\{x_n\}$ in X such that $x_{n+1} = Tx_n$, for each $n \geq 0$. If there exists a positive integer k such that $x_k = x_{k+1}$, then x_k is a fixed point of T. Hence, we shall assume that $x_n \neq x_{n+1}$, for all $n \geq 0$. Now, from (2.1)

$$\psi(M(x_{n+1}, x_{n+2}, t)) = \psi(M(Tx_n, Tx_{n+1}, t))$$
$$\leq \psi(\min\{M(x_n, x_{n+1}, t), M(x_n, x_{n+1}, t), M(x_{n+1}, x_{n+2}, t)\})$$
$$-\phi(\min\{M(x_n, x_{n+1}, t), M(x_{n+1}, x_{n+2}, t)\}). \qquad (2.2)$$

Suppose that $M(x_n, x_{n+1}, t) > M(x_{n+1}, x_{n+2}, t)$, for some positive integer n. Then from (2.2), we have
$\psi(M(x_{n+1}, x_{n+2}, t)) \leq \psi(M(x_{n+1}, x_{n+2}, t)) - \phi(M(x_{n+1}, x_{n+2}, t))$, that is, $\phi(M(x_{n+1}, x_{n+2}, t)) \leq 0$, which implies that $M(x_{n+1}, x_{n+2}, t) = 1$. This gives that $x_{n+1} = x_{n+2}$, which is a contradiction.

Therefore, $M(x_{n+1}, x_{n+2}, t) \leq M(x_n, x_{n+1}, t)$ for all $n \geq 0$, and $\{M(x_n, x_{n+1}, t)\}$ is a monotone increasing sequence of non-negative real numbers. Hence, there exists an $r > 0$ such that $\lim\limits_{n \to \infty} M(x_n, x_{n+1}, t) = r$.

In view of the above facts, from (2.2), we have

$$\psi(M(x_{n+1}, x_{n+2}, t)) \leq \psi(M(x_n, x_{n+1}, t)) - \phi(M(x_n, x_{n+1}, t)), \text{ for all } n \geq 0,$$

Taking the limit as $n \to \infty$ in the above inequality and using the continuities of ϕ and ψ we have $\psi(r) \leq \psi(r) - \phi(r)$, which is a contradiction unless $r = 1$. Hence

$$M(x_n, x_{n+1}, t) \to 1 \quad \text{as } n \to \infty. \qquad (2.3)$$

Next, we show that $\{x_n\}$ is Cauchy sequence. If otherwise, there exist λ, $\epsilon > 0$ with $\lambda \in (0, 1)$ such that for each integer k, there are two integers $l(k)$ and $m(k)$ such that $m(k) > l(k) \geq k$ and

$$M(x_{l(k)}, x_{m(k)}, \epsilon) \leq 1 - \lambda, \quad \text{for all } k > 0. \qquad (2.4)$$

By choosing $m(k)$ to be the smallest integer exceeding $l(k)$ for which (2.4) holds, then for all $k > 0$, we have

$$M(x_{l(k)}, x_{m(k)-1}, \epsilon) > 1 - \lambda.$$

Now, by triangle inequality, for any s with $0 < s < \frac{\epsilon}{2}$, for all $k > 0$, we have

$$1 - \lambda \geq M(x_{l(k)}, x_{m(k)}, \epsilon)$$
$$\geq M(x_{l(k)}, x_{l(k)+1}, s) * M(x_{l(k)+1}, x_{m(k)+1}, \epsilon - 2s) * M(x_{m(k)+1}, x_{m(k)}, s).$$
$$(2.5)$$

For $t > 0$, we define the function $h_1(t) = \varlimsup\limits_{n \to \infty} M(x_{l(k)+1}, x_{m(k)+1}, t)$.

Taking $\lim\sup$ on both the sides of (2.5), using (2.3) and the continuity property of $*$, by Lemma (1.2), we conclude that

$$1 - \lambda \geq 1 * \varlimsup_{k\to\infty} M(x_{l(k)+1}, x_{m(k)+1}, \epsilon - 2s) * 1$$
$$= \varlimsup_{k\to\infty} M(x_{l(k)+1}, x_{m(k)+1}, \epsilon - 2s)$$
$$= h_1(\epsilon - 2s).$$

By an application of Lemma (1.3), h_1 is left continuous.

Letting limit as $s \to 0$ in the above inequality, we obtain

$$h_1(\epsilon) = \varlimsup_{k\to\infty} M(x_{l(k)+1}, x_{m(k)+1}, \epsilon) \leq 1 - \lambda. \qquad (2.6)$$

Next, for all $t > 0$, we define the function
$$h_2(t) = \varliminf_{k\to\infty} M\left(x_{l(k)+1}, x_{m(k)+1}, t\right).$$
In above similar process, we can prove that

$$h_2(\epsilon) = \varliminf_{k\to\infty} M\left(x_{l(k)+1}, x_{m(k)+1}, \epsilon\right) \geq 1 - \lambda. \qquad (2.7)$$

Combining (2.6) and (2.7), we get
$$\varlimsup_{k\to\infty} M(x_{l(k)+1}, x_{m(k)+1}, \epsilon) \leq 1 - \lambda \leq \varliminf_{k\to\infty} M(x_{l(k)+1}, x_{m(k)+1}, \epsilon).$$
This implies that

$$\lim_{n\to\infty} M(x_{l(k)+1}, x_{m(k)+1}, t) = 1 - \lambda. \qquad (2.8)$$

Again by (2.6),
$$\varlimsup_{k\to\infty} M(x_{l(k)}, x_{m(k)}, \epsilon) \leq 1 - \lambda.$$

For $t > 0$, we define the function

$$h_3(t) = \varliminf_{k\to\infty} M(x_{l(k)}, x_{m(k)}, \epsilon). \qquad (2.9)$$

Now for $s > 0$,
$$M(x_{l(k)}, x_{m(k)}, \epsilon + 2s) \geq M(x_{l(k)}, x_{l(k)+1}, s) * M(x_{l(k)+1}, x_{m(k)+1}, \epsilon) * M(x_{m(k)+1}, x_{m(k)}, s).$$
Taking $\lim\inf$ both the sides, we have
$$\varliminf_{k\to\infty} M(x_{l(k)}, x_{m(k)}, \epsilon + 2s) \geq 1 * \varliminf_{k\to\infty} M(x_{l(k)+1}, x_{m(k)+1}, \epsilon) * 1 = 1 - \lambda.$$
Thus,

$$h_3(\epsilon + 2s) \geq 1 - \lambda. \qquad (2.10)$$

Taking limit as $s \to 0$, we get $h_3(\epsilon) \geq 1 - \lambda$. Combining (2.9) and (2.10) we obtain

$$\lim_{n\to\infty} M(x_{l(k)}, x_{m(k)}, \epsilon) = 1.$$

Now,

$$\psi(M(x_{l(k)+1}, x_{m(k)+1}, \epsilon)) \leq \psi(\min(M(x_{l(k)}, x_{m(k)}, \epsilon),$$
$$M(x_{l(k)}, x_{l(k)+1}, \epsilon)), M(x_{m(k)}, x_{m(k)+1}, \epsilon))$$
$$-\phi(\min(M(x_{l(k)}, x_{m(k)}, \epsilon), M(x_{m(k)}, x_{m(k)+1}, \epsilon))).$$

Taking limit as $k \to \infty$, we get

$\psi(1 - \lambda) \leq \psi(1 - \lambda) - \phi(1 - \lambda)$, which is a contradiction.

Thus, $\{x_n\}$ is Cauchy sequence. Since X is complete, there exists $p \in X$ such that $x_n \to p$ as $n \to \infty$. Now,

$$\psi(M(x_{n+1}, Tp, t)) = \psi(M(Tx_n, Tp, t))$$
$$\leq \psi(\min\{M(x_n, p, t), M(x_n, x_{n+1}, t), M(p, Tp, t)\})$$
$$-\phi(\min\{M(x_n, p, t), M(p, Tp, t)\}).$$

Taking limit as $n \to \infty$, we get

$\psi(M(p, Tp, t)) \leq \psi(M(p, Tp, t)) - \phi(M(p, Tp, t))$,

which implies that $\phi(M(p, Tp, t)) = 0$, that is,

$M(p, Tp, t) = 1$ or $p = Tp$.

We next establish that fixed point is unique. Let p and q be two fixed points of T.

Putting $x = p$ and $y = q$ in (2.1),

$\psi(M(Tp, Tq, t)) \leq \psi(\min M(p, q, t), M(p, Tp, t), M(q, Tq, t)) - \phi(\min M(p, q, t), M(q, Tq, t))$ or, $\psi(M(p, q, t)) \leq \psi(\min M(p, q, t), M(p, p, t), M(q, q, t)) - \phi(\min M(p, q, t), M(q, q, t))$ or, $\psi(M(p, q, t)) \leq \psi(M(p, q, t)) - \phi(M(p, q, t))$ or, $\phi(M(p, q, t)) \leq 0$, or, equivalently, $M(p, q, t) = 1$, that is, $p = q$.

The following example is in support of Theorem 2.1.

Example 2.1. *Let $X = [0, 1]$. Let*

$$M(x, y, t) = e^{-\frac{|x-y|}{t}},$$

*for all $x, y \in X$ and $t > 0$, then $(X, M, *)$ is a complete fuzzy metric space, where '$*$' is product t-norm. Let $\psi, \phi : (0, 1] \to [0, \infty)$ be defined by $\psi(s) = \frac{1}{s} - 1$ and $\phi(s) = \frac{1}{s} - \frac{1}{\sqrt{s}}$. Then ψ and ϕ satisfy all the conditions of Theorem (2.1). Let the mapping $T : X \to X$ be defined by $Tx = \frac{x}{2}$, for all $x \in X$.
Now, we will show that*

$$\psi(M(Tx, Ty, t)) \leq \psi(M(x, y)) - \phi(N(x, y)), \tag{2.11}$$

where $M(x, y) = \min\{M(x, y, t), M(x, Tx, t), M(y, Ty, t)\}$ and $N(x, y) = \phi(\min\{M(x, y, t), M(y, Ty, t)\})$. Herein;

$$\max\{|x - y|, \frac{x}{2}, \frac{y}{2}\} = \begin{cases} x - y & 0 \leq y \leq \frac{x}{2} \\ \frac{x}{2} & \frac{x}{2} < y \leq x \\ \frac{y}{2} & x < y \leq 2x \\ y - x & 2x < y \leq 1 \end{cases}$$

and

$$\max\{|x - y|, \frac{y}{2}\} = \begin{cases} x - y & 0 \le y \le \frac{2x}{3} \\ \frac{y}{2} & \frac{2x}{3} < y \le 2x \\ y - x & 2x < y \le 1. \end{cases}$$

Case (1): When $0 \le y \le \frac{x}{2}$ or $2x < y \le 1$, then

$$\psi(M(Tx, Ty, t)) = \psi(e^{-|\frac{x-y}{2t}|}) = e^{|\frac{x-y}{2t}|} - 1$$

and

$$\psi(M(x, y)) - \phi(N(x, y)) = \psi(e^{-\frac{|x-y|}{t}}) - \phi(e^{-\frac{|x-y|}{t}}) = e^{|\frac{x-y}{2t}|} - 1.$$

Obviously, in this case, (2.11) is satisfied.

Case (2): When $\frac{x}{2} < y \le \frac{2x}{3}$, then

$$\psi(M(Tx, Ty, t)) = \psi(e^{-\frac{x-y}{2t}}) = e^{\frac{x-y}{2t}} - 1$$

and

$$\psi(M(x, y)) - \phi(N(x, y)) = \psi(e^{-\frac{x}{2t}}) - \phi(e^{-\frac{x-y}{t}}) = e^{\frac{x}{2t}} - 1 - e^{\frac{x-y}{t}} + e^{\frac{x-y}{2t}}.$$

In this case, $\frac{x}{2} \ge x - y$ and exponential function is an increasing function. Therefore, $e^{\frac{x-y}{2t}} \le e^{\frac{x}{2t}} - e^{\frac{x-y}{t}} + e^{\frac{x-y}{2t}}$ and hence (2.11) is satisfied.

Case (3): When $\frac{2x}{3} < y \le x$, then

$$\psi(M(Tx, Ty, t)) = \psi(e^{-\frac{x-y}{2t}}) = e^{\frac{x-y}{2t}} - 1$$

and

$$\psi(M(x, y)) - \phi(N(x, y)) = \psi(e^{-\frac{x}{2t}}) - \phi(e^{-\frac{y}{2t}}) = e^{\frac{x}{2t}} - 1 - e^{\frac{y}{2t}} + e^{\frac{y}{4t}}.$$

Since, in this case, $\frac{x-y}{2} \le \frac{y}{4}$ and $\frac{x}{2} \ge \frac{y}{2}$, (2.11) is satisfied.

Case (4): $x < y \le 2x$, then

$$\psi(M(Tx, Ty, t)) = \psi(e^{-\frac{y-x}{2t}}) = e^{\frac{y-x}{2t}} - 1$$

and

$$\psi(M(x, y)) - \phi(N(x, y)) = \psi(e^{-\frac{y}{2t}}) - \phi(e^{-\frac{y}{2t}}) = e^{\frac{x}{4t}} - 1.$$

Since, in this case, $\frac{y}{2} \ge y - x$, (2.11) is satisfied. Hence, all the conditions of Theorem (2.1) are satisfied. Thus, 0 is the unique fixed point of T.

References

1. Alber, Y.I., Guerre-Delabriere, S.: Principle of weak contractive mapes in Hilbert space. In: New Results in Operator Theory and its Applications, vol. 98. Birkhuser, Basel, pp. 7–22 (1997)
2. Kramosil, O., Michalek, J.: Fuzzy metric and statistical metric spaces. Kybernetica **11**, 326–334 (1975)
3. George, A., Veeramani, P.: On some results in fuzzy metric space. Fuzzy Sets Syst. **64**, 395–399 (1994)
4. Schweizer, B., Sklar, A.: Statistical metric spaces. Pac. J. Math. **10**, 313–334 (1960)
5. Grabice, M.: Fixed points in fuzzy metric spaces. Fuzzy Sets Syst. **27**, 385–389 (1988)
6. Rhoades, B.E.: Some theorems on weakly contractive maps. Nonlinear Anal. **47**, 2683–2693 (2001)
7. Berinde, V.: Approximating fixed points of weak ϕ-contractions. Fixed Point Theor. **4**, 131–142 (2003)
8. Zhang, Q., Song, Y.: Fixed point theory for generalized ϕ-weak contractions. Appl. Math. Lett. **22**, 75–78 (2009)
9. Dutta, P.N., Choudhury, B.S.: A generalization of contraction principle in metric spaces. Fixed Point Theor. Appl. 1–8 (2008). Article ID 406368
10. Som, T., Choudhury, B.S., Das, K.P.: Two common fixed point results in fuzzy metric spaces. Internat. Rev. Fuzzy. Math. **6**, 21–32 (2011)
11. Doric, D.: Common fixed point for generalized $(\psi\text{-}\phi)$-weak contractions. Appl. Math. Lett. **22**, 1896–1900 (2009)
12. Popescu, O.: Fixed point for $(\psi\text{-}\phi)$-weak contractions. Appl. Math. Lett. **24**, 1–4 (2011)
13. Karapinar, E.: Fixed point theory for cyclic weak ϕ-contraction. Appl. Math. Lett. **24**, 822–825 (2011)
14. Rouhani, B., Moradi, S.: Common fixed point of multivalued generalized ϕ-weak contractive mappings. Fixed Point Theor. Appl. 1–13 (2010). Article ID 708984
15. Samet, B., Vetro, C., Vetro, P.: Fixed point theorems for $(\alpha\text{-}\psi)$-contractive type mappings. Nonlinear Anal. **75**, 2154–2165 (2012)
16. Choudhury, B.S., Konar, P., Rhoades, B.E., Metiya, N.: Fixed point theorems for generalized weakly contractive mappings. Nonlin. Anal. **74**, 2116–2126 (2011)
17. Moradi, S., Farajzadeh, A.: On the fixed point of $(\psi\text{-}\phi)$-weak and generalized $(\psi\text{-}\phi)$-weak contraction mappings. Appl. Math. Lett. **25**, 1257–1262 (2012)
18. Jamala, N., Sarwara, M., Imdad, M.: Fixed point results for generalized $(\psi\text{-}\phi)$-weak contractions with an application to system of non-linear integral equations. Trans. A Razm. Math. Inst. **171**, 182–194 (2017)
19. Aydi, H., Karapnar, E., Shatanawi, W.: Coupled fixed point results for $(\psi\text{-}\phi)$-weakly contractive condition in ordered partial metric spaces. Comput. Math. Appl. **62**, 4449–4460 (2011)
20. Aydi, H., Postolache, M., Shatanawi, W.: Coupled fixed point results for $(\psi\text{-}\phi)$-weakly contractive mappings in ordered G-metric spaces. Comput. Math. Appl. **63**, 298–309 (2012)
21. An, T.V., Chi, K.P., Karapnar, E., Thanh, T.D.: An extension of generalized $(\psi\text{-}\phi)$-weak contractions. Int. J. Math. Math. Sci. 1–11 (2012). Article ID 431872
22. Latif, A., Mongkolkeha, C., Sintunavarat, W.: Fixed point theorems for generalized $\alpha\text{-}\beta$-weakly contraction mappings in metric spaces and applications. Sci. World J. 1–14 (2014). Article ID 784207

23. Jha, K., Abbas, M., Beg, I., Pant, R.P., Imdad, M.: Common fixed point theorem for $(\psi\text{-}\phi)$-weak contraction in suzzy metric space. Bull. Math. Anal. Appl. **3**, 149–158 (2011)
24. Luo, T.: Fuzzy $(\psi\text{-}\phi)$-contractive mapping and fixed point theorem. Appl. Math. Sci. **8**(148), 7375–7381 (2016)
25. Saha, P., Choudhury, B.S., Das, P.: Weak coupled coincidence point results having a partially ordering in fuzzy metric space. Fuzzy. Inf. Eng. **7**, 1–18 (2016)

Identifying Individuals Using Fourier and Discriminant Analysis of Electrocardiogram

Ranjeet Srivastva[1(✉)] and Yogendra Narain Singh[2]

[1] Department of Information Technology, Babu Banarasi Das Northern India
Institute of Technology, Lucknow, U.P., India
ranjeetbbdit@gmail.com
[2] Department of Computer Science and Engineering,
Institute of Engineering and Technology, Lucknow, U.P., India
singhyn@gmail.com

Abstract. From the last one and a half decades, the electrocardiogram
(ECG) has emerged as a new modality for human identification. The
research shows that the people heartbeats recorded using diagnostic
method called ECG exhibit discriminatory features that can distinguish
themselves. The ECG as a biometric inherently provides liveness detection and robustness against falsification. This paper presents a novel
method of ECG analysis for human identification using Fourier and linear discriminant analysis, which does not require detection of fiducial
points of ECG wave. The method utilizes autocorrelation coefficients of
filtered ECG signal, to extract significant features of it. The performance
of the proposed method is evaluated on MIT-BIH arrhythmia and QT
database of physionet. The experimental results show the equal error rate
(EER) of 0.17% and 0.03% on MIT-BIH arrhythmia and QT database,
respectively that outperform the other methods on these databases.

Keywords: Individual identification · Electrocardiogram
Fourier transform · Discriminant analysis

1 Introduction

The emerging technology that recognizes people based on their unique physiological and behavioral characteristics, termed as biometrics. These days, biometric
traits are used in a wide variety of applications such as in access control, financial and business transactions, health care and other applications [1]. Automatic
and accurate identification of an individual is critical along with reducing the
probability of intruders getting access to an authentication system [2]. As the
proliferation of computer and internet, identity theft becomes the major concern
of the modern society [3]. Traditional personal authentication systems based on
passwords, PIN numbers and ID cards are unable to fulfil the requirement of
high security applications and they are more susceptible to identity theft [4].

© Springer Nature Singapore Pte Ltd. 2018
D. Ghosh et al. (Eds.): ICMC 2018, CCIS 834, pp. 286–295, 2018.
https://doi.org/10.1007/978-981-13-0023-3_27

Biometrics has emerged as a potential tool for accurate and efficient authentication of an individual but there are some challenging issues such as confidentiality and vitalityness making the system more prone to spoofing attacks [5].

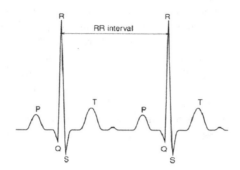

Fig. 1. ECG waveform features.

In order to address these issues one of the state-of-the-art biometrics electrocardiogram (ECG) is a better alternative to the conventional biometrics. ECG is generated from a complex self regulatory system of the heart. It is highly secure, confidential and impossible to mimic. It is universally present in all living individuals thus provides real-time vitality testing [6]. The basic elements of a single heartbeat of ECG consist of P-QRS-T waveforms are shown in Fig. 1.

Beil *et al.* have demonstrated the use of ECG to discriminate 20 subjects using a set of temporal and amplitude features [7]. They have achieved 100% identification rate by multivariate analysis of ECG features. Shen *et al.* have used the appearance and time domain features of the heartbeat and achieved classification accuracy of 95% and 80% for template matching and decision based neural network approaches, respectively [8]. Israel *et al.* have investigated the timing characteristics of ECG signal, from the heartbeat of 29 individuals using linear discriminant analysis (LDA) [9].

Wang *et al.* used analytical feature extraction with discrete cosine transform (DCT) of autocorrelated heartbeat signals [10]. Singh and Gupta have used signal processing methods to delineate ECG wave fiducials from each heartbeat and achieved 98% classification accuracy for 50 subjects [11]. Plataniotis *et al.* have developed an ECG biometric system based on classification of DCT coefficients of the autocorrelated ECG data segment [12]. Agrafioti and Hatzinakos have demonstrated an autocorrelation based feature extraction approach in conjunction with DCT or LDA [13]. In a recent study, Srivastva and Singh have introduced a new method for ECG analysis used in biometric recognition [14,15]. They have reported 97% identification performance using Walsh Hadamard transform and LDA [14]. The authentication performance achieved by DCT and LDA have minimum EER of 0.06% [15].

The major concerns of most of the studies include detection accuracy of fiducial points, selection of features those are insensitive to change in physiology of the heart, variations of heart rate, age and time. The individuality of ECG over a large population is yet to be explored. To address the issues related to ECG biometrics, the paper advocates the use of proposed method. It does not require specific fiducial points of the ECG waveforms and thus not requires pulse synchronization. Therefore, the method is computationally efficient and exhibits better identification performance. The proposed method utilizes the autocorrelation (AC) coefficients, calculated from the filtered ECG signals. The Fourier analysis of autocorrelated ECG segments is performed to form a feature vector. The dimensionality of the feature vector is reduced using LDA before calculating match score for classification. The rest of the paper is outlined as follows: Sect. 2 presents the novel method of ECG waveform analysis and its characterization that is used for the biometric applications. The experimental results are presented in Sect. 3. Finally, the conclusion is noted in Sect. 4.

2   Methodology

Human recognition is essentially a pattern recognition process involves preprocessing, feature extraction, feature normalization, and classification. The proposed biometric system is depicted in Fig. 2. Preprocessing involves noise and artifact removal step. Features are extracted from an ECG data by autocorrelation followed by Fourier transform of ECG window. The LDA is used for

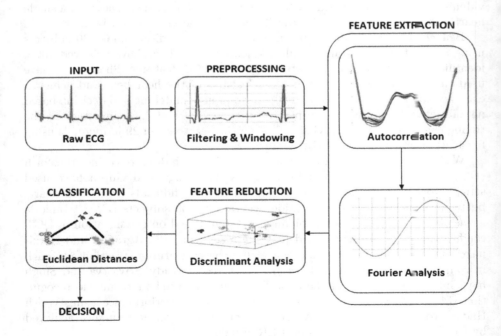

Fig. 2. Proposed ECG biometric system.

dimensionality reduction and the last step of the identification process is classification based on similarity scores of the subjects.

Normally different type of noises contaminate ECG signals. These include low-frequency noise components resulted from baseline oscillations, respiration or body movements and high frequency noise components from power line interferences. The combination of low pass and high pass filters is used to eliminate the effects of noise with the following difference equations, respectively [16].

$$y_n = 2y_{n-1} - y_{n-2} + x_n - 2x_{n-6} + x_{n-12} \tag{1}$$

$$y_n = 32x_{n-16} - (y_{n-1} + x_n - x_{n-32}) \tag{2}$$

The cutoff frequency of low pass filter and high pass filter is about 11 Hz and 5 Hz respectively, which has been chosen considering that the frequency band of normal ECG signal lies within this range.

The filtered ECG signals are segmented into non-overlapping segments. The only restriction regarding the division of ECG data is that the segments have to be longer than the normal cardiac cycle to include at least two or more heartbeats. The length of the window can be chosen heuristically and varies with the sampling frequency of data. For this experiment, all the records are re-sampled at the sampling rate of 200 Hz, and the data window of 50 s and 10 s are selected for MIT-BIH arrhythmia database and QT database, respectively.

ECG is highly repetitive signal that exhibits distinctive characteristics in a population. ECG analysis based on its dominant fiducials require pulse synchronization, and exact localization of wave boundaries. To extract features from ECG data without fiducial detectors, autocorrelation is applied on windowed ECG, that blend samples into a sequence of sums of products. The AC provides an automatic, shift invariant representation of similarity features over multiple cardiac cycles. The normalized AC ($\widehat{R}_{xx}[m]$) of filtered ECG signal, $x[i]$ of length N is computed using the following formula,

$$\widehat{R}_{xx}[m] = \sum_{i=0}^{N-|m|-1} \frac{x[i] * x[i+m]}{\widehat{R}_{xx}[0]} \tag{3}$$

where $x[i + m]$ is the time shifted version of the windowed ECG with a time lag of $m = 0, 1, \ldots.(M - 1); M << N$.

The discrete Fourier transform (DFT) coefficients are calculated from autocorrelated ECG signals. It maximizes the inter-class variability and intra-class similarity. The DFT is frequency domain representation of the original input sequence in the time domain. Let $x_0, x_1 \ldots\ldots\ldots x_{N-1}$ be the sequence of N complex numbers. It can be transformed into an N-periodic sequence of complex numbers by the following formula,

$$X_k = \sum_{n=0}^{N-1} x_n e^{\frac{-2\pi ikn}{N}}, \quad k = 0, 1 \ldots.N - 1 \tag{4}$$

Here each X_k is a complex number, that encodes both amplitude and phase of a complex sinusoidal component $(e^{2\pi ikn/N})$ of function x_n. The sinusoid's frequency is k cycles per N samples.

The LDA is a known method of dimensionality reduction and feature extraction. It preserves the class specific discriminability by linearly transforming the feature characteristics into a low dimension space. More formally, for a given training set $Z = \{Z_i\}_{i=1}^C$ containing the patterns of C classes. Each class $Z_i = \{Z_{ij}\}_{j=1}^{C_i}$ has a number of windows Z_{ij} and a set of K feature basis vectors $\{\psi_m\}_{m=1}^K$ is estimated by maximizing Fisher's ratio. This ratio is defined as the between-class to within class scatter matrix. The maximization is equivalent to the solution of the following eigenvalue problem:

$$\psi = argmax \left(\frac{|\psi^T S_b \psi|}{|\psi^T S_w \psi|} \right) \tag{5}$$

where $\psi = [\psi_1, \ldots, \psi_K]$, and S_b and S_w are the between and within class scatter matrices, respectively defined as,

$$S_b = \frac{1}{N} \sum_{i=1}^C C_i (Z_i - \overline{Z})(Z_i - \overline{Z})^T \tag{6}$$

$$S_w = \frac{1}{N} \sum_{i=1}^C \sum_{j=1}^{C_i} (Z_{ij} - \overline{Z_i})(Z_{ij} - \overline{Z_i})^T \tag{7}$$

where $\overline{Z_i} = \frac{1}{C_i} \sum_{j=1}^{C_i} Z_{ij}$ is the mean of class Z_i and N is the total number of training windows and $N = \sum_{i=1}^C C_i$. The LDA finds ψ as the K most significant eigenvectors of $(S_w)^{-1} S_b$ that correspond to the first K largest eigenvalues. Using these basis vectors, a test input window Z is subjected to the linear projection $y = \psi^T Z$.

3   Experimental Results

The performance of the identification system is analyzed through equal error rate (EER) [17]. The EER is an error rate where the frequency of false acceptance (FAR) and the frequency of false rejection (FRR) assume the same value. In order to confirm the benefit of the combined system the receiver operating characteristics (ROC) curve of the authentication process has also been considered. The ROC curve is a two-dimensional measure of classification performance that plots the likelihood of false acceptance (FAR) against the likelihood of genuine acceptance (GAR) [5]. The accuracy of the identification system can be defined as,

$$Accuracy(\%) = 100 - EER(\%) \tag{8}$$

The performance of the proposed method is tested on MIT-BIH arrhythmia database and QT database of physionet [22]. Both databases include ECG recordings of normal subjects and arrhythmia patients (men and women) of age between 20 and 84 years. Forty-eight ECG recordings of MIT-BIH arrhythmia database and thirty-nine records of QT database are used in this study. The original sampling rate is 360 Hz and 250 Hz for MIT-BIH arrhythmia and QT database, respectively. All these records are re-sampled at 200 Hz for this experiment. After preprocessing, eleven windows of 50 s (10000 samples) and 10 s (2000 samples) in length are chosen from preprocessed ECG signal of MIT-BIH arrhythmia database and QT database, respectively. The windows exclude the 10 s samples from start and end of the recording to avoid sensor and body stabilization effects. To extract features a data set of $528(48 \times 11) \times 10000$ for MIT-BIH arrhythmia database and of $429(39 \times 11) \times 2000$ for QT database are formed.

Autocorrelation is applied to these data set which forms a feature vector of 528×180 and 429×180 for MIT-BIH arrhythmia database and QT database, respectively. The autocorrelation time lag can be set to different settings for maximum correlation between samples. For this experiment, it is set to 180 samples due to the fact that a normal heart rate for adults ranges from 60 to 100 beats a minute. The Fourier analysis of these feature vectors is performed in order to minimize the intrasubject variations and to maximize the intersubject variations. The LDA is used for dimensionality reduction of feature vectors to different dimensions. The intrasubject variability and intersubject similarity on first three dimensions as achieved by LDA for ten subjects from each database is shown in Fig. 3.

The results of EER at different dimensions on different databases are presented in Table 1. On MIT-BIH arrhythmia database the EER value is found to be 10% at dimension 1, and it decreases to 0.17% at dimension 10. The EER is linearly increasing above the dimension 10. On QT database the EER values are found to be 12%, 1.9%, 0.35%, 0.2%, 0.35%, 0.04% 0.04% and 0.03% at dimensions 1, 2, 4, 5, 7, 10, 13 and 15, respectively. The EER value increases above dimension 15. The lowest values of EER are reported to 0.17% and 0.03%, respectively on MIT-BIH arrhythmia database and QT database at dimension 10 and 15, respectively. The ROC curves represent the ratio of GAR and FAR at different dimensions are shown in Fig. 4. The identification results on MIT-BIH arrhythmia database achieve 100% GAR on FAR of 0.75%, 0.22%, 0.71%, 0.35% and 0.35% at the dimensions 5, 10, 15, 20 and 25, respectively that are shown in Fig. 4(a). Similarly, the performance on QT database achieves 100% GAR on FAR of 0.2%, 0.13%, 0.07%, 0.07% and 0.2% at the dimensions 5, 10, 15, 20 and 25, respectively that are shown in Fig. 4(b).

The highest identification accuracy on both databases is found to be about 100% which is better than all known approaches tested on these databases. For example, when we compare the proposed method with fiducial based identification methods, it's performance is better than [18]. Although [7,8] achieve 100% identification accuracy, these methods were tested on only a group of 20 subjects. The result of proposed method can also be compared with non-fiducial based ECG identification methods [10,12,19–21]. Among these, the methods

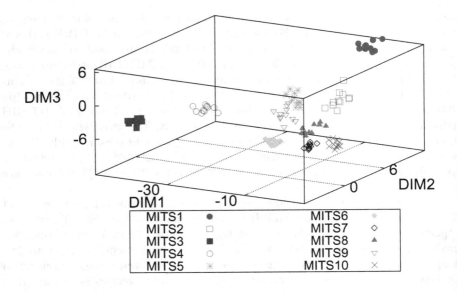

(a)

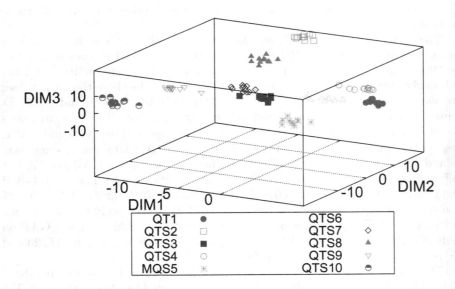

(b)

Fig. 3. Intrasubject similarity and intersubject variability represented by first three dimensions as shown by DIM1, DIM2 and DIM3 for ten different subjects of (a) MIT-BIH Arrhythmia database and (b) QT database.

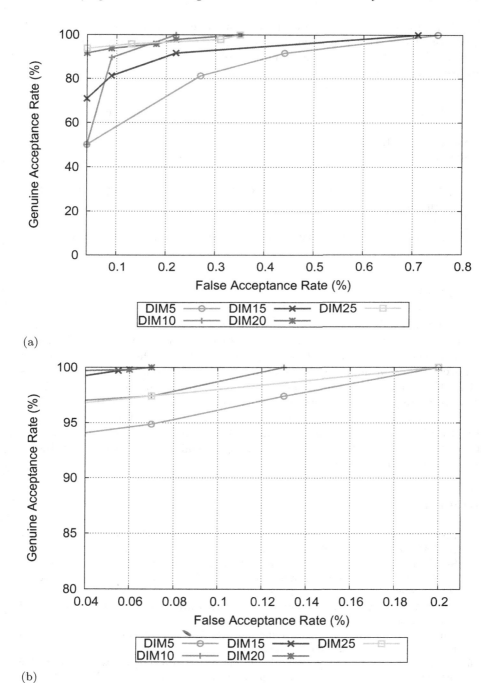

Fig. 4. ROC curve for (a) MIT-BIH Arrhythmia database, and (b) QT database.

[10,12,19,21] reports better performance but these methods were tested only on small set of subjects. The issues like sensitivity to the accurate localization of fiducial points of ECG wave and individuality of ECG over larger population are resolved by applying the proposed method.

Table 1. Equal error rates for different databases at different dimensions

Number of dimensions	Equal error rate (%)											
	1	2	4	5	7	10	13	15	18	20	22	25
MIT-BIH Arrhythmia database	10	5	0.84	0.79	0.37	**0.17**	0.21	0.37	0.52	0.59	0.69	0.73
QT database	12	1.9	0.35	0.2	0.35	0.04	0.04	**0.03**	0.035	0.035	0.04	0.2

4   Conclusion

The conventional biometrics are susceptible to the falsification and spoofing attacks. The ECG has the strong potential to overcome these issues of conventional biometrics. It is proven to be a liveliness indicator. The paper has proposed a novel method of human identification using Fourier and discriminant analysis of the ECG. The method need not to require any fiducial point detection of ECG waveforms rather it has inherently explored the significant points of the ECG signals. Fourier analysis is used to represent the discriminatory features of the ECG while LDA is used to preserve them. The proposed method is proved to be robust as it has reported higher accuracy to normal subjects as well as subjects suffering from severe arrhythmia.

References

1. Pouryayevali, S.: ECG biometrics: new algorithm and multimodal biometric system. Master of Applied Science thesis, University of Toronto (2015)
2. Singh, Y.N., Gupta, P.: ECG to individual identification. In: Proceedings of 2nd IEEE International Conference on Biometrics: Theory, Applications and Systems, BTAS 2008, pp. 1–8, October 2008
3. Singh, Y.N., Gupta, P.: Correlation based classification of heartbeats for individual identification. Soft Comput. **15**(3), 449–460 (2009)
4. Singh, Y.N., Singh, S.K.: Identifying individuals using eigenbeat features of electrocardiogram. J. Eng. **2013**, 1–8 (2013)
5. Singh, Y.N., Singh, S.K., Gupta, P.: Fusion of electrocardiogram with unobtrusive biometrics: an efficient individual authentication system. Pattern Recognit. Lett. **33**(14), 1932–1941 (2012)
6. Singh, Y.N.: Human recognition using Fisher's discriminant analysis of heartbeat interval features and ECG morphology. Neurocomputing **167**, 322–335 (2015)

7. Biel, L., Pettersson, O., Philipson, L., Wide, P.: ECG analysis: a new approach in human identification. IEEE Trans. Instrum. Meas. **50**(3), 808–812 (2001)

8. Shen, T.W., Tompkins, W.J., Hu, Y.H.: One-lead ECG for identity verification. In: 2nd Joint Conference of the IEEE Engineering in Medicine and Biology Society and the Biomedical Engineering Society, Houston, pp. 62–63 (2002)

9. Israel, S.A., Irvine, J.M., Andrew, C., Mark, D.W., Brenda, K.W.: ECG to identify individuals. Pattern Recognit. **38**(1), 133–142 (2005)

10. Wang, Y., Agrafioti, F., Hatzinakos, D., Plataniotis, K.N.: Analysis of human electrocardiogram for biometric recognition. EURASIP J. Adv. Signal Process. **2008**, 1–11 (2008)

11. Singh, Y.N., Gupta, P.: Biometrics method for human identification using electrocardiogram. In: Tistarelli, M., Nixon, M.S. (eds.) ICB 2009. LNCS, vol. 5558, pp. 1270–1279. Springer, Heidelberg (2009). https://doi.org/10.1007/978-3-642-01793-3_128

12. Plataniotis, K., Hatzinakos, D., Lee, J.: ECG biometric recognition without fiducial detection. In: Proceedings of Biometrics Symposiums, BSYM, Baltimore, Maryland, USA (2006)

13. Agrafioti, F., Hatzinakos, D.,: ECG based recognition using second order statistics. In: IEEE Sixth Annual Communication Networks and Services Research Conference, Canada, pp. 82–87 (2008)

14. Srivastva, R., Singh, Y.N.: ECG biometric analysis using Walsh-Hadamard transform. In: Kolhe, M.L., et al. (eds.) Advances in Data and Information Sciences. LNNS, vol. 38. Springer (2017). https://doi.org/10.1007/978-981-10-8360-0_19

15. Srivastva, R., Singh, Y.N.: Human recognition using discrete cosine transform and discriminant analysis of ECG. In: Proceedings of IEEE 2017 Fourth International Conference on Image Information Processing, JUIT, Solan, pp. 368–372 (2017)

16. Pan, J., Tompkins, W.J.: A real-time QRS detection algorithm. IEEE Trans. Biomed. Eng. **32**(3), 230–236 (1985)

17. Duda, R.O., Hart, P.E., Stork, D.G.: Pattern Classification, 2nd edn. Wiley, India (2000)

18. Singh, Y.N.: Individual identification using linear projection of heartbeat features. Appl. Comput. Intell. Soft Comput. **2014**, 1–14 (2014)

19. Wubbeler, G., Stavridis, M., Kreiseler, D., Bousseljot, R.D., Elster, C.: Verification of humans using the electrocardiogram. Pattern Recognit. Lett. **28**, 1172–1175 (2007)

20. Chan, A.D.C., Hamdy, M.M., Badre, A., Badee, V.: Wavelet distance measure for person identification using electrocardiograms. IEEE Trans. Instrum. Meas. **57**(2), 248–253 (2008)

21. Li, M., Narayanan, S.: Robust ECG biometrics by fusing temporal and cepstral information. In: 2010 20th International Conference Pattern Recognition, ICPR, pp. 1326–1329, August 2010

22. PhysioNet: PhysioBank archives. Massachusetts Institute of Technology Cambridge. http://www.physionet.org/physiobank/database/#ecg

Generalized Statistical Convergence for Sequences of Function in Random 2-Normed Spaces

Ekrem Savaş[1] and Mehmet Gürdal[2(✉)] ⓘ

[1] Department of Mathematics, Istanbul Ticaret University, Üsküdar-Istanbul, Turkey
ekremsavas@yahoo.com
[2] Department of Mathematics, Suleyman Demirel University, 32260 Isparta, Turkey
gurdalmehmet@sdu.edu.tr

Abstract. In this paper, we introduce a new type of convergence for a sequence of function, namely, λ-statistically convergent sequences of functions in random 2-normed space, which is a natural generalization of convergence in random 2-normed space. In particular, following the line of recent work of Karakaya et al. [12], we introduce the concepts of uniform λ-statistical convergence and pointwise λ-statistical convergence in the topology induced by random 2-normed spaces. We define the λ-statistical analog of the Cauchy convergence criterion for pointwise and uniform λ-statistical convergence in a random 2-normed space and give some basic properties of these concepts. In addition, the preservation of continuity by pointwise and uniform λ-statistical convergence is proven.

Keywords: λ-statistical convergence · Random 2-normed space
The sequences of functions

1 Introduction and Preliminaries

Our aim is to propose some new variants of statistical convergence (and more general λ-statistical convergence) for sequences of functions in random 2-normed spaces. We put special attention on functions in random 2-normed spaces, in a sense extending original ideas of Balcerzak et al. [3] and Karakaya et al. [12].

The theory of probabilistic normed (PN) spaces is important area of research in functional analysis. Lots of work have been done by this theory and it has many important applications in real life situations. PN spaces are the vector spaces in which the norms of the vectors are uncertain due to randomness. A PN space is a generalization of an ordinary normed linear space. In a PN space, the norms of the vectors are represented by probability distribution functions instead of non-negative real numbers. If x is an element of a PN space, then its norm is denoted by F_x, and the value $F_x(t)$ is interpreted as the probability that the norm of x is smaller than t. The probabilistic metric space was introduced by Menger [13] which is an interesting and an important generalization of the notion

© Springer Nature Singapore Pte Ltd. 2018
D. Ghosh et al. (Eds.): ICMC 2018, CCIS 834, pp. 296–308, 2018.
https://doi.org/10.1007/978-981-13-0023-3_28

of a metric space. The theory of probabilistic normed (or metric) space was initiated and developed in [1, 19–21]; further it was extended to random/probabilistic 2-normed spaces by Golet [9] using the concept of 2-norm which is defined by Gähler (see [7]); and Gürdal and Pehlivan [10] studied statistical convergence in 2-Banach spaces.

In order to extend the notion of convergence of sequences, statistical convergence of sequences was introduced by Fast [5]. A lot of developments have been made in this areas after the work of Fridy [6]. Over the years and under different names, statistical convergence has been discussed in the theory of Fourier analysis, summability theory and number theory. Recently, Mursaleen [14] studied λ-statistical convergence as a generalization of the statistical convergence, and in [15] he considered the concept of statistical convergence of sequences in random 2-normed spaces. Quite recently, Savaş and Mohiuddine [18] defined λ-statistical convergence for double sequences in probabilistic normed spaces, and also Savaş [16] studied generalized statistical convergence in random 2-normed space (also see [17]).

In another direction the idea of statistical convergence of sequences of real functions was studied in [3], and some important results and references on statistical convergence and function sequences can be found in [4, 8]. Recently, in [12], Karakaya et al. studied the statistical convergence of sequences of functions with respect to the intuitionistic fuzzy normed spaces. Also in [11], Karakaya et al. introduced the concept of λ-statistical convergence of sequences of functions in the intuitionistic fuzzy normed spaces.

The notion of λ-statistical convergence of sequences of functions has not been studied previously in the setting of random 2-normed spaces. Motivated by this fact, in this paper, as a variant of statistical convergence, the notion of λ-statistical convergence of sequences of functions is introduced in a random 2-normed space. In Sect. 2, we prove some results concerning to convergence in pointwise λ-statistical convergence and uniform λ-statistical convergence of sequences of functions in a random 2-normed spaces. We demonstrate the λ-statistical analog of the Cauchy convergence criterion for pointwise and uniform λ-statistical convergence in a random 2-normed space and give some basic properties of these concepts. Finally, we prove that pointwise and uniform λ-statistical convergence preserves continuity.

First we recall some of the basic concepts, that will be used in this paper.

The notion of statistical convergence depends on the density of subsets of $\mathbb{N}$, the set of natural numbers. Let K be a subset of $\mathbb{N}$. Then the asymptotic density of the set K denoted by $\delta(K)$ is defined as

$$\delta(K) = \lim_{n \to \infty} \frac{1}{n} |\{k \le n : k \in K\}|,$$

where the vertical bars denote the cardinality of the enclosed set. A number sequence $x = (x_k)_{k \in \mathbb{N}}$ is said to be statistically convergent to a point L if for every $\varepsilon > 0$, the set $K(\varepsilon) = \{k \le n : |x_k - L| \ge \varepsilon\}$ has asymptotic density zero, i.e.,

$$\lim_{n \to \infty} \frac{1}{n} |\{k \le n : |x_k - L| \ge \varepsilon\}| = 0.$$

In this case we write S-$\lim x = L$ or $x_k \to L\,(S)$ (see [5,6]).

The following definitions are due to Mursaleen [14].

Definition 1. *Let K be a subset of $\mathbb{N}$ and $\lambda = (\lambda_n)$ be a non-decreasing sequences of positive real numbers tending to ∞ and such that*

$$\lambda_{n+1} \le \lambda_n + 1, \quad \lambda_1 = 0.$$

Let K be a subset of $\mathbb{N}$, the set of natural numbers. The number

$$\delta_\lambda (K) = \lim_n \frac{1}{\lambda_n} |\{k \in K : n - \lambda_n + 1 \le k \le n\}|,$$

is said to be the λ-density of K.

Definition 2. *A sequence $x = (x_k)$ in X is said to be λ-statistically convergent to $L \in X$ and is denoted by S_λ-$\lim x = L$, if, for every $\varepsilon > 0$, the set $K(\varepsilon)$ has λ-density zero, i.e.,*

$$\lim_n \frac{1}{\lambda_n} |K_n(\varepsilon)| = 0,$$

where $K_n(\varepsilon) = \{k \in I_n : |x_k - L| \ge \varepsilon\}$ and $I_n = [n - \lambda_n + 1, n]$.

Definition 3 ([7]). *Let X be a real vector space of dimension d, where $2 \le d < \infty$. A 2-norm on X is a function $\|\cdot, \cdot\| : X \times X \to \mathbb{R}$ which satisfies* (i) *$\|x, y\| = 0$ if and only if x and y are linearly dependent;* (ii) *$\|x, y\| = \|y, x\|$;* (iii) *$\|\alpha x, y\| = |\alpha| \|x, y\|$, $\alpha \in \mathbb{R}$;* (iv) *$\|x, y + z\| \le \|x, y\| + \|x, z\|$. The pair $(X, \|\cdot, \cdot\|)$ is then called a 2-normed space.*

As an example of a 2-normed space we may take $X = \mathbb{R}^2$ being equipped with the 2-norm $\|x, y\| :=$ the area of the parallelogram spanned by the vectors x and y, which may be given explicitly by the formula

$$\|x, y\| = |x_1 y_2 - x_2 y_1|, \quad x = (x_1, x_2), \quad y = (y_1, y_2).$$

All the concepts listed below are studied in depth in the fundamental book by Schweizer and Sklar [19].

Definition 4. *Let $\mathbb{R}$ denote the set of real numbers, $\mathbb{R}_+ = \{x \in \mathbb{R} : x \ge 0\}$ and $S = [0, 1]$ the closed unit interval. A mapping $f : \mathbb{R} \to S$ is called a distribution function if, it is non-decreasing and left continuous with $\inf_{t \in \mathbb{R}} f(t) = 0$ and $\sup_{t \in \mathbb{R}} f(t) = 1$.*

We denote the set of all distribution functions by D^+ such that $f(0) = 0$. If $a \in \mathbb{R}_+$, then $H_a \in D^+$, where

$$H_a(t) = \begin{cases} 1 \text{ if } t > a, \\ 0 \text{ if } t \le a. \end{cases}$$

It is obvious that $H_0 \ge f$ for all $f \in D^+$.

Definition 5. *A triangular norm (t-norm) is a continuous mapping* $* : S \times S \to S$ *be such that* $(S, *)$ *is an abelian monoid with unit one and* $c * d \leq a * b$ *if* $c \leq a$ *and* $d \leq b$ *for all* $a, b, c, d \in S$. *A triangle function* τ *is a binary operation on* D^+ *which is commutative, associative and* $\tau(f, H_0) = f$ *for every* $f \in D^+$.

Definition 6. *Let* X *be a linear space of dimension greater than one,* τ *be a triangle function, and* $F : X \times X \to D^+$. *Then* F *is called a probabilistic 2-norm and* (X, F, τ) *a probabilistic 2-normed space if the following conditions are satisfied:*

(i) $F(x, y; t) = H_0(t)$ *if* x *and* y *are linearly dependent, where* $F(x, y; t)$ *denotes the value of* $F(x, y)$ *at* $t \in \mathbb{R}$,
(ii) $F(x, y; t) \neq H_0(t)$ *if* x *and* y *are linearly independent,*
(iii) $F(x, y; t) = F(y, x; t)$ *for all* $x, y \in X$,
(iv) $F(\alpha x, y; t) = F(x, y; \frac{t}{|\alpha|})$ *for every* $t > 0$, $\alpha \neq 0$ *and* $x, y \in X$,
(v) $F(x + y, z; t) \geq \tau(F(x, z; t), F(y, z; t))$ *whenever* $x, y, z \in X$, *and* $t > 0$. *If* *(v) is replaced by*
(vi) $F(x+y, z; t_1 + t_2) \geq F(x, z; t_1) * F(y, z; t_2)$ *for all* $x, y, z \in X$ *and* $t_1, t_2 \in \mathbb{R}_+$;
then $(X, F, *)$ *is called a random 2-normed space (for short, RTNS).*

We provide the following example.

Example 1. Let $(X, \|., .\|)$ be a 2-normed space, and let $a * b = ab$ for all $a, b \in S$. For all $x, y \in X$ and every $t > 0$, consider

$$F(x, y; t) = \frac{t}{t + \|x, y\|}.$$

Clearly $(X, F, *)$ is a random 2-normed space.

Let $(X, F, *)$ be a RTN space. Since $*$ is a continuous t-norm, the system of (ε, η)-neighbourhoods of θ (the null vector in X)

$$\left\{ \mathcal{N}_{(\theta, z)}(\varepsilon, \eta) : \varepsilon > 0, \ \eta \in (0, 1), \ z \in X \right\},$$

where

$$\mathcal{N}_{(\theta, z)}(\varepsilon, \eta) = \left\{ (x, z) \in X \times X : F_{(x, z)}(\varepsilon) > 1 - \eta \right\}.$$

determines a first countable Hausdorff topology on $X \times X$, called the F-topology. Thus, the F-topology can be completely specified by means of F-convergence of sequences. It is clear that $x - y \in \mathcal{N}_{(\theta, z)}$ means $y \in \mathcal{N}_{(x, z)}$ and vice-versa.

A sequence $x = (x_k)$ in X is said to be F-convergent to $L \in X$ if for every $\varepsilon > 0$, $\eta \in (0, 1)$ and for each non-zero $z \in X$ there exists a positive integer N such that;

$$(x_k, z - L) \in \mathcal{N}_{(\theta, z)}(\varepsilon, \eta) \text{ for each } k \geq N$$

or equivalently,

$$(x_k, z) \in \mathcal{N}_{(L, z)}(\varepsilon, \eta) \text{ for each } k \geq N.$$

In this case we write $F\text{-}\lim (x_k, z) = L$.

We also recall that the concept of convergence and Cauchy sequence in a random 2-normed space is studied in [2].

Definition 7. *Let* $(X, F, *)$ *be a RN space. Then, a sequence* $x = \{x_k\}$ *is said to be convergent to* $L \in X$ *with respect to the random norm* F *if, for every* $\varepsilon > 0$ *and* $\eta \in (0, 1)$, *there exists* $k_0 \in \mathbb{N}$ *such that* $F_{(x_k - L, z)}(\varepsilon) > 1 - \eta$ *whenever* $k \geq k_0$. *It is denoted by* F-$\lim x = L$ *or* $x_k \to_F L$ *as* $k \to \infty$.

Definition 8. *Let* $(X, F, *)$ *be a RN space. Then, a sequence* $x = \{x_k\}$ *is called a Cauchy sequence with respect to the random norm* F *if, for every* $\varepsilon > 0$ *and* $\eta \in (0, 1)$, *there exists* $k_0 \in \mathbb{N}$ *such that* $F_{(x_k - x_m, z)}(\varepsilon) > 1 - \eta$ *for all* $k, m \geq k_0$.

Definition 9. *Let* $(X, F, *)$ *be a RN space. Then, a sequence* $x = \{x_k\}$ *is said to be* λ-*statistically convergent to* $L \in X$ *with respect to the* F-*topology if for every* $\varepsilon > 0$, $\eta \in (0, 1)$ *and each non-zero* $z \in X$ *such that;*

$$\delta_\lambda \left(\{k \in \mathbb{N} : F_{(x_k - L, z)}(\varepsilon) \leq 1 - \eta\} \right) = 0$$

or equivalently

$$\delta_\lambda \left(\{k \in \mathbb{N} : F_{(x_k - L, z)}(\varepsilon) > 1 - \eta\} \right) = 1.$$

In this case we write S_λ^{R2N}-$\lim x = L$ *or* $x_k \to L \left(S_\lambda^{R2N} \right)$.

If $\lambda_n = n$ for every n then every λ-statistically convergent sequences in random 2-normed space $(X, F, *)$ reduce to statistically convergent sequences in random 2-normed space $(X, F, *)$.

Definition 10. *Let* $(X, F, *)$ *be a RN space. Then, a sequence* $x = \{x_k\}$ *is said to be* λ-*statistical Cauchy to* $L \in X$ *with respect to the* F-*topology if, for every* $\varepsilon > 0$, $\eta \in (0, 1)$ *and each non-zero* $z \in X$ *there exists a positive integer* $N = N(\varepsilon)$ *such that*

$$\delta_\lambda \left(\{k \in \mathbb{N} : F_{(x_k - x_N, z)}(\varepsilon) \leq 1 - \eta\} \right) = 0$$

or equivalently

$$\delta_\lambda \left(\{k \in \mathbb{N} : F_{(x_k - x_N, z)}(\varepsilon) > 1 - \eta\} \right) = 1.$$

In this case we write S_λ^{R2N}-$\lim x = L$ *or* $x_k \to L \left(S_\lambda^{R2N} \right)$.

2 Kinds of λ-Statistical Convergence for Functions in RTNS

In this section we are concerned with convergence in pointwise λ-statistical convergence and uniform λ-statistical convergence of sequences of functions in a random 2-normed spaces. Particularly, we introduce the λ-statistical analog of the Cauchy convergence criterion for pointwise and uniform λ-statistical convergence in a random 2-normed space. Finally, we prove that pointwise and uniform λ-statistical convergence preserves continuity.

2.1   Pointwise λ-Statistical Convergence in RTNS

Fix a random 2-normed space $(Y, F', *)$. Assume that $(X, F, *)$ is a RTN space and that $\mathcal{N}'_{(\theta, z)}(\varepsilon, \eta) = \left\{ x, z \in X \times X : F'_{(x, z)}(\varepsilon) > 1 - \eta \right\}$, called the F'-topology, is given.

Let $f_k : (X, F, *) \to (Y, F', *)$, $k \in \mathbb{N}$, be a sequence of functions. A sequences of functions $(f_k)_{k \in \mathbb{N}}$ (on X) is said to be F-convergence to f (on X) if for every $\varepsilon > 0$, $\eta \in (0, 1)$ and for each non-zero $z \in X$, there exists a positive integer $N = N(\varepsilon, \eta, x)$ such that

$$(f_k(x) - f(x), z) \in \mathcal{N}'_{\theta, z}(\varepsilon, \eta) = \left\{ (x, z) \in X \times X : F'_{((f_k(x) - f(x)), z)}(\varepsilon) > 1 - \eta \right\}$$

for each $k \geq N$ and for each $x \in X$ or equivalently,

$$(f_k(x), z) \in \mathcal{N}'_{(f(x), z)}(\varepsilon, \eta) \text{ for each } k \geq N \text{ and for each } x \in X.$$

In this case we write $f_k \to_{F^2} f$.

First let us define pointwise λ-statistical convergence in a random 2-normed space.

Definition 11. *Let $f_k : (X, F, *) \to (Y, F', *)$, $k \in \mathbb{N}$, be a sequence of functions. $(f_k)_{k \in \mathbb{N}}$ is said to be pointwise λ-statistical convergence to a function f (on X) with respect to F-topology if, for every $x \in X$, $\varepsilon > 0$, $\eta \in (0, 1)$ and each non-zero $z \in X$ the set*

$$\delta_\lambda \left(\left\{ k \in \mathbb{N} : (f_k(x), z) \notin \mathcal{N}'_{(f(x), z)}(\varepsilon, \eta) \right\} \right) = 0,$$

or equivalently

$$\delta_\lambda \left(\left\{ k \in \mathbb{N} : (f_k(x), z) \in \mathcal{N}'_{(f(x), z)}(\varepsilon, \eta) \right\} \right) = 1.$$

In this case we write $f_k \to f \left(S_\lambda^{RTN} \right)$.

Theorem 1. *Let $(X, F, *)$, $(Y, F', *)$ be RTN spaces. Assume that $(f_k)_{k \in \mathbb{N}}$ is pointwise convergent (on X) with respect to F-topology where $f_k : (X, F, *) \to (Y, F', *)$, $k \in \mathbb{N}$. Then $f_k \to f \left(S_\lambda^{RTN} \right)$ (on X). However the converse of this is not true.*

Proof. Let $\varepsilon > 0$ and $\eta \in (0, 1)$. Suppose $(f_k)_{k \in \mathbb{N}}$ is F-convergent on X. In this case the sequence $(f_k(x))$ is convergent with respect to F'-topology for each $x \in X$. Then, there exists a number $k_0 = k_0(\varepsilon) \in \mathbb{N}$ such that $(f_k(x), z) \in \mathcal{N}'_{(f(x), z)}(\varepsilon, \eta)$ for every $k \geq k_0$, every non-zero $z \in X$ and for each $x \in X$. This implies that the set

$$A(\varepsilon, \eta) = \left\{ k \in \mathbb{N} : (f_k(x), z) \notin \mathcal{N}'_{f(x), z}(\varepsilon, \eta) \right\} \subseteq \{1, 2, 3, ..., k_0 - 1\}.$$

Since finite subset of $\mathbb{N}$ has λ-density 0, we have $\delta_\lambda(A(\varepsilon, \eta)) = 0$. That is, $f_k \to f \left(S_\lambda^{RTN} \right)$ (on X).

Example 2. Considering X as in Example 1, we have $(X, F, *)$ as a RTN space induced by the random 2-norm $F_{(x,y)}(\varepsilon) = \frac{\varepsilon}{\varepsilon + \|x,y\|}$. Define a sequence of functions $f_k : [0, 1] \to \mathbb{R}$ via

$$f_k(x) = \begin{cases} x^k + 1 & \text{if } n - \sqrt{\lambda_n} + 1 \le k \le n \text{ and } x \in [0, \tfrac{1}{2}) \\ 0 & \text{if otherwise and } x \in [0, \tfrac{1}{2}) \\ x^k + \tfrac{1}{2} & \text{if } n - \sqrt{\lambda_n} + 1 \le k \le n \text{ and } x \in [\tfrac{1}{2}, 1) \\ 1 & \text{if otherwise and } x \in [\tfrac{1}{2}, 1) \\ 2 & \text{if } x = 1. \end{cases}$$

Then, for every $\varepsilon > 0$, $\eta \in (0, 1)$, $x \in [0, \tfrac{1}{2})$ and each non-zero $z \in X$, let $A_n(\varepsilon, \eta) = \left\{ k \in \mathbb{N} : (f_k(x), z) \notin \mathcal{N}'_{(f(x),z)}(\varepsilon, \lambda) \right\}$. We observe that;

$$A_n(\varepsilon, \eta) = \left\{ k \in I_n : \frac{\varepsilon}{\varepsilon + \|f_k(x), z\|} \le 1 - \eta) \right\}$$

$$= \left\{ k \in I_n : \|f_k(x), z\| \ge \frac{\varepsilon \eta}{1 - \varepsilon} \right\}$$

$$= \left\{ k \in I_n : f_k(x) = x^k + 1 \right\}.$$

and $|A_n(\varepsilon, \lambda)| \le \sqrt{\lambda_n}$. Thus, for each $x \in [0, \tfrac{1}{2})$, since

$$\delta_\lambda(A_n(\varepsilon, \eta)) = \lim_{n \to \infty} \frac{|A_n(\varepsilon, \eta)|}{\lambda_n} = \lim_{n \to \infty} \frac{\sqrt{\lambda_n}}{\lambda_n} = 0$$

$(f_k)_{k \in \mathbb{N}}$ is λ-statistically convergent to 0 with respect to F-topology. Similarly, if we take $x \in [\tfrac{1}{2}, 1)$ and $x = 1$, it can be easily seen that $(f_k)_{k \in \mathbb{N}}$ is λ-statistical convergence to $\tfrac{1}{2}$ and 2 with respect to F-topology, respectively. Hence $(f_k)_{k \in \mathbb{N}}$ is pointwise λ-statistical convergent with respect to F-topology (on X).

Theorem 2. *Let $(X, F, *)$, $(Y, F', *)$ be RTN spaces and let $f_k : (X, F, *) \to (Y, F', *)$, $k \in \mathbb{N}$, be a sequence of functions. Then the following statements are equivalent:*

(i) $f_k \to f \left(S_\lambda^{RTN} \right)$.

(ii) $\delta_\lambda \left(\left\{ k \in \mathbb{N} : (f_k(x), z) \notin \mathcal{N}'_{(f(x),z)}(\varepsilon, \eta) \right\} \right) = 0$ *for every $\varepsilon > 0$, $\eta \in (0, 1)$, for each $x \in X$ and each non-zero $z \in X$.*

(iii) $\delta_\lambda \left(\left\{ k \in \mathbb{N} : (f_k(x), z) \in \mathcal{N}'_{(f(x),z)}(\varepsilon, \eta) \right\} \right) = 1$ *for every $\varepsilon > 0$, $\eta \in (0, 1)$, for each $x \in X$ and each non-zero $z \in X$.*

(iv) $S_\lambda\text{-}\lim F'_{(f_k(x) - f(x), z)}(\varepsilon) = 1$ *for every $x \in X$ and each non-zero $z \in X$.*

Proof is standard.

Theorem 3. *Let $(f_k)_{k \in \mathbb{N}}$ and $(g_k)_{k \in \mathbb{N}}$ be two sequences of functions from $(X, F, *)$ to $(Y, F', *)$ with $a * a > a$ for every $a \in (0, 1)$. If $f_k \to f \left(S_\lambda^{RTN} \right)$ and $g_k \to g \left(S_\lambda^{RTN} \right)$, then $(\alpha f_k + \beta g_k) \to (\alpha f + \beta g) \left(S_\lambda^{RTN} \right)$ where $\alpha, \beta \in \mathbb{R}$ (or $\mathbb{C}$).*

Proof. Let $\varepsilon > 0$ and $\eta \in (0,1)$. Since $f_k \to f\left(S_\lambda^{RTN}\right)$ and $g_k \to g\left(S_\lambda^{RTN}\right)$ for each $x \in X$, if we define

$$A_1 = \left\{k \in \mathbb{N} : (f_k(x), z) \notin \mathcal{N}'_{(f(x),z)}(\tfrac{\varepsilon}{2}, \eta)\right\} \text{ and } A_2 = \left\{k \in \mathbb{N} : (g_k(x), z) \notin \mathcal{N}'_{(g(x),z)}(\tfrac{\varepsilon}{2}, \eta)\right\}$$

then $\delta_\lambda(A_1) = 0$ and $\delta_\lambda(A_2) = 0$. Since $\delta_\lambda(A_1) = 0$ and $\delta_\lambda(A_2) = 0$, if we represent A by $(A_1 \cup A_2)$ then $\delta_\lambda(A) = 0$. Hence $A_1 \cup A_2 \neq \mathbb{N}$ and there exists $\exists k_0 \in \mathbb{N}$ such that;

$$(f_{k_0}(x), z) \in \mathcal{N}'_{(f(x),z)}(\tfrac{\varepsilon}{2}, \eta) \text{ and } (g_{k_0}(x), z) \in \mathcal{N}'_{(g(x),z)}(\tfrac{\varepsilon}{2}, \eta)$$

Let

$$B = \left\{k \in \mathbb{N} : ((\alpha f_k(x) + \beta g_k(x)), z) \notin \mathcal{N}'_{((\alpha f(x) + \beta g(x)),z)}(\varepsilon, \eta)\right\}.$$

We shall show that $A^c \subset B$ for each $x \in X$. Let $k_0 \in A^c$. In this case,

$$(f_{k_0}(x), z) \in \mathcal{N}'_{(f(x),z)}(\tfrac{\varepsilon}{2}, \eta) \text{ and } (g_{k_0}(x), z) \in \mathcal{N}'_{(g(x),z)}(\tfrac{\varepsilon}{2}, \eta).$$

From the above expressions, we have

$$F'_{((\alpha f_k(x) + \beta g_k(x) - \alpha f(x) + \beta g(x)),z)}(\varepsilon) \geq F'_{((\alpha f_k(x) - \alpha f(x)),z)}\left(\tfrac{\varepsilon}{2}\right) * F'_{((\beta g_k(x) - \beta g(x)),z)}\left(\tfrac{\varepsilon}{2}\right)$$

$$= F'_{((f_k(x) - f(x)),z)}\left(\tfrac{\varepsilon}{2\alpha}\right) * F'_{((g_k(x) - g(x)),z)}\left(\tfrac{\varepsilon}{2\beta}\right)$$

$$> (1 - \eta) * (1 - \eta)$$

$$> 1 - \eta.$$

This implies $A^c \subset B$. Since $B^c \subset A$ and $\delta_\lambda(A) = 0$, hence $\delta_\lambda(B^c) = 0$. That is

$$\delta_\lambda\left(\left\{k \in \mathbb{N} : ((\alpha f_k(x) + \beta g_k(x)), z) \notin \mathcal{N}'_{((\alpha f(x) + \beta g(x)),z)}(\varepsilon, \eta)\right\}\right) = 0.$$

Definition 12. Let $(X, F, *)$, $(Y, F', *)$ be RTN spaces and let $f_k : (X, F, *) \to (Y, F', *)$, $k \in \mathbb{N}$, be a sequence of functions. A sequence $(f_k)_{k \in \mathbb{N}}$ is called pointwise λ-statistical Cauchy sequence in RTN space if, for every $\varepsilon > 0$, $\eta \in (0,1)$ and each non-zero $z \in X$ there exists $M = M(\varepsilon, \eta, x) \in \mathbb{N}$ such that;

$$\delta_\lambda\left(\left\{k \in \mathbb{N} : (f_k(x) - f_M(x), z) \notin \mathcal{N}'_{(\theta,z)}(\varepsilon, \eta)\right\}\right) = 0.$$

Theorem 4. Let $(X, F, *)$, $(Y, F', *)$ be RTN spaces such that $a * a > a$ for every $a \in (0,1)$ and let $f_k : (X, F, *) \to (Y, F', *)$, $k \in \mathbb{N}$, be a sequence of functions. If $(f_k)_{k \in \mathbb{N}}$ is a pointwise λ-statistical convergent sequence with respect to F-topology, then $(f_k)_{k \in \mathbb{N}}$ is a pointwise λ-statistical Cauchy sequence with respect to F-topology. However the converse of this is not true.

Proof. Suppose that $(f_k)_{k \in \mathbb{N}}$ is a pointwise λ-statistical convergent to f with respect to F-topology. Let $\varepsilon > 0$ and $\eta \in (0,1)$ be given. If we state A and A^c by

$$A = \left\{k \in \mathbb{N} : (f_k(x), z) \notin \mathcal{N}'_{(f(x),z)}(\tfrac{\varepsilon}{2}, \eta)\right\} \text{ and } A^c = \left\{k \in \mathbb{N} : (f_k(x), z) \in \mathcal{N}'_{(f(x),z)}(\tfrac{\varepsilon}{2}, \eta)\right\},$$

then $\delta_\lambda(A) = 0$ and $\delta_\lambda(A^c) = 1$. Now, for every $k, m \in A^c$,

$$F'_{(f_k(x)-f_m(x),z)}(\varepsilon) \geq F'_{(f_k(x)-f(x),z)}\left(\frac{\varepsilon}{2}\right) * F'_{(f_m(x)-f(x),z)}\left(\frac{\varepsilon}{2}\right)$$
$$> (1-\eta) * (1-\eta)$$
$$> 1 - \eta.$$

So, $\delta_\lambda\left(\left\{k \in \mathbb{N} : (f_k(x) - f_m(x), z) \in \mathcal{N}'_{(\theta,z)}(\varepsilon, \eta)\right\}\right) = 1$. Therefore

$$\delta_\lambda\left(\left\{k \in \mathbb{N} : (f_k(x) - f_m(x), z) \notin \mathcal{N}'_{(\theta,z)}(\varepsilon, \eta)\right\}\right) = 0,$$

i.e., $(f_k)_{k\in\mathbb{N}}$ is a pointwise λ-statistical Cauchy sequence with respect to F-topology.

The next result is a modification of a well-known result.

Theorem 5. *Let $(X, F, *)$, $(Y, F', *)$ be a RTN spaces such that $a * a > a$ for every $a \in (0,1)$. Assume that $f_k \to f \left(S_\lambda^{RTN}\right)$ (on X) where functions $f_k : (X, F, *) \to (Y, F', *)$, $k \in \mathbb{N}$, are equi-continuous (on X) and $f : (X, F, *) \to (Y, F', *)$. Then f is continuous (on X) with respect to F-topology.*

Proof. We prove that f is continuous with respect to F-topology. Let $x_0 \in X$ and $(x - x_0, z) \in \mathcal{N}_{\theta,z}(\varepsilon, \eta)$ be fixed. By the equi-continuity of f_k's, for every $\varepsilon > 0$ and each non-zero $z \in X$, there exists a $\gamma \in (0,1)$ with $\gamma < \eta$ such that $(f_k(x) - f_k(x_0), z) \in \mathcal{N}'_{(\theta,z)}(\frac{\varepsilon}{3}, \gamma)$ for every $k \in \mathbb{N}$. Since $f_k \to f\left(S_\lambda^{RTN}\right)$, if we state respectively A and B by the sets $A = \left\{k \in \mathbb{N} : (f_k(x_0), z) \notin \mathcal{N}'_{(f(x_0),z)}(\frac{\varepsilon}{3}, \gamma)\right\}$ and $B = \left\{k \in \mathbb{N} : (f_k(x), z) \notin \mathcal{N}'_{(f(x),z)}(\frac{\varepsilon}{3}, \gamma)\right\}$, then $\delta_\lambda(A) = 0$ and $\delta_\lambda(B) = 0$. Therefore, $\delta_\lambda(A \cup B) = 0$ and $A \cup B$ is different from $\mathbb{N}$. So, there exists $k \in \mathbb{N}$ such that $(f_k(x_0), z) \in \mathcal{N}'_{(f(x_0),z)}(\frac{\varepsilon}{3}, \gamma)$ and $(f_k(x), z) \in \mathcal{N}'_{(f(x),z)}(\frac{\varepsilon}{3}, \gamma)$. We have

$$F'_{(f(x_0)-f(x),z)}(\varepsilon) \geq F'_{(f(x_0)-f_k(x_0),z)}\left(\frac{\varepsilon}{3}\right) * \left[F'_{(f_k(x_0)-f_k(x),z)}\left(\frac{\varepsilon}{3}\right) * F'_{(f_k(x)-f(x),z)}\left(\frac{\varepsilon}{3}\right)\right]$$
$$> (1-\gamma) * [(1-\gamma) * (1-\gamma)]$$
$$> (1-\gamma) * (1-\gamma)$$
$$> 1 - \gamma$$
$$> 1 - \eta$$

and the continuity of f with respect to F-topology is proved.

2.2   Uniformly λ-Statistical Convergence in RTNS

Let us define uniform λ-statistical convergence in a random 2-normed space.

Definition 13. *Let* $(X, F, *)$, $(Y, F', *)$ *be RTN spaces. We say that a sequence of functions* $f_k : (X, F, *) \to (Y, F', *)$, $k \in \mathbb{N}$, *is uniform* λ-*statistically convergent to a function* f *(on* X*) with respect to* F-*topology if and only if* $\forall \varepsilon > 0$, $\exists M \subset \mathbb{N}$, $\delta_\lambda(M) = 1$, $\exists k_0 = k_0(\varepsilon, \eta, x) \in M \ni \forall k > k_0$, $k \in M$, $\forall z \in X$ *and* $\forall x \in X, \eta \in (0, 1)$ $(f_k(x), z) \in \mathcal{N}'_{(f(x), z)}(\varepsilon, \eta)$.

In this case we write $f_k \rightrightarrows f\left(S_\lambda^{RTN}\right)$.

We state the following result without proof, which can be established using standard technique.

Theorem 6. *Let* $(X, F, *)$, $(Y, F', *)$ *be RTN spaces and let* $f_k : (X, F, *) \to (Y, F', *)$, $k \in \mathbb{N}$, *be a sequence of functions. Then for every* $\varepsilon > 0$ *and* $\eta \in (0, 1)$, *the following statements are equivalent:*

(i) $f_k \rightrightarrows f\left(S_\lambda^{RTN}\right)$.

(ii) $\delta_\lambda\left(\left\{k \in \mathbb{N} : (f_k(x), z) \notin \mathcal{N}'_{(f(x), z)}(\varepsilon, \eta)\right\}\right) = 0$ *for every* $x \in X$ *and each non-zero* $z \in X$.

(iii) $\delta_\lambda\left(\left\{k \in \mathbb{N} : (f_k(x), z) \in \mathcal{N}'_{(f(x), z)}(\varepsilon, \eta)\right\}\right) = 1$ *for every* $x \in X$ *and each non-zero* $z \in X$.

(iv) S_λ–$\lim F'_{(f_k(x)-f(x), z)}(\varepsilon) = 1$ *for every* $x \in X$ *and each non-zero* $z \in X$.

Definition 14. *Let* $(X, F, *)$ *be a RTN space. A subset* Y *of* X *is said to be bounded on RTN spaces if for every* $\eta \in (0, 1)$ *there exists* $\varepsilon > 0$ *such that* $(x, z) \in \mathcal{N}_{(\theta, z)}(\varepsilon, \eta)$ *for all* $x \in Y$ *and every non-zero* $z \in X$.

Definition 15. *Let* $(X, F, *)$, $(Y, F', *)$ *be RTN spaces and let* $f_k : (X, F, *) \to (Y, F', *)$, $k \in \mathbb{N}$, *and* $f : (X, F, *) \to (Y, F', *)$ *be bounded functions. Then* $f_k \rightrightarrows f\left(S_\lambda^{RTN}\right)$ *if and only if* S_λ-$\lim\left(\inf_{x \in X} F'_{(f_k(x)-f(x), z)}(\varepsilon)\right) = 1$.

Example 3. Let $(X, F, *)$ be as considered in Example 1. Define a sequence of functions $f_k : [0, 1) \to \mathbb{R}$ via

$$f_k(x) = \begin{cases} x^k + 1 & \text{if } n - \sqrt{\lambda_n} + 1 \leq k \leq n \\ 2 & \text{otherwise.} \end{cases}$$

Then, for every $\varepsilon > 0$, $\eta \in (0, 1)$ and each non-zero $z \in X$, let $A_n(\varepsilon, \eta) = \left\{k \in I_n : (f_k(x), z) \notin \mathcal{N}'_{(1, z)}(\varepsilon, \lambda)\right\}$. For all $x \in X$, we have $\delta_\lambda(A_n(\varepsilon, \lambda)) = 0$. Since $f_k \to 1\left(S_\lambda^{RTN}\right)$ for all $x \in X$, $f_k \rightrightarrows 1\left(S_\lambda^{RTN}\right)$ (on $[0, 1)$).

Remark 1. If $f_k \rightrightarrows f\left(S_\lambda^{RTN}\right)$, then $f_k \to f\left(S_\lambda^{RTN}\right)$. But not necessarily conversely.

We establish the above remark providing the following example.

Example 4. Define the sequence of functions

$$f_k(x) = \begin{cases} 0 & \text{if } n - \sqrt{\lambda_n} + 1 \leq k \leq n \\ \frac{k^2 x}{1 + k^3 x^2} & \text{otherwise} \end{cases}$$

on $[0,1]$. Since $f_k\left(\frac{1}{k}\right) \to 1\left(S_\lambda^{RTN}\right)$ and $f_k(0) \to 0\left(S_\lambda^{RTN}\right)$, this sequence of functions is pointwise λ-statistically convergence to 0 with respect to F-topology. But by Definition 11, it is not uniform λ-statistical convergence with respect to F-topology.

Theorem 7. *Let $(X, F, *)$, $(Y, F', *)$ be RTN spaces. Assume that $(f_k)_{k \in \mathbb{N}}$ is uniformly convergent (on X) with respect to F-topology where $f_k : (X, F, *) \to (Y, F', *)$, $k \in \mathbb{N}$. Then $f_k \rightrightarrows f\left(S_\lambda^{RTN}\right)$ (on X). However the converse of this is not true.*

Proof. Assume that $(f_k)_{k \in \mathbb{N}}$ is uniformly convergent to f on X with respect to F-topology. In this case, for every $\varepsilon > 0$, $\eta \in (0, 1)$ and every non-zero $z \in X$, there exists a positive integer $k_0 = k_0(\varepsilon, \eta)$ such that $\forall x \in X$ and $\forall k > k_0$, $(f_k(x), z) \in \mathcal{N}'_{(f(x),z)}(\varepsilon, \eta)$. That is, for $k \le k_0$

$$A(\varepsilon, \eta) = \left\{k \in \mathbb{N} : (f_k(x), z) \notin \mathcal{N}'_{(f(x),z)}(\varepsilon, \eta)\right\} \subseteq \{1, 2, 3, ..., k_0\}.$$

Since finite subset of $\mathbb{N}$ has λ-density 0, we have $\delta_\lambda(A(\varepsilon, \eta)) = 0$. That is, $f_k \rightrightarrows f\left(S_\lambda^{RTN}\right)$ (on X).

Definition 16. *Let $(X, F, *)$, $(Y, F', *)$ be RTN spaces and let $f_k : (X, F, *) \to (Y, F', *)$, $k \in \mathbb{N}$, be a sequence of functions. Then a sequence $(f_k)_{k \in \mathbb{N}}$ is called uniform λ-statistical Cauchy sequence in RTN space if for every $\varepsilon > 0$, $\eta \in (0, 1)$ and each non-zero $z \in X$ there exists $N = N(\varepsilon, \eta) \in \mathbb{N}$ such that*

$$\delta_\lambda\left(\left\{k \in \mathbb{N} : (f_k(x) - f_N(x), z) \notin \mathcal{N}'_{(\theta,z)}(\varepsilon, \eta)\right\}\right) = 0.$$

Theorem 8. *Let $(X, F, *)$, $(Y, F', *)$ be RTN spaces such that $a * a > a$ for every $a \in (0, 1)$ and let $f_k : (X, F, *) \to (Y, F', *)$, $k \in \mathbb{N}$, be a sequence of functions. If $(f_k)_{k \in \mathbb{N}}$ is a uniform λ-statistical convergence sequence with respect to F-topology, then $(f_k)_{k \in \mathbb{N}}$ is a uniform λ-statistical Cauchy sequence with respect to F-topology. However the converse of this is not true.*

Proof. Suppose that $f_k \rightrightarrows f\left(S_\lambda^{RTN}\right)$. Let $A = \left\{k \in \mathbb{N} : (f_k(x), z) \in \mathcal{N}'_{f(x),z}\right.$ $\left.(\varepsilon, \eta)\right\}$. By Definition 9, for every $\varepsilon > 0$, $\eta \in (0, 1)$ and each non-zero $z \in X$, there exists $A \subset \mathbb{N}$, $\delta_\lambda(A) = 0$ and $\exists k_0 = k_0(\varepsilon, \eta) \in A$ such that $\forall k > k_0$, $k \in A$ and $\forall x \in X$, $(f_k(x), z) \in \mathcal{N}'_{(f(x),z)}(\frac{\varepsilon}{2}, \eta)$. Choose $N = N(\varepsilon, \eta) \in A$, $N > k_0$. So, $(f_N(x), z) \in \mathcal{N}'_{(f(x),z)}(\frac{\varepsilon}{2}, \eta)$. For every $k \in A$, we have

$$F'_{(f_k(x)-f_N(x),z)}(\varepsilon) \ge F'_{(f_k(x)-f(x),z)}\left(\frac{\varepsilon}{2}\right) * F'_{(f(x)-f_N(x),z)}\left(\frac{\varepsilon}{2}\right)$$
$$> (1 - \eta) * (1 - \eta)$$
$$> 1 - \eta.$$

Hence, $\delta_\lambda \left(\left\{ k \in \mathbb{N} : (f_k(x) - f_N(x), z) \in \mathcal{N}'_{(\theta,z)}(\varepsilon, \eta) \right\} \right) = 1$. Therefore

$$\delta_\lambda \left(\left\{ k \in \mathbb{N} : (f_k(x) - f_N(x), z) \notin \mathcal{N}'_{(\theta,z)}(\varepsilon, \eta) \right\} \right) = 0,$$

i.e., (f_k) is an uniformly λ-statistical Cauchy sequence in RTN space.

The next result is a modification of a well-known result.

Theorem 9. *Let $(X, F, *)$, $(Y, F', *)$ be RTN spaces such that $a*a > a$ for every $a \in (0,1)$ and the map $f_k : (X, F, *) \to (Y, F', *)$, $k \in \mathbb{N}$, be continuous (on X) with respect to F-topology. If $f_k \rightrightarrows f \left(S_\lambda^{RTN} \right)$ (on X) then $f : (X, F, *) \to (Y, F', *)$ is continuous (on X) with respect to F-topology. However the converse of this is not true.*

Proof. Let $x_0 \in X$ and $(x_0 - x, z) \in \mathcal{N}_{(\theta,z)}(\varepsilon, \eta)$ be fixed. By F-continuity of f_k's, for every $\varepsilon > 0$ and each non-zero $z \in X$, there exists a $\gamma \in (0,1)$ with $\gamma < \eta$ such that $(f_k(x_0) - f_k(x), z) \in \mathcal{N}'_{(\theta,z)}(\frac{\varepsilon}{3}, \gamma)$ for every $k \in \mathbb{N}$. Since $f_k \rightrightarrows f \left(S_\lambda^{RTN} \right)$, for all $x \in X$, if we state respectively $A(\varepsilon, \eta)$ and $B(\varepsilon, \eta)$ by the sets $A = \left\{ k \in \mathbb{N} : (f_k(x_0), z) \notin \mathcal{N}'_{(f(x_0),z)}(\frac{\varepsilon}{3}, \gamma) \right\}$ and $B = \left\{ k \in \mathbb{N} : (f_k(x), z) \notin \mathcal{N}'_{(f(x),z)}(\frac{\varepsilon}{3}, \gamma) \right\}$, then $\delta_\lambda(A) = 0$ and $\delta_\lambda(B) = 0$. Therefore, $\delta_\lambda(A \cup B) = 0$ and $A \cup B$ is different from $\mathbb{N}$. So, there exists $k \in \mathbb{N}$ such that $(f_k(x_0), z) \in \mathcal{N}'_{(f(x_0),z)}(\frac{\varepsilon}{3}, \gamma)$ and $(f_k(x), z) \in \mathcal{N}'_{(f(x),z)}(\frac{\varepsilon}{3}, \gamma)$. It follows that

$$F'_{(f(x)-f(x_0),z)}(\varepsilon) \geq F'_{(f(x)-f_m(x),z)}\left(\frac{\varepsilon}{3}\right) * \left[F'_{(f_m(x_0)-f_m(x_0),z)}\left(\frac{\varepsilon}{3}\right) * F'_{(f_m(x_0)-f(x_0),z)}\left(\frac{\varepsilon}{3}\right) \right]$$
$$> (1 - \gamma) * [(1 - \gamma) * (1 - \gamma)]$$
$$> (1 - \gamma) * (1 - \gamma)$$
$$> 1 - \gamma$$
$$> 1 - \eta.$$

This implies that f is continuous (on X) with respect to F-topology.

References

1. Alsina, C., Schweizer, B., Sklar, A.: On the definition of a probabilistic normed space. Aequationes Math. **46**, 91–98 (1993)
2. Asadollah, A., Nourouz, K.: Convex sets in probabilistic normed spaces. Chaos, Solitons Fractals **36**, 322–328 (2008)
3. Balcerzak, M., Dems, K., Komisarski, A.: Statistical convergence and ideal convergence for sequences of functions. J. Math. Anal. Appl. **328**, 715–729 (2007)
4. Caserta, A., Giuseppe, D., Kočinac, L.: Statistical convergence in function spaces. Abstr. Appl. Anal. **2011**, 1–11 (2011)
5. Fast, H.: Sur la convergence statistique. Colloquium Math. **2**, 241–244 (1951)
6. Fridy, J.: On statistical convergence. Analysis **5**, 301–314 (1985)
7. Gähler, S.: Lineare 2-normietre räume. Math. Nachr. **28**, 1–43 (1964)

8. Giuseppe, D., Kočinac, L.: Statistical convergence in topology. Topol. Appl. **156**, 28–45 (2008)

9. Golet, I.: On probabilistic 2-normed spaces. Novi Sad J. Math. **35**, 95–102 (2006)

10. Gürdal, M., Pehlivan, S.: The statistical convergence in 2-banach spaces. Thai J. Math. **2**, 107–113 (2004)

11. Karakaya, V., Şimşek, N., Ertürk, M., Gürsoy, F.: λ-statistical convergence of sequences of functions with respect to the intuitionistic fuzzy normed spaces. J. Funct. Spaces Appl. **2012**, 1–14 (2012)

12. Karakaya, V., Şimşek, N., Ertürk, M., Gürsoy, F.: Statistical convergence of sequences of functions with respect to the intuitionistic fuzzy normed spaces. Abstr. Appl. Anal. **2012**, 1–19 (2012)

13. Menger, K.: Statistical metrics. Proc. Nat. Acad. Sci. **28**, 535–537 (1942)

14. Mursaleen, M.: λ-statistical convergence. Math. Slovaca **50**, 111–115 (2000)

15. Mursaleen, M.: On statistical convergence in random 2-normed spaces. Acta Sci. Math. (Szeged) **76**, 101–109 (2010)

16. Savaş, E.: On generalized statistical convergence in random 2-normed space. Iran. J. Sci. Technol. **A4**, 417–423 (2012)

17. Savaş, E., Gürdal, M.: Certain summability methods in intuitionistic fuzzy normed spaces. J. Intell. Fuzzy Syst. **27**, 1621–1629 (2014)

18. Savaş, E., Mohiuddine, S.: λ-statistically convergent double sequences in probabilistic normed spaces. Math. Slovaca **62**, 99–108 (2012)

19. Schweizer, B., Sklar, A.: Probabilistic Metric Spaces. Elsevier Science Publishing Co., New York (1983)

20. Sempi, C.: A short and partial history of probabilistic normed spaces. Mediterr. J. Math. **3**, 283–300 (2006)

21. Serstnev, A.: On the notion of a random normed space. Dokl. Akad. Nauk SSSR **149**, 280–283 (1963)

On Linear Theory of Thermoelasticity for an Anisotropic Medium Under a Recent Exact Heat Conduction Model

Manushi Gupta$^{(\boxtimes)}$ and Santwana Mukhopadhyay

Department of Mathematical Sciences, Indian Institute of Technology (BHU),
Varanasi 221005, India
manushig.rs.mat16@iitbhu.ac.in

Abstract. The aim of this paper is to discuss about a new thermoelasticity theory for a homogeneous and anisotropic medium in the context of a recent heat conduction model proposed by Quintanilla (2011). The coupled thermoelasticity being the branch of science that deals with the mutual interactions between temperature and strain in an elastic medium had become the interest of researchers since 1956. Quintanilla (2011) have introduced a new model of heat conduction in order to reformulate the heat conduction law with three phase-lags and established mathematical consistency in this new model as compared to the three phase-lag model. This model has also been extended to thermoelasticity theory. Various Taylor's expansion of this model has gained the interest of many researchers in recent times. Hence, we considered the model's backward time expansion of Taylor's series upto second-order and establish some important theorems. Firstly, uniqueness theorem of a mixed type boundary and initial value problem is proved using specific internal energy function. Later, we give the alternative formulation of the problem using convolution which incorporates the initial conditions into the field equations. Using this formulation, the convolution type variational theorem is proved. Further, we establish a reciprocal relation for the model.

Keywords: Non-Fourier heat conduction model
Generalized thermoelasticiy · Uniqueness · Variational principle
Reciprocity theorem

1 Introduction

The infinite speed of propagation of thermal signal proposed by Fourier's law violates Einstein's relativity theory which motivated the researchers to work in the direction of eliminating this apparent physical drawback. Many modifications had been done to develop new theories which tried to eliminate the infinite behaviour of heat propagation. These modified heat conduction laws are subsequently referred to as non-Fourier heat conduction models. Classical coupled dynamical theory of thermoelasticity was introduced by Biot [1] which was

© Springer Nature Singapore Pte Ltd. 2018
D. Ghosh et al. (Eds.): ICMC 2018, CCIS 834, pp. 309–324, 2018.
https://doi.org/10.1007/978-981-13-0023-3_29

based on the Fourier's law of heat conduction and hence suffered from the similar drawback. Consequently, thermoelasticity theories are developed on the basis of non-Fourier heat conduction models. Gradual and systematic development of the heat conduction and thermoelasticity theory is observed which had been recorded in many review articles and books [2–10].

The theories of Lord and Shulman [11] and Green and Lindsay [12] provided us the innovative theories of thermoelasticity which described the finite speed of the thermal signal. The first theory was based on Cattaneo–Vernotte heat conduction model [13–15] which deals with one thermal relaxation time parameter i.e.

$$\vec{q}(x, t + \tau) = -k\vec{\nabla}T \tag{i}$$

where $\vec{q}$ is the heat flux, k is the thermal material, T is the temperature and τ is relaxation time. Then, later Green and Naghdi [16–18] developed completely new thermoelasticity theory by modifying the equation of heat propagation by incorporating the thermal pulse transmission. Green and Naghdi's theory is categorized into three parts, namely, GN-I, GN-II, and GN-III, in which thermal displacement (ν) and temperature (T) are considered as the constitutive variables with $\dot{\nu} = T$. GN-III represents the more general form with equation as

$$\vec{q} = -[k\vec{\nabla}T + k^*\vec{\nabla}\nu] \tag{ii}$$

where newly introduced k* is positive constant called as rate of thermal conductivity of the material. In 1995, Tzou [19,20] introduced the effect of microstructural interactions in the fast transient process of the heat transport phenomenon and expressed it as

$$\vec{q}(x, t + \tau_q) = -[k\vec{\nabla}T(x, t + \tau_T)] \tag{iii}$$

where t_q and t_T are positive delay parameters. It is called a dual-phase-lag heat conduction model as it involves two delay times, known as phase lags. Later, Roychoudhuri [21] presented an extension of the dual-phase-lag by including an extra delay time, τ_ν, which is termed as the phase lag of thermal displacement gradient. The corresponding equation for heat flux is

$$\vec{q}(x, t + \tau_q) = -[k\vec{\nabla}T(x, t + \tau_T) + k^*\vec{\nabla}\nu(x, t + \tau_\nu)]. \tag{iv}$$

These theories have dragged the interest of many researchers in recent years and encouraged them to discuss their various features. For example, Dreher et al. [22] analysed dual-phase-lag and three-phase-lag heat conduction model and shown that this constitutive equation along with the energy equation

$$-\nabla\vec{q}(x, t) = c\dot{T}(x, t), \tag{v}$$

gives a sequence of eigenvalues in the point spectrum such that its real parts tend to infinity which further proved the ill-posedness of the problem in the Hadamard sense.

The taylor's approximation of these heat conduction equation have become the new area of interest in the sense of research [23–32]. Lately, Quintanilla [33] worked on the three-phase-lag model in a different way and examined the stability and spatial behavior of the new proposed model by taking $\tau_\nu > \tau_q = \tau_T$. Subsequently, the study by Leseduarte and Quintanilla [34] on this model, presented a Phragmen–Lindelof type alternative which demonstrated that the solutions either decay in an exponential way or amplify at infinity in an exponential way. This result further extended to the Taylor series approximation of the equation of heat conduction to the delay term and derived the forward and backward in time equations in the forms as folows:

(i)

$$\vec{q}(x,t) = -k\vec{\nabla}T - k^*(\vec{\nabla}\nu(x,t) + \tau\vec{\nabla}\dot{\nu} + \frac{\tau^2}{2}\vec{\nabla}\ddot{\nu}), \qquad (vi)$$

where $\tau = \tau_\nu - \tau_q > 0$.

(ii)

$$\vec{q}(x,t) = -k\vec{\nabla}T - k^*(\vec{\nabla}\nu(x,t) - \tau\vec{\nabla}\dot{\nu} + \frac{\tau^2}{2}\vec{\nabla}\ddot{\nu}), \qquad (vii)$$

where, $\tau = \tau_q - \tau_\nu > 0$.

They further obtained Phragmen–Lindelof type alternatives for the solutions of the heat conduction equations corresponding to both the heat conduction laws as given above. Subsequently, Kumari and Mukhopadhyay [54] studied some theorems related to model (vi). Also, Quintanilla [35] checked the uniqueness and stability of the model considering (vi) in an alternative way.

Some pioneering work on thermoelasticity theory have been reported by eminent researchers like, Nickell and Sackman [40], Iesan [38,39], Ignaczak [36], Gurtin [37], etc. and it has been shown that the state of dynamics of a thermoelastic system can be determined by using the variational method which describes it as the extremum of functional or function. Ignaczak [36] and Gurtin [37] explained the variational principle for the initial-boundary value problem by incorporating the initial condtions into the field equations. With the help of this alternative formulation, Iesan [38,39] and then Nickell and Sackman [40], established a convolution type variational principle for the linear coupled thermoelasticity. Later, the first variational theorem of Gurtin type for solids with micro-structure was presented by Iesan [41].

The reciprocity theorem is used to derive various methods of integrating the elasticity equations in terms of Green's function and it has significant practical applications in the solution of engineering problems [42]. Maysel [43] developed the Betti–Maxwell reciprocity theorem for the static problems in theory of thermoelasticity. Later, the reciprocity theorem was extended to uncoupled thermoelasticity, coupled thermoelasticity and coupled thermoelasticity for anisotropic homogeneous material by Predeleanu [44], Ionescu-Cazimir [45] and Nowacki [42], respectively. Iesan [41] presented the first reciprocal relation without using the Laplace transform. Convolution type reciprocity theorems were also derived by Iesan [38,39]. Scalia [46] used a method to deduce reciprocity relations without using the Laplace transform and without incorporation of the initial data in the field equations. An exhaustive treatment of the variational principles in thermoelasticity is available in the books by Lebon [47], Carlson [48], Hetnarski and Ignaczak [49] and Hetnarski and Eslami [50]. Recently, the convolution type variational principles and reciprocal relations on different theories of thermoelasticity were given by Chirita and Ciarletta [51], Mukhopadhyay and Prasad [52], Kothari and Mukhopadhyay [53] and Kumari and Mukhopadhyay [54].

Here, we have considered the second-order Taylor's approximation of the proposed model of the form (vii) given by Leseduarte and Quintanilla [33] with a single delay term in which the micro-structural effects in the heat transport phenomenon is considered. We tried to prove some theorems for this model for homogeneous and anisotropic medium which have many applications. We start with describing the basic equations with respect to the considered model and consider a mixed initial-boundary value problem with non-homogeneous initial conditions. Then, we work in the direction to prove the uniqueness by using the specific internal energy function. Next, we present the alternative formulation of the mixed initial-boundary value problem using convolution. The benefit of this formulation is that it incorporates the initial conditions into the field equations, due to which there is no need to consider the initial conditions separately. Lastly, using this formulation, we have presented the variational principle of convolution type and a reciprocity theorem. The present model has not yet been studied in this direction by any researcher with the best of our knowledge. Hence, it is believed that the theorems established in the present paper will be useful for further study in this area.

2 Basic Equations and Problem Formulation

Following Leseduarte and Quintanilla [34], we consider the constitutive relation for heat flux, temperature gradient and thermal displacement in the form given in equation (vii). Hence, we consider the basic governing and the constitutive relations in context of this Quintanilla model under linear theory of thermoelasticity for a homogeneous and anisotropic material as follows:

The Equation of Motion:

$$\sigma_{ij,j} + \rho H_i = \rho \ddot{u}_i. \tag{1}$$

The Equation of Energy:

$$\rho T_0 \dot{S} = -q_{i,i} + \rho h. \tag{2}$$

The Constitutive Equations:

$$\sigma_{ij} = C_{ijkl} e_{kl} - \alpha_{ij}\theta; \tag{3}$$

$$\rho S = \rho c_E \frac{\theta}{T_0} + \alpha_{ij} e_{ij}; \tag{4}$$

$$\dot{q}_i = -\{K_{ij}\frac{\partial}{\partial t} + K_{ij}^*(1 - \tau\frac{\partial}{\partial t} + \frac{\tau^2}{2}\frac{\partial^2}{\partial t^2})\}\eta_j. \tag{5}$$

The Geometrical Relations:

$$\eta_j = \theta_{,j}; \tag{6}$$

$$e_{ij} = \frac{1}{2}(u_{i,j} + u_{j,i}) = u_{(i,j)}. \tag{7}$$

In this system of equations, we use a rectangular coordinate system x_k in three dimensional Euclidean space with usual indicial notations and used the following notations:

σ_{ij}-the components of stress tensor, u_i-the displacement components, e_{ij}-the components of strain tensor, H_i-the components of body force vector per unit mass, q_i-the components of heat flux vector, θ-the temperature variation from the uniform reference temperature T_0, η_i-the components of temperature gradient vector, h-the heat source per unit mass, S-the entropy per unit mass, ρ - the mass density and c_E-specific heat at constant strain. The comma in the subscript is used to represent the partial derivatives with respect to the space variables and the over-headed dots denote the differentiation with respect to the time variable t. C_{ijkl}, α_{ij} and K_{ij} and K_{ij}^* denote the elasticity tensor, thermoelasticity tensor, thermal conductivity tensor and rate of thermal conductivity tensor respectively. The subscripts i, j, k and l take the values 1, 2, 3 and the summation is represented by repetition of index.

Mixed Initial Boundary Value Problem

Now, we consider $\overline{V}$ as the closure of an open, bounded, connected domain with boundary, ∂V, enclosing an homogeneous and anisotropic thermoelastic material. Let V denote the interior of $\overline{V}$ and n_i be the components of an outward drawn unit normal to ∂V. Let B_i, (i = 1, 2, 3, 4) be the subsets of ∂V such that $B_1 \cup B_2 = B_3 \cup B_4 = \partial V$ and $B_1 \cap B_2 = B_3 \cap B_4 = \phi$. The motion relative to an undistorted stress free reference state is considered for the present study.

For a mixed initial and boundary value problem, we consider the field equations and constitutive relations given by Eqs. (1–7) defined in $V \times [0, \infty)$ together with the following initial conditions and boundary conditions:

Initial Conditions: On V

$$\left.\begin{array}{c} u_i(x,0) = u_{0_i}(x), \ \dot{u}_i(x,0) = v_i(x), \\ \theta(x,0) = \theta_0(x), \ \dot{\theta}(x,0) = \theta_1(x), \ q_i(x,0) = q_{0_i}(x). \end{array}\right\} \tag{8}$$

Boundary Conditions:

$$\left.\begin{array}{c} u_i = \tilde{u}_i(x,t) \ on \ B_1 \times [0, \infty), \\ \sigma_i = \sigma_{ij}n_j = \tilde{\sigma}_i(x,t) \ on \ B_2 \times [0, \infty), \\ q = q_i n_i = \tilde{q}(x,t) \ on \ B_3 \times [0, \infty), \\ \theta = \tilde{\theta}(x,t) \ on \ B_4 \times [0, \infty). \end{array}\right\} \tag{9}$$

Here, u_{0_i}, v_i, θ_0, θ_1, q_{0_i} represent the specified initial displacement component, velocity component, temperature, rate of temperature and heat flux, respectively together with $\tilde{u}_i$, $\tilde{\sigma}_i$, $\tilde{\theta}$, $\tilde{q}$, which denote the known surface displacement component, component of traction vector, temperature and normal heat flux, respectively. The smoothness requirements and other regularity assumptions on the ascribable functions are also considered as hypotheses on data. Also, we assume that u_{0_i}, v_i, θ_0, θ_1, q_{0_i} are continuous on $\overline{V}$, H_i and h are continuously differentiable on $\overline{V} \times [0, \infty)$. $\tilde{q}$ and $\tilde{\sigma}$ are piecewise continuous on $B_3 \times [0, \infty)$ and $B_2 \times [0, \infty)$, respectively. $\tilde{u}_i$ and $\tilde{\theta}$ are continuous on $B_1 \times [0, \infty)$ and $B_4 \times [0, \infty)$, respectively.

Further, we assume that the C_{ijkl}, α_{ij}, K_{ij} and K_{ij}^* are smooth on $\overline{V}$ and satisfy

$$C_{ijkl} = C_{klij} = C_{jikl} = C_{ijlk}, \qquad \alpha_{ij} = \alpha_{ji}, \qquad K_{ij} = K_{ji}, \qquad K_{ij}^* = K_{ji}^*, \qquad (10)$$

$$C_{ijkl} e_{ij} e_{kl} > 0, \ \ for \ all \ e_{ij} \ on \ \overline{V} \times [0, \infty), \tag{11}$$

$$K_{ij} \varphi_i \varphi_j > 0 \ for \ any \ real \ \varphi_i \ on \ \overline{V} \times [0, \infty), \tag{12}$$

$$K_{ij}^* \psi_i \psi_j > 0 \ for \ any \ real \ \psi_i \ on \ \overline{V} \times [0, \infty), \tag{13}$$

The material constants and delay time parameters satisfy the following inequalities:

$$\rho > 0, \ c_E > 0, \ T_0 > 0, \ \tau > 0, \ K_{ij} - \tau K_{ij}^* > 0 \ \ on \ V. \tag{14}$$

Now we define an admissible state as $R = \{u_i, \theta, \eta_i, e_{ij}, \sigma_{ij}, q_i, S\}$, which is an ordered array of functions $u_i, \theta, \eta_i, e_{ij}, \sigma_{ij}, q_i, S$ defined on $\overline{V} \times [0, \infty)$ with the properties that $u_i \in C^{2,2}$, $\theta \in C^{1,2}$, $\eta_i \in C^{0,2}$, $\sigma_{ij} \in C^{1,0}$, $q_i \in C^{1,1}$, $S \in C^{0,1}$ and $e_{ij} = e_{ji}$, $\sigma_{ij} = \sigma_{ji}$ on $\overline{V} \times [0, \infty)$. We further define two operations, addition of two admissible states and multiplication of an admissible state with a scalar as follows:

$$R + R' = \{u_i + u_i, \theta + \theta', \ldots\ldots, S + S'\},$$
$$\lambda^* R' = \{\lambda^* u_i, \lambda^* \theta, \ldots\ldots, \lambda^* S\},$$

where λ^* is any scalar. Then the set of all admissible states is clearly a linear space.

Further, an admissible state is the solution of the present mixed problem if it satisfies all the field Eqs. (1–7), the initial conditions (8) and the boundary conditions (9).

3   Uniqueness of Solution

For the uniqueness of solution, we consider the specific internal energy for the present initial-boundary value problem which is in the form

$$E = \frac{1}{2} C_{ijkl} \dot{e}_{kl} \dot{e}_{ij} + \frac{\rho c_E}{2 T_0} \dot{\theta}^2, \tag{15}$$

Clearly, from Eqs. (11) and (14), we can say that the specific internal energy (Eq. 15) is positive definite and using Eqs. (3), (4) and (10), we get

$$\dot{E} = \dot{\sigma}_{ij} \ddot{e}_{ij} + \rho \ddot{S} \dot{\theta}. \tag{16}$$

Now, by using relations (2), (5), (6) and (7), we obtain

$$\dot{E} = \dot{\sigma}_{ij} \ddot{u}_{i,j} - \frac{1}{T_0} \dot{q}_{i,i} \dot{\theta} + \frac{\rho h}{T_0} \dot{\theta}$$

$$= (\dot{\sigma}_{ij} \ddot{u}_i)_{,j} - \dot{\sigma}_{ij,j} \ddot{u}_i - \frac{1}{T_0} (\dot{q}_i \dot{\theta})_i + \frac{1}{T_0} (\dot{q}_i \eta_i) + \frac{\rho h}{T_0} \dot{\theta}$$

$$= (\dot{\sigma}_{ij} \ddot{u}_i)_{,j} - \frac{1}{T_0} (\dot{q}_i \dot{\theta})_{,i} + \rho \dot{H}_i \ddot{u}_i + \frac{\rho h}{T_0} \dot{\theta} - \rho \ddot{u}_i \ddot{u}_i$$

$$- \frac{\dot{\eta}_i}{T_0} \{ K_{ij} \dot{\eta}_j + K_{ij}^* \eta_j - \tau K_{ij}^* \dot{\eta}_j + \frac{\tau^2}{2} K_{ij}^* \ddot{\eta}_j \}. \tag{17}$$

Integrating both sides of Eq. (17) over V, using divergence theorem and by using (1), we get

$$\frac{\partial}{\partial t} \int_V (E + \frac{\rho}{2} \ddot{u}_i \ddot{u}_i + \frac{K_{ij}^*}{2T_0} \eta_i \eta_j + \frac{\tau^2 K_{ij}^*}{4T_0} \dot{\eta}_i \dot{\eta}_j) dV + \frac{1}{T_0} \int_V (K_{ij} - \tau K_{ij}^*) \dot{\eta}_i \dot{\eta}_j dV$$

$$= \int_V (\rho \dot{H}_i \ddot{u}_i + \frac{\rho h \dot{\theta}}{T_0}) dV + \int_A (\dot{\bar{\sigma}}_i \ddot{u}_i - \frac{1}{T_0} \dot{\theta} \dot{\bar{q}}) dA. \tag{18}$$

We will now establish the uniqueness of solution of the present mixed initial-boundary value problem by the following uniqueness theorem.

Theorem 3.1 (Uniqueness theorem):
Statement: The mixed initial-boundary value problem given by Eqs. (1)–(7), which satisfies the initial conditions (8) and boundary conditions (9) has at most one solution.

Proof: We assume that there are two sets of solutions $u_i^{(\gamma)}$, $\theta^{(\gamma)}$, $e_{ij}^{(\gamma)}$, $\sigma_{ij}^{(\gamma)}$, $q_i^{(\gamma)}$, $S^{(\gamma)}$ for $\gamma = 1, 2$. Then, we will construct the difference between these two sets of functions as

$$\bar{u}_i = u_i^{(1)} - u_i^{(2)}, \bar{\theta} = \theta^{(1)} - \theta^{(2)}, \ldots \ldots \ldots \ldots \ldots, \bar{S} = S^{(1)} - S^{(2)}. \tag{19}$$

Since, the set of all admissible states is a linear space, so the difference functions defined by (19) also satisfy the Eqs. (1–7) with zero body forces and heat source, the initial conditions (8) and the boundary conditions (9) in their homogeneous form and hence Eq. (18) too. Therefore from Eq. (18), we obtain

$$\frac{\partial}{\partial t} \int_V (\bar{E} + \frac{\rho}{2} \ddot{\bar{u}}_i \ddot{\bar{u}}_i + \frac{K_{ij}^*}{2T_0} \bar{\eta}_i \bar{\eta}_j + \frac{\tau^2 K_{ij}^*}{4T_0} \dot{\bar{\eta}}_i \dot{\bar{\eta}}_j) dV + \frac{1}{T_0} \int_V (K_{ij} - \tau K_{ij}^*) \dot{\bar{\eta}}_i \dot{\bar{\eta}}_j dV = 0. \tag{20}$$

Interchanging the variable t with ξ and integrating above equation over time interval $(0, t)$ and using the homogeneous initial conditions for difference functions, we obtain

$$\int_V (\overline{E} + \frac{\rho}{2}\dot{\overline{u}}_i\dot{\overline{u}}_i + \frac{K_{ij}^*}{2T_0}\overline{\eta}_i\overline{\eta}_j + \frac{\tau^2 K_{ij}^*}{4T_0}\dot{\overline{\eta}}_i\dot{\overline{\eta}}_j)dV + \frac{1}{T_0}\int_0^t\int_V (K_{ij} - \tau K_{ij}^*)\dot{\overline{\eta}}_i\dot{\overline{\eta}}_j dV\,d\xi = 0.$$

(21)

From Eqs. (11), (12), (13) and (14) we observe that the component in each term present on the left hand side of Eq. (21) is non-negative. Thus we conclude that each term in Eq. (21) must be zero which implies that

$$\ddot{\overline{u}}_i = 0, \quad \dot{\overline{\theta}} = 0 \quad on\ \overline{V} \times [0, \infty).$$

(22)

From (16), we get

$$\frac{\partial^2 \overline{u}_i}{\partial t^2} = 0, \quad \frac{\partial \overline{\theta}}{\partial t} = 0, \ on\ \overline{V} \times [0, \infty).$$

(23)

Therefore, in view of the initial conditions $\overline{u}_i(x, 0) = 0, \dot{\overline{u}}_i(x, 0) = 0$ and $\overline{\theta}(x, 0) = 0$, we get from Eq. (23) that

$$\overline{u}_i = 0, \quad \overline{\theta} = 0 \quad on\ \overline{V} \times [0, \infty),$$

i.e.,

$$u_i^{(1)} = u_i^{(2)}, \quad \theta^{(1)} = \theta^{(2)} \quad on\ \overline{V} \times [0, \infty).$$

This completes the proof of the uniqueness theorem.

4   Alternative Formulation of Mixed Problem

This section discusses the alternative formulation of the above mixed initial-boundary value problem in which the initial conditions are combined into the field equations (Gurtin [37]). For this purpose, we proceed as follows:

Let ϕ and ψ be two functions defined on $\overline{V} \times [0, \infty)$ such that both are continuous on $[0, \infty)$ for each $x \in V$. Then the convolution $\phi * \psi$ of ϕ and ψ is defined as

$$[\phi * \psi](x, t) = \int_0^t \phi(x, t - \tau)\psi(x, \tau)d\tau, \quad (x, t) \in \overline{V} \times [0, \infty).$$

We will use the commutativity, associativity and distributivity of convolution and the property that

$$\phi * \psi = 0 \Rightarrow \phi = 0\ or\ \psi = 0$$

(24)

Now, we define the functions g and l on $[0, \infty)$ as

$$g(t) = t, \quad l(t) = 1. \tag{25}$$

Also, let functions f_i and W be defined on $\overline{V} \times [0, \infty)$ as

$$f_i = g * \rho H_i + \rho(t v_i + u_{0_i}), \tag{26}$$

$$W = l * \frac{\rho h}{T_0} + \rho c_E \frac{\theta_0}{T_0} + \alpha_{ij} u_{0_{i,j}}, \tag{27}$$

and let

$$N_i = l * (t q_{0_i} + t \theta_{0,j} K_{ij} - t \tau \theta_{0,j} K_{ij}^* + t \theta_{1,j} \frac{\tau^2}{2} K_{ij}^* + \theta_{0,j} \frac{\tau^2}{2} K_{ij}^*), \tag{28}$$

Consider $p(x,t)$ and $\dot{p}(x,t)$, two functions defined on $\overline{V} \times [0, \infty)$ such that both are continuous and differentiable on $[0, \infty)$. Then the following results hold clearly:

$$g * \ddot{p}(x,t) = p(x,t) - [t\dot{p}(x,0) + p(x,0)], \tag{29}$$

$$l * \dot{p}(x,t) = p(x,t) - p(x,0), \tag{30}$$

$$g * \dot{p}(x,t) = l * (l * \dot{p}(x,t)) = l * [p(x,t) - p(x,0)] = l * p(x,t) - tp(x,0). \tag{31}$$

By this formulation we, therefore, obtain the following theorem that characterises our mixed problem in an alternative way.

Theorem 4.1:
Statement: The function u_i, θ, η_i, e_{ij}, σ_{ij}, q_i, S satisfy Eqs. (1), (2) and (5) and the initial conditions (8) if and only if

$$g * \sigma_{ij,j} + f_i = \rho u_i, \tag{32}$$

$$\rho S = -l * \frac{q_{i,i}}{T_0} + W, \tag{33}$$

$$L_1 * q_i = -L_1 * K_{ij} \eta_j - L_2 * K_{ij}^* \eta_j + N_i, \tag{34}$$

where $L_1 = l * l$ and $L_2 = l * (g + \tau l + \frac{\tau^2}{2})$, f_i, W and N_i are given by Eqs. (26), (27), and (28), respectively.

Proof: Firstly, assuming the governing Eqs. (1), (2) and (5) and initial conditions (8) hold good. Then, taking the convolution of Eq. (1) with g and using the results from Eqs. (29) and (8), we arrive at the Eq. (32). Similarly, taking the convolution of the Eq. (2) with l and using (30), (4) and (8) we obtain the Eq. (33). Again, taking the convolution of Eq. (5) with $l*g$, and using the relation from (29), (31) and (8) we get the Eq. (34).

Similarly, we can prove the converse of the above theorem, by reverse arguments. Hence, finally we get the following theorem.

Theorem 4.2: Let $R = \{u_i, \theta, \eta_i, e_{ij}, \sigma_{ij}, q_i, S\}$ be an admissible state. Then R is a solution of the mixed problem if and only if it satisfies the Eqs. (32)–(34), (3), (4), (6), (7) and the boundary conditions (9).

5   Variational Theorem

Using the alternative formulation and the theorem established in the previous section, we will formulate a variational principle on linear theory of thermoelasticity for anisotropic and homogeneous medium under the present heat conduction model (using backward expansion of Taylor's series) given by Leseduarte and Quintanilla [34].

Theorem 5.1:
Statement: Let Λ be a linear space of all admissible states with addition and scalar multiplication as describe in Sect. 2. If for each $t \in [0, \infty)$ and for every $\Gamma = \{u_i, \theta, \eta_j, e_{ij}, \sigma_{ij}, q_i, S\} \in \Lambda$, we define a functional $F_t\{\Gamma\}$ on Λ by

$$F_t\{\Gamma\} = \int_V [\frac{1}{2}L_1 * g * C_{ijkl}e_{kl} * e_{ij} - \frac{1}{2}L_1 * \rho u_i * u_i$$

$$- L_1 * g * \sigma_{ij} * e_{ij} - L_1 * g * l * \frac{1}{T_0}q_i * \eta_i$$

$$+ L_1 * u_i * (\rho u_i - g * \sigma_{ij,j} - f_i) - L_1 * g * \theta * (\rho S + l * \frac{q_{i,i}}{T_0} - W)$$

$$+ g * l * \frac{1}{T_0}(-L_1 * \frac{1}{2}K_{ij}\eta_j - L_2 * \frac{1}{2}K_{ij}^*\eta_j + N_i) * \eta_i$$

$$+ \frac{T_0}{2\rho c_E}L_1 * g * (\rho S - \alpha_{rs}e_{rs}) * (\rho S - \alpha_{ij}e_{ij})]dV + \int_{B_1} L_1 * g * \tilde{u}_i * \sigma_i dA$$

$$+ \int_{B_2} L_1 * g * (\sigma_i - \tilde{\sigma}_i) * u_i dA + \frac{1}{T_0} \int_{B_3} L_1 * g * l * q * \tilde{\theta}dA$$

$$+ \frac{1}{T_0} \int_{B_4} M_1 * g * l * (q - \tilde{q}) * \theta dA, \tag{35}$$

then the variation of this functional,

$$\delta F_t\{\Gamma\} = 0, \quad t \in [0, \infty), \tag{36}$$

if and only if, Γ is a solution of the mixed initial-boundary value problem given by Eqs. (1)–(7) with the initial conditions (8) and the boundary conditions (9).

Proof: Let $\Gamma' = \{u_i', \theta', \eta_i', e_{ij}', \sigma_{ij}', q_i', S'\} \in \Lambda$, which implies that $\Gamma + \lambda\Gamma' \in \Lambda$, for every real λ. Then Eq. (35) together with properties of convolution, the definition of variation and the divergence theorem, implies

$$\delta_{\Gamma'}\Omega_t\{\Gamma\} = \int_V [L_1 * g * \{C_{ijkl}e_{kl} - \frac{T_0\alpha_{ij}}{\rho c_E}(\rho S - \alpha_{rs}e_{rs}) - \sigma_{ij}\} * e_{ij}'$$

$$+ L_1 * g * \{\frac{T_0}{\rho c_E}(\rho S - \alpha_{rs}e_{rs}) - \theta\} * \rho S'$$

$$+ g * l * \frac{1}{T_0}(-L_1 * K_{ij}\eta_j - L_2 * K_{ij}^*\eta_j + N_i - L_1 * q_i) * \eta_i'] dV$$

$$- \int_V [L_1 * (g * \sigma_{ij,j} + f_i - \rho u_i) * u_i' + L_1 * g * (\rho S + l * \frac{q_{i,i}}{T_0} - W) * \theta'] dV$$

$$- \int_V [L_1 * g * (e_{ij} - u_{(i,j)}) * \sigma_{ij}' - L_1 * g * l * \frac{1}{T_0}(\theta_{,i} - \eta_i) * q_i'] dV$$

$$+ \int_{B_1} L_1 * g * (\tilde{u}_i - u_i) * \sigma_i' dA + \int_{B_2} L_1 * g * (\sigma_i - \tilde{\sigma}_i) * u_i' dA$$

$$+ \frac{1}{T_0} \int_{B_3} L_1 * g * l * (\tilde{\theta} - \theta) * q' dA + \frac{1}{T_0} \int_{B_4} L_1 * g * l * (q - \tilde{q}) * \theta' dA,$$

$$(37)$$

for all $t \in [0, \infty)$.

Firstly, assuming that Γ is a solution of the mixed initial-boundary value problem, then from Theorem 4.1, the relations (32) to (34) and the boundary conditions (9) gives

$$\delta_{\Gamma'} \Omega_t\{\Gamma\} = 0, \ t \in [0, \infty) \tag{38}$$

for every $\Gamma' = \{u_i', \theta', \gamma_i', e_{ij}', \sigma_{ij}', q_i', S'\} \in \Lambda$, and therefore we get (36). This completes the proof of the necessary part of the Theorem 5.1.

Conversely, let (36) holds true and hence (38) holds for every $\Gamma' = \{u_i', \theta', \eta_i', e_{ij}', \sigma_{ij}', q_i', S'\} \in \Lambda$. Then, we have to show that Γ is a solution of mixed initial-boundary value problem.

Since (38) holds for every $\Gamma' \in \Lambda$, we choose $\Gamma' = \{u_i', 0, 0, 0, 0, 0, 0\}$ and let u_i', along with all the space derivatives, vanish on $\partial V \times [0, \infty)$. Therefore, we deduce from Eqs. (37) and (38)

$$\int_V (g * \sigma_{ij,j} + f_i - \rho u_i) * u_i' dV = 0 \quad for \ t \in [0, \infty). \tag{39}$$

By using Lemma-1 (see Gurtin [37]) and convolution properties, we find that Eq. (32) holds.

Similarly, by substituting appropriate choices of Γ' into (37), we can prove with the help of three Lemmas (1–3) (Gurtin [37]) that Γ also satisfies the Eqs. (33), (34), (3), (4), (6), (7) and the boundary conditions (9). Therefore, from Theorem 4.2, Γ is the solution of the present mixed problem. The proof of the above theorem is therefore complete.

6 Reciprocity Theorem

Now, we consider two different systems of thermoelastic loadings

$$L^\beta = (H_i^{(\beta)}, h^{(\beta)}, \tilde{u}_i^{(\beta)}, \tilde{\theta}^{(\beta)}, \tilde{q}_i^{(\beta)}, \tilde{\sigma}_i^{(\beta)}, u_{0_i}^{(\beta)}, v_i^{(\beta)}, \theta_0^{(\beta)}, \theta_1^{(\beta)}, q_{0_i}^{(\beta)}), \ \beta = 1, 2$$

$$(40)$$

The corresponding thermoelastic configurations are denoted as

$$I^\beta = (u_i^{(\beta)}, \theta^{(\beta)}) \tag{41}$$

that satisfy (32)–(34), (3), (4), (6), (7) and (9).

We aim to establish a reciprocity theorem that states the relation between these two sets of thermoelastic loadings and thermoelastic configurations. For this, we use the following notations:

$$f_i^{(\beta)} = \rho(g * H_i^{(\beta)} + tv_i^{(\beta)} + u_{0_i}^{(\beta)}), \tag{42}$$

$$W^{(\beta)} = l * \frac{\rho h^{(\beta)}}{T_0} + \rho c_E \frac{\theta_0^{(\beta)}}{T_0} + \alpha_{ij} u_{0_{i,j}}^{(\beta)}, \tag{43}$$

$$N_i^{(\beta)} = l * (t q_{0_i}^{(\beta)} + t K_{ij}\theta_{0,j}^\beta - t\tau K_{ij}^*\theta_{0,j}^\beta + t\frac{\tau^2}{2} K_{ij}^*\theta_{1,j}^\beta + \frac{\tau^2}{2} K_{ij}^*\theta_{0,j}^\beta), \tag{44}$$

for $\beta = 1, 2$. Then, we have the reciprocity theorem as given below.

Theorem 6.1 (Reciprocity theorem): If a thermoelastic solid is associated with two different systems of thermoelastic loadings, L^β, $(\beta = 1, 2)$ and I^β, $(\beta = 1, 2)$ are the corresponding thermoelastic configurations, then the following reciprocity relation holds:

$$\int_V L_1 * [f_i^{(1)} * u_i^{(2)} - g * W^{(1)} * \theta^{(2)}]dV + \int_A L_1 * g * \left[\sigma_i^{(1)} * u_i^{(2)} + \frac{1}{T_0}l * q^{(1)} * \theta^{(2)}\right] dA$$

$$- \int_V g * l * \left[\frac{1}{T_0}N_i^{(1)} * \eta_i^{(2)}\right] dV = \int_V L_1 * \left[f_i^{(2)} * u_i^{(1)} - g * W^{(2)} * \theta^{(1)}\right] dV$$

$$+ \int_A L_1 * g * \left[\sigma_i^{(2)} * u_i^{(1)} + \frac{1}{T_0}l * q^{(2)} * \theta^{(1)}\right] dA - \int_V g * l * \left[\frac{1}{T_0}N_i^{(2)} * \eta_i^{(1)}\right] dV, \tag{45}$$

where, $f_i^{(\beta)}$, $W^{(\beta)}$, $N_i^{(\beta)}$ $(\beta = 1, 2)$ associated with two systems are given by Eqs. (42), (43), (44) respectively.

Proof: From Eq. (3) we have

$$\sigma_{ij}^{(\beta)} = C_{ijkl}e_{kl}^{(\beta)} - \alpha_{ij}\theta^{(\beta)}, \tag{46}$$

Now, taking convolution of Eq. (46) for $\beta = 1$ with $e_{ij}^{(2)}$ and for $\beta = 2$ with $e_{ij}^{(1)}$ and then subtracting the results, we get

$$(\sigma_{ij}^{(1)} + \alpha_{ij}\theta^{(1)}) * e_{ij}^{(2)} = (\sigma_{ij}^{(2)} + \alpha_{ij}\theta^{(2)}) * e_{ij}^{(1)} + C_{ijkl}(e_{kl}^{(1)} * e_{ij}^{(2)} - e_{kl}^{(2)} * e_{ij}^{(1)}).$$

Hence due to the symmetry properties of C_{ijkl}, we have

$$C_{ijkl}(e_{kl}^{(1)} * e_{ij}^{(2)} - e_{kl}^{(2)} * e_{ij}^{(1)}) = C_{ijkl}e_{kl}^{(1)} * e_{ij}^{(2)} - C_{klij}e_{kl}^{(1)} * e_{ij}^{(2)} = 0 \qquad (47)$$

Therefore,

$$(\sigma_{ij}^{(1)} + \alpha_{ij}\theta^{(1)}) * e_{ij}^{(2)} = (\sigma_{ij}^{(2)} + \alpha_{ij}\theta^{(2)}) * e_{ij}^{(1)}. \qquad (48)$$

Again from Eq. (4), we can write

$$\rho S^{(\beta)} - \alpha_{ij}e_{ij}^{(\beta)} = \rho c_E \frac{\theta^{(\beta)}}{T_0}, \qquad \beta = 1, 2 \qquad (49)$$

Taking convolution of Eq. (49) for $\beta = 1$ with $\theta^{(2)}$ and for $\beta = 2$ with $\theta^{(1)}$ and subtracting, we get

$$(\rho S^{(1)} - \alpha_{ij}e_{ij}^{(1)}) * \theta^{(2)} = (\rho S^{(2)} - \alpha_{ij}e_{ij}^{(2)}) * \theta^{(1)}. \qquad (50)$$

Equations (48) and (50) yield

$$(\sigma_{ij}^{(1)} * e_{ij}^{(2)} - \rho S^{(1)} * \theta^{(2)}) = (\sigma_{ij}^{(2)} * e_{ij}^{(1)} - \rho S^{(2)} * \theta^{(1)}). \qquad (51)$$

Now, we introduce the notation

$$L_{\alpha\beta} = \int_V L_1 * g * \left[\sigma_{ij}^{(\alpha)} * e_{ij}^{(\beta)} - \rho S^{(\alpha)} * \theta^{(\beta)} \right] dV, \qquad \alpha, \beta = 1, 2. \qquad (52)$$

Now, from Eqs. (7) and (32)–(34), we get

$$L_1 * g * (\sigma_{ij}^{(\alpha)} * e_{ij}^{(\beta)} - \rho S^{(\alpha)} * \theta^{(\beta)})$$

$$= L_1 * g * \sigma_{ij}^{(\alpha)} * u_{i,j}^{\beta} - L_1 * g * (-l * \frac{q_{i,i}^{(\alpha)}}{T_0} + W^{(\alpha)}) * \theta^{(\beta)}$$

$$= L_1 * g * (\sigma_{ij}^{(\alpha)} * u_i^{\beta})_{,j} - L_1 * g * (\sigma_{ij,j}^{(\alpha)} * u_i^{\beta})$$

$$+ \frac{1}{T_0}L_1 * g * (l * q_i^{(\alpha)} * \theta^{(\beta)})_{,i} - \frac{1}{T_0}L_1 * g * l * q_i^{(\alpha)} * \eta_i^{(\beta)}$$

$$- L_1 * g * W^{(\alpha)} * \theta^{(\beta)}$$

$$L_1 * g * (\sigma_{ij}^{(\alpha)} * e_{ij}^{(\beta)} - \rho S^{(\alpha)} * \theta^{(\beta)})$$

$$= L_1 * g * (\sigma_{ij}^{(\alpha)} * u_i^{(\beta)})_{,j} - L_1 * \rho u_i^{(\alpha)} * u_i^{(\beta)} + L_1 * f_i^{(\alpha)} * u_i^{(\beta)}$$

$$+ \frac{1}{T_0}L_1 * g * l * (q_i^{(\alpha)} * \theta^{(\beta)})_{,i} + \frac{1}{T_0}g * l * (L_1 * K_{ij}\eta_j^{(\alpha)}$$

$$+ L_2 * K_{ij}^*\eta_j^{(\alpha)}) * \eta_i^{(\beta)} - \frac{1}{T_0}g * l * N_i^{(\alpha)} * \eta_i^{(\beta)}$$

$$- L_1 * g * W^{(\alpha)} * \theta^{(\beta)}. \qquad (53)$$

From Eqs. (52) and (53), we therefore obtain

$$
L_{\alpha\beta} = \int_V L_1 * \left[f_i^{(\alpha)} * u_i^{(\beta)} - g * W^{(\alpha)} * \theta^{(\beta)} \right] dV
$$

$$
+ \int_{\partial V} L_1 * g * \left[\sigma_i^{(\alpha)} * u_i^{\beta} + \frac{1}{T_0} l * q_i^{(\alpha)} * \theta^{(\beta)} \right] dA
$$

$$
- \int_V \left[L_1 * \rho u_i^{(\alpha)} * u_i^{(\beta)} - \frac{1}{T_0} g * l * L_1 * K_{ij} \eta_j^{(\alpha)} * \eta_i^{(\beta)} \right.
$$

$$
\left. - \frac{1}{T_0} g * l * L_2 * K_{ij}^* \eta_j^{(\alpha)} * \eta_i^{(\beta)} \right] dV
$$

$$
- \int_V \left[\frac{1}{T_0} g * l * N_i^{(\alpha)} * \eta_i^{(\beta)} \right] dV. \tag{54}
$$

Clearly, from Eqs. (51) and (52), we have

$$
L_{12} = L_{21}. \tag{55}
$$

Hence, Eqs. (54) and (55) prove the reciprocity relation (45), which completes the proof of the Theorem 6.1.

References

1. Biot, M.A.: Thermoelasticity and irreversible thermodynamics. J. Appl. Phys. **27**, 240–253 (1956)
2. Chandrasekharaiah, D.S.: Thermoelasticity with second sound: a review. Appl. Mech. Rev. **39**(3), 355–376 (1986)
3. Chandrasekharaiah, D.S.: Hyperbolic thermoelasticity: a review of recent literature. Appl. Mech. Rev. **51**(12), 705–729 (1998)
4. Joseph, D.D., Preziosi, L.: Heat waves. Rev. Mod. Phys. **61**, 41–73 (1989)
5. Hetnarski, R.B., Ignaczak, J.: Generalized thermoelasticity. J. Therm. Stresses **22**, 451–476 (1999)
6. Dreyer, W., Struchtrup, H.: Heat pulse experiments revisited. Continuum Mech. Therm. **5**, 3–50 (1993)
7. Ozisik, M.N., Tzou, D.Y.: On the wave theory of heat conduction. ASME J. Heat Transfer **116**, 526–535 (1994)
8. Ignaczak, J., Ostoja-Starzewski, M.: Thermoelasticity With Finite Wave Speeds. Oxford University Press, New York (2010)
9. Muller, I., Ruggeri, T.: Extended Thermodynamics. Springer Tracts on Natural Philosophy. Springer, New York (1993). https://doi.org/10.1007/978-1-4684-0447-0
10. Marín, E.: Does Fourier's law of heat conduction contradict the theory of relativity? Latin-American J. Phys. Edu. **5**, 402–405 (2011)
11. Lord, H.W., Shulman, Y.A.: Generalized dynamical theory of thermoelasticity. J. Mech. Phys. Solids **15**(5), 299–309 (1967)
12. Green, A.E., Lindasy, K.A.: Thermoelasticity. J. Elast. **2**, 1–7 (1972)

13. Cattaneo, C.: A form of heat conduction equation which eliminates the paradox of instantaneous propagation. Compte Rendus **247**, 431–433 (1958)
14. Vernotte, P.: Les paradoxes de la theorie continue de l'equation de la chaleur. Compte Rendus **246**, 3154–3155 (1958)
15. Vernotte, P.: Some possible complications in the phenomena of thermal conduction. Compte Rendus **252**, 2190–2191 (1961)
16. Green, A.E., Naghdi, P.M.: A re-examination of the base postulates of thermoemechanics. Proc. R. Soc. Lond. A **432**, 171–194 (1991)
17. Green, A.E., Naghdi, P.M.: On undamped heat waves in an elastic solid. J. Therm. Stresses **15**, 253–264 (1992)
18. Green, A.E., Naghdi, P.M.: Thermoelasticity without energy dissipation. J. Elast. **31**, 189–208 (1993)
19. Tzou, D.Y.: A unified field approach for heat conduction from macro to micro scales. ASME J. Heat Transfer **117**, 8–16 (1995)
20. Tzou, D.Y.: The generalized lagging response in small-scale and high-rate heating. Int. J. Heat Mass Transfer **38**(17), 3231–3240 (1995)
21. Roychoudhuri, S.K.: On a thermoelastic three-phase-lag model. J. Therm. Stresses **30**, 231–238 (2007)
22. Dreher, M., Quintanilla, R., Racke, R.: Ill-posed problems in thermomechanics. Appl. Math. Lett. **22**, 1374–1379 (2009)
23. Quintanilla, R.: Exponential stability in the dual-phase-lag heat conduction theory. J. Non-Equilib. Thermodyn. **27**, 217–227 (2002)
24. Horgan, C.O., Quintanilla, R.: Spatial behaviour of solutions of the dual-phase-lag heat equation. Math. Methods Appl. Sci. **28**, 43–57 (2005)
25. Kumar, R., Mukhopadhyay, S.: Analysis of the effects of phase-lags on propagation of harmonic plane waves in thermoelastic media. Comput. Methods Sci. Tech. **16**(1), 19–28 (2010)
26. Mukhopadhyay, S., Kumar, R.: Analysis of phase-lag effects on wave propagation in a thick plate under axisymmetric temperature distribution. Acta Mech. **210**, 331–344 (2010)
27. Mukhopadhyay, S., Kothari, S., Kumar, R.: On the representation of solutions for the theory of generalized thermoelasticity with three phase-lags. Acta Mech. **214**, 305–314 (2010)
28. Quintanilla, R.: A condition on the delay parameters in the one-dimensional dual-phase-lag thermoelastic theory. J. Therm. Stresses **26**, 713–721 (2003)
29. Quintanilla, R., Racke, R.: A note on stability of dual-phase-lag heat conduction. Int. J. Heat Mass Transfer **49**, 1209–1213 (2006)
30. Quintanilla, R., Racke, R.: Qualitative aspects in dual-phase-lag thermoelasticity. SIAM J. Appl. Math. **66**, 977–1001 (2006)
31. Quintanilla, R., Racke, R.: Qualitative aspects in dual-phase-lag heat conduction. Proc. R. Soc. Lond. A **463**, 659–674 (2007)
32. Quintanilla, R., Racke, R.: A note on stability in three-phase-lag heat conduction. Int. J. Heat Mass Transfer **51**, 24–29 (2008)
33. Quintanilla, R.: Some solutions for a family of exact phase-lag heat conduction problems. Mech. Res. Commun. **38**, 355–360 (2011)
34. Leseduarte, M.C., Quintanilla, R.: Phragman-Lindelof alternative for an exact heat conduction equation with delay. Commun. Pure Appl. Math. **12**(3), 1221–1235 (2013)
35. Quintanilla, R.: On uniqueness and stability for a thermoelastic theory. Math. Mech. Solids **22**(6), 1387–1396 (2017)

36. Ignaczak, J.: A completeness problem for stress equations of motion in the linear elasticity theory. Arch. Mech. Stos **15**, 225–234 (1963)

37. Gurtin, M.E.: Variational principles for linear Elastodynamics. Arch. Ration. Mech. Anal. **16**, 34–50 (1964)

38. Iesan, D.: Principes variationnels dans la theorie de la thermoelasticite couplee. Ann. Sci. Univ. 'Al. I. Cuza' Iasi Mathematica **12**, 439–456 (1966)

39. Iesan, D.: On some reciprocity theorems and variational theorems in linear dynamic theories of continuum mechanics. Memorie dell'Accad. Sci. Torino. Cl. Sci. Fis. Mat. Nat. Ser. **4**(17), 17–37 (1974)

40. Nickell, R., Sackman, J.: Variational principles for linear coupled thermoelasticity. Quart. Appl. Math. **26**, 11–26 (1968)

41. Iesan, D.: Sur la théorie de la thermoélasticité micropolaire couplée. C. Rend. Acad. Sci. Paris **265**, 271–275 (1967)

42. Nowacki, W.: Fundamental relations and equations of thermoelasticity. In: Francis, P.H., Hetnarski, R.B. (eds.) Dynamic Problems of Thermoelasticity (English Edition). Noordhoff Internationa Publishing, Leyden (1975)

43. Maysel, V.M.: The Temperature Problem of the Theory of Elasticity. Kiev (1951). (in Russian)

44. Predeleanu, P.M.: On thermal stresses in viscoelastic bodies. Bull. Math. Soc. Sci. Math. Phys. **3**(51), 223–228 (1959)

45. Ionescu-Cazimir, V.: Problem of linear thermoelasticity: theorems on reciprocity I. Bull. Acad. Polon. Sci. Ser. Sci. Tech. **12**, 473–480 (1964)

46. Scalia, A.: On some theorems in the theory of micropolar thermoelasticity. Int. J. Eng. Sci. **28**, 181–189 (1990)

47. Lebon, G.: Variational Principles in Thermomechanics. Springer-Wien, New York (1980). https://doi.org/10.1007/978-3-540-88467-5

48. Carlson, D.E.: Linear thermoelasticity. In: Truesdell, C. (ed.) Flugge's Handbuch der Physik, vol. VI a/2, pp. 297–345. Springer, Heidelberg (1973). https://doi.org/10.1007/978-3-662-39776-3_2

49. Hetnarski, R.B., Ignaczak, J.: Mathematical Theory of Elasticity. Taylor and Francis, New York (2004)

50. Hetnarski, R.B., Eslami, M.R.: Thermal Stresses: Advanced Theory and Applications. In: Gladwell, G.M.L. (ed.) Solid Mechanics and Its Applications, vol. 158. Springer, Dordrecht (2010). https://doi.org/10.1007/978-1-4020-9247-3

51. Chirita, S., Ciarletta, M.: Reciprocal and variational principles in linear thermoelasticity without energy dissipation. Mech. Res. Commun. **37**, 271–275 (2010)

52. Mukhopadhyay, S., Prasad, R.: Variational and reciprocal principles in linear theory of type-III thermoelasticity. Math. Mech. Solids **16**, 435–444 (2011)

53. Kothari, S., Mukhopadhyay, S.: Some theorems in linear thermoelasticity with dual phase-lags for an Anisotropic Medium. J. Therm. Stresses **36**, 985–1000 (2013)

54. Kumari, B., Mukhopadhyay, S.: Some theorems on linear theory of thermoelasticity for an anisotropic medium under an exact heat conduction model with a delay. Math. Mech. Solids **22**(5), 1177–1189 (2016, 2017)

Author Index

Printed in the United States
By Bookmasters